Wave Energy Potential, Behavior and Extraction

Wave Energy Potential, Behavior and Extraction

Special Issue Editor

Hua Li

MDPI • Basel • Beijing • Wuhan • Barcelona • Belgrade • Manchester • Tokyo • Cluj • Tianjin

Special Issue Editor
Hua Li
Texas A&M University-Kingsville
USA

Editorial Office
MDPI
St. Alban-Anlage 66
4052 Basel, Switzerland

This is a reprint of articles from the Special Issue published online in the open access journal *Energies* (ISSN 1996-1073) (available at: http://www.mdpi.com/journal/energies/special_issues/wave_energy).

For citation purposes, cite each article independently as indicated on the article page online and as indicated below:

LastName, A.A.; LastName, B.B.; LastName, C.C. Article Title. *Journal Name* **Year**, *Article Number*, Page Range.

ISBN 978-3-03928-396-5 (Pbk)
ISBN 978-3-03928-397-2 (PDF)

Contents

About the Special Issue Editor

Hua Li (Dr.) received his Ph.D. degree in Industrial Engineering from Texas Tech University and his B.S. degree from Tsinghua University. He is an Associate Professor and Graduate Coordinator in Mechanical and Industrial Engineering Department at Texas A&M University-Kingsville. He has extensive research experience in renewable energy, especially wind and wave energy, energy-saving product design, optimization, simulation, and engineering education. He has received more than $5M grants as PI and Co-PI from different federal agencies, and he has supervised more than 20 Ph.D. and M.S. students in Texas A&M University-Kingsville.

Preface to "Wave Energy Potential, Behavior and Extraction"

This book comprises successful submissions to a Special Issue of Energies on the subject "Wave Energy Potential, Behavior and Extraction". The topics covered in this Special Issue include wave energy potential analysis and wave energy converter design in the form of original research and review articles. Of the 13 papers that were submitted, 10 were accepted for publication. The average processing time was about 38.5 days. The authors represent diverse geographical locations: China (3); USA (3); UK (2); Sweden (2); Denmark (1); Australia (1); and Italy (1). Papers published in this Special Issue covered a wide range of topics, including wave energy utilization for offshore oil platform, wave energy converter design, economic analysis of wave energy converter, as well as thorough reviews on the status of wave energy development. I found the task of editing and selecting papers for this collection to be both stimulating and rewarding. I would also like to thank the staff and reviewers of Energies for their time, efforts, and input.

Hua Li
Special Issue Editor

Review

Ocean Wave Energy Converters: Status and Challenges

Tunde Aderinto [1] and Hua Li [2,*]

[1] Sustainable Energy Systems Engineering, Texas A&M University-Kingsville, Kingsville, TX 78363, USA; tunde.aderinto@students.tamuk.edu

[2] Mechanical and Industrial Engineering Department, Texas A&M University-Kingsville, Kingsville, TX 78363, USA

[*] Correspondence: hua.li@tamuk.edu

Received: 1 April 2018; Accepted: 9 May 2018; Published: 14 May 2018

Abstract: Wave energy is substantial as a resource, and its potential to significantly contribute to the existing energy mix has been identified. However, the commercial utilization of wave energy is still very low. This paper reviewed the background of wave energy harvesting technology, its evolution, and the present status of the industry. By covering the theoretical formulations, wave resource characterization methods, hydrodynamics of wave interaction with the wave energy converter, and the power take-off and electrical systems, different challenges were identified and discussed. Solutions were suggested while discussing the challenges in order to increase awareness and investment in wave energy industry as a whole.

Keywords: wave energy; wave energy converters; design; challenges

---

1. Introduction

The concept of harvesting energy from the ocean is not a recent idea. A patent of a wave energy converter (WEC) was filed as at 1799 [1], and several hundreds of patents related to wave energy conversion have been in existence by the late 20th century [2–4]. The first reported practical form of wave energy usage was by Y. Masuda when ocean waves were used to power navigation buoys [5]. The resurgence in wave energy research in the 1970s–1980s was partially as the result of the so-called oil crises of 1973 since the crises created the awareness of governments and policy makers to the temporal and spatial nature of fossil fuel reserves [6,7]. Hence, a lot of research and development efforts were spearheaded by governments [8,9] and inter government associations [10]. Another reason was the recognition of fossil fuel especially petroleum-based fuel as a 'dirty' fuel because it releases carbon dioxide and other greenhouse gases into the atmosphere causing global warming and other adverse effects on the environment [11].

Renewable energy sources as choices of alternative forms of energy supply are attractive since they are inexhaustible and cleaner during operation. Ocean wave energy has the second largest potential among all ocean renewable energy sources [12,13]. Although opinion varies on the amount that can be successfully exploited out of the total wave resource, some studies have conservatively estimated this amount to be up to 10–20% of the total potential [14–17]. At this amount, it is a substantial part of the current total world power consumption [18]. A study in 2010 estimated the world total theoretical potential of ocean energy resource about 29,500 TWh/yr [19]. This estimation did not consider technical, geographical, and economic constraints that would make the actual recoverable resource to be less. Many countries and regions having considerable exposure to the oceans within and along their borders have realized the potential of wave energy contributing to their energy needs. Related studies have been performed to evaluate the amount of exploitable ocean wave energy resource in those areas. Some countries have actively conducted studies to estimate the amount of ocean energy available within their countries for exploitation. For example, the United States with a substantial exposure to the ocean along its borders has a good potential of ocean energy exploitation. A comprehensive study conducted in the U.S. estimated the wave energy resource along its coast to be 2640 TWh/yr with the recoverable energy up to 1170 TWh/yr [20]. The total recoverable wave energy resource in the United Kingdom has been estimated around 50 TWh/yr [21] while another study [22] estimated it around 69 TWh/yr. Regions such as Australia, South America and Southern Coast of Africa have been recognized as high density wave energy areas with peak level of up to 90 kW/m along their coast lines [23].

Many countries have seen some development in the planning, installation, and operation of wave energy converters. Although the amount is still low compared to other renewable energy sources, such as solar and wind, the progress shows that interests and awareness in the ocean wave energy as a viable source of energy are increasing. The information shown in Table 1 is valid as of the end of 2016. It should be noted that all of these WEC installations included in Table 1 were not in commercialization stage yet, and WECs at prototype design or conceptualization stage were not included.

Table 1. Wave Energy Converters Installations (kW) around the World [24].

Country	Planned	Installed	Operational	Total
Canada	0	0	11	11
New Zealand	0	20	0	20
Denmark	39	12	1	52
Italy	0	150	0	150
Mexico	200	0	0	200
Ghana	0	0	450	450
Spain	0	230	296	526
Korea	0	0	665	665
China	0	400	300	700
Portugal	350	0	400	750
United States	1335	500	30	1865
Sweden	0	0	3200	3200
Ireland	5000	0	0	5000

There have been many concepts and designs of wave energy converters over the years. Many of them remained at the design stage, laboratory test stage, or prototype testing stage [25,26], while only a few have been deployed [27]. Despite all these studies focused on wave energy conversion technology and the amount of identified resource potential, the commercial utilization of ocean wave energy is still low compared to other renewable sources, such as solar energy and wind energy. One of the issues affecting the commercial utilization of wave energy is the difficulty in integrating power from large WECs into the electricity grid due to the high variability of the wave properties [28]. Another reason is that there is little or nonexistence of grid facilities where the wave resources are located. Even when the coastlines have facilities for grid connections, the coastlines may have other competing uses such as recreation, which may make getting permits for building large scale WECs closed to coastlines a challenge. Extreme weather conditions are also factors to be considered that make the construction, operation, and maintenance of WEC systems challenging. Lastly, wave properties (height, period and direction) have very high variabilities in both space and time, which make it difficult to plan and harvest the energy from ocean waves.

There were reported successes in testing of models and prototypes of WECs [25,26,29]. These models had capacities ranging from a few hundred watts to up to 500 kW [30]. Transforming these laboratory and prototype models to large scale commercial successes have been one major challenge hindering the commercialization of WECs as the focus being on integrating large scale WECs into the electricity grid. Compared to other renewable energy systems that are successful in both technology development and commercialization, such as wind and solar, wave energy is still at a nascent stage. There are very few grid-connected commercialized WEC systems operating anywhere in the world [24]. For example, a company in United States had to abandon its WEC program in 2010 because the program would not be profitable [31].

In this paper, the authors reviewed different concepts of wave energy conversion while covering some historical perspectives and backgrounds, discussed the ocean wave energy resource assessment, and presented the evolution of the conversion technologies while analyzing the hydrodynamics of wave energy as the WECs interacting with the ocean environment. The general status of the emerging wave energy conversion industry was studied with focus on the challenges facing the industry and suggested solutions.

2. Wave Energy Resource Assessment

To effectively plan either small or large scale WECs, it is important to have an estimate of both the total resource potential and the exploitable resource. The change in ocean wave properties when it travels from the deep offshore to coastal areas is critical since it affects the wave energy resource. The characterization of the ocean wave properties would ensure that the appropriate technology to extract the optimum amount of energy is applied. The wave energy harvesting should start with the proper characterization of available wave energy resources in a given area. The available wave energy resource depends mainly on the wave height and wave period [32–35]. Many organizations at both the government and private levels have conducted various studies to estimate the wave energy resource potential at both global [36,37] and regional levels [38–41]. For example, National Renewable Energy Laboratory, a United States government agency, has estimated the gross and recoverable wave energy resource potentials in United States [20]. Besides those studies sponsored at government level, numerous researchers have estimated the wave energy resource potential at national, regional and sub-regional levels. Some of these studies can be found in China [42–45], Ireland [32,46], Spain [39,47], the Mediterranean Sea [48–50], the Aegean Sea [51,52], Black Sea [40], India Ocean [53], etc. Table 2 shows estimations of the wave energy resource potential in different regions of the world. This data provides general overview. However, it is lacking in technical and economic studies, as well as being limited by geographical constraints between countries in these regions.

Table 2. Theoretical Potential of Wave Energy Resource in Different Regions of the World [19].

Regions	Wave Energy Potential (TWh/y)
Mediterranean Sea and Atlantic Archipelagos (Azores, Cape Verde, Canaries)	1300
Central America	1500
Western and Northern Europe	2800
Africa	3500
North America and Greenland	4000
South America	4600
Australia, New Zealand and Pacific Islands	5600
Asia	6200
TOTAL	29,500

Most of these estimations' resolutions are yearly average, and only offer an overview on wave energy potential. Actual location specified data will have to be either measured or forecasted for the particular location if detailed WEC design, installation, and operation are to be undertaken. The work done by [54,55] analyzed wave data in the Gulf of Mexico focusing on the assessment of wave energy potential considering both the spatial and temporal variation of the wave energy resource using the energy event method. A lot of different data bases are available with ocean wave properties in many parts of the world. However, these ocean wave properties are usually for providing information for other ocean related activities, including seafaring vessels, offshore structures for oil and gas exploitation activities, weather stations, and even for recreational activities such as surfing. While data from these data bases can be beneficial, they are not tailored for WEC systems since some vital information may be missing from them. For example, data that gives significant wave height, wave period and direction may not have good enough grid resolution to be adequately used for implementing a WEC design and planning.

Ocean wave property data are normally collected from weather stations, ocean buoys and the satellite. Most of these data have grid resolutions of about 20 km by 20 km that makes them suitable for wave energy resource assessment but not sufficient for specific WEC design. Meanwhile, the wave properties vary when the bathymetry changes rapidly from the offshore regions to the coastal areas [56–58]. For the wave energy resource assessment in the nearshore areas, two factors are important: (1) the location of the wave properties includes the effect of wave reaction and shoaling in the nearshore areas forming hotspots with high energy potential as waves travel from offshore to coastal areas; (2) the change in the wave properties per unit area becomes pronounced, therefore, more data points should be required in this region in order to accurately assess the wave energy resource in the area. Some work has been done in specific nearshore areas in order to identify the hot spots best suited for wave energy extraction, the qualification of the wave energy potential, and the frequency of occurrence of the wave properties (wave height and period) [58].

Some computer applications have been developed to simulate and forecast ocean wave properties, and to improve the grid resolution of wave data to make them suitable for specific needs. These applications were validated with data collected from weather stations and ocean buoys. For example, the wave energy resource of the Iberian Peninsula was studied [47], in which the wave properties were simulated using a coupled SWAN-WAVEWATCH III approach. Each of the two applications provided the boundary condition for one another. WAVEWATCH III was used for the deep offshore areas while SWAN was used for the coastal areas. They were able to achieve a resolution of approximately 5.5 km by 11.1 km in the offshore region and 920 m by 920 m in the coastal areas. Resolutions of 20 km by 20 km, 10 km by 10 km, and 2 km by 2 km were achieved for forecasting and simulating wave properties using WAVEWATCH III in the Mediterranean Sea [48]. WAVEWATCH III was also used by [46] for both offshore and nearshore areas, in which the grid resolution ranged from 10 km by 10 km in the offshore region to 225 m by 225 m in the coastal areas. Other scholars have used WAVEWATCH III, SWAN, WAMIT, and combination of these applications [53,59,60] to simulate and forecast ocean wave properties that have shown a very good accuracy by comparing with actual measured data. Therefore, these computer applications have proven over time to be reliable methods for the simulation and forecast of ocean wave properties. Recent advancement in the assessment of ocean wave energy resources have progressed from merely predicting ocean wave properties for calculating the potential, to computer tools to better evaluate the temporal and spatial variability of the resource [55]. The use of machine learning tools, optimization methods, and artificial neural networks [61–63] has increased the accuracy of wave energy prediction. Furthermore, the combination of optimization methods with data obtained from wave property predicting tools has allowed further improving of the accuracy in the wave energy assessment [64,65].

These ocean energy assessment methods have provided very good framework for WEC designers in the design and planning stage. However, the spatial resolution of the wave energy assessment is one major limitation especially for nearshore areas, where the wave properties vary rapidly with change in bathymetry [56–58]. The finest spatial resolutions are in the range of 200 m, so in situ investigation will still be required to obtain accurate wave data for detailed WEC design.

3. Wave Energy Extraction Technologies

Although there are many different patents filed for WECs, with each claiming to be unique, different ways are used to classify wave energy converters. WECs can be classified based on their working principles, i.e., the mode of capturing energy. WECs can also be classified based on their ways of converting the energy from the waves to electrical energy. Another classification method is based on water depth of the ocean where the WEC is sited (Shoreline, Nearshore and Offshore). WECs can also be classified based on the ratio of the magnitude of the wavelength over the interacting part of the

WEC. If the WEC absorber is smaller than the wavelength of the incoming waves, it is called a point absorber [66–71]. For terminators [72], the dominant wave direction is perpendicular to the structural extension of the WEC, while it is called an attenuator if the structural extension is parallel to the incoming wave direction [73]. In this paper, the WECs were broadly classified under three categories (Figure 1): Oscillating Water Columns (OWC), Oscillating Body Systems, and Overtopping Converters.

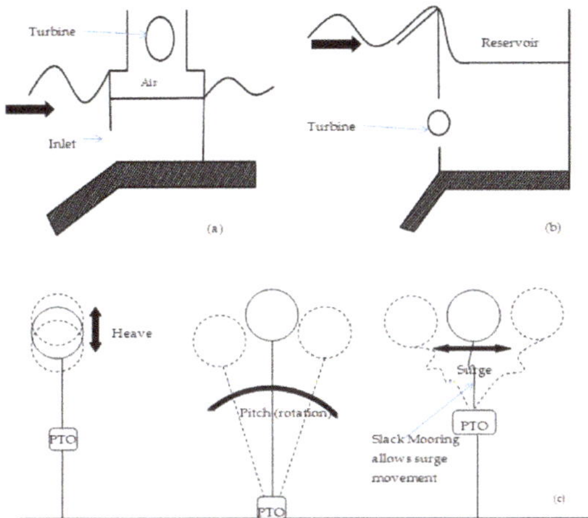

Figure 1. Classification of Wave Energy Converter (WEC) extraction technologies: (**a**) oscillating water column, (**b**) overtopping devices, and (**c**) oscillating bodies [13].

3.1. Oscillating Water Column (OWC)

The Oscillating Water Column is one of the first identified methods by which energy can be extracted from ocean wave. The earliest OWCs were located on fixed structures on the shoreline or very close to the shore [29] or on naturally occurring or manmade rigid structures, such as breakwaters or rock cliffs on the sea [26]. Applications as a floating device are also possible, and these are suitable for offshore locations [74]. The operating mode of OWC shown in Figure 2 is a simplified one. A partially submerged structure with an opening in the underwater section is made to trap air above the free air surface. The air inside the internal free surface moves according to the oscillatory motion of the water column to drive a turbine placed across. Many prototypes based on OWC have been built and tested. A lot of configurations and modifications have been adapted in order to increase the efficiency. For example, a backward bent duct buoy was used to capture surge and pitch motion of waves [75], and a spar buoy with a long vertical tube opened at the bottom to make the water column moving only in the heave direction to produce pneumatic power through the relative motion between the water and vertical column. Comparing these basic technologies of energy capture mechanism of the OWC, it can be concluded that the performance of any of these devices will be influenced by the water depth and wave direction in addition to the wave height and wave period, which are used to estimate the resource potential. For example, a spar buoy will be more suitable for offshore locations where the water depth can accommodate the draft of the elongated vertical tube with more powerful waves, while the spar buoy is not sensitive to wave direction [30]. The sloped configuration will be more suitable for regions closed to the shore with considerable water depth so that it can float and capture both the surging and heaving waves. Meanwhile, since the sloped buoy is sensitive to wave direction, nearshore areas with more unidirectional waves will make sloped buoy perform better.

Most of the fixed OWC wave energy converters were located on either the coastline or the nearshore areas. Fixed structures are more practical and convenient in these areas because ease of installation, operation, and maintenance leads to lower costs. In addition, power transmission facilities for such devices are easier to be installed compared to devices located in the deep offshore areas. The main disadvantage in these low depth regions is that the wave possesses less power due to wave breaking. The structure (Figure 1a) is designed in such a way that an opening is submerged below the water surface in which the incident waves oscillate the air trapped in the oscillating column. Prototypes of the fixed OWC device have been installed in Norway [25], Portugal [76], Iceland [77], Italy [78], Spain [79], Japan [26], etc. One of the major advantages as discussed earlier is that such systems can be integrated into existing or planned coastal structures, such as breakwater and retaining walls. So construction costs can be shared since the civil design and construction costs take the greatest portion of fixed OWC's cost besides the well turbine.

Although the energy conversion efficiency was very low, buoys were first deployed in Japan to capture wave energy [30]. As the name implies, these devices are usually allowed to float on the sea surface, and they are moored loosely to a fixed point to allow free oscillations. Similar to the fixed OWC devices, the opening of the water column was normally submerged below the sea surface during operation. Experimentally modifications were made to increase the amount of power capture, e.g., a backward bent duct buoy (Figure 2) in which the opening of the oscillating column faced backward in relatively to the direction of the incident wave [74]. The spar buoy was also a modification of the floating OWC wave energy converter that can be in sloped or straight configuration [30].

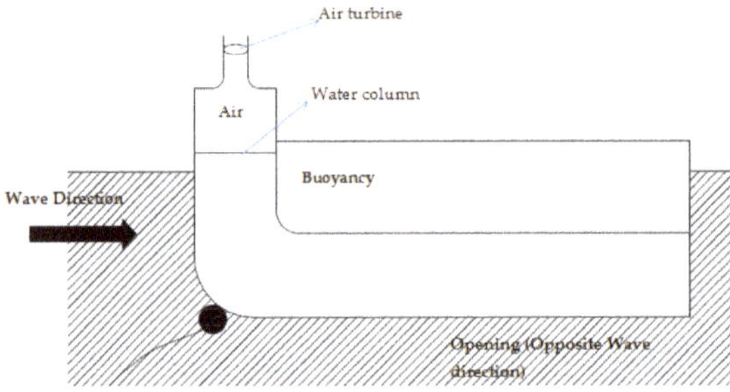

Figure 2. Schematic Representation of a Backward Bent Duct Buoy (Floating Oscillating Water Column (OWC)) [75].

When the first OWC wave energy converter was deployed by Masuda in the 1940s [30], the theoretical knowledge of the OWC's hydrodynamics was still limited that caused very low power absorbing efficiency. Since then, various modifications have been made in order to improve the power capturing efficiency of the OWC device. The spar buoy [74] and BBDB [75] were examples of shape modifications that were made to improve the power capturing efficiency. With the increased knowledge about the power capture of an oscillating system through its interaction with the ocean waves (more discussion in Section 4.2) and the increased availability and capability of computational fluid dynamics tools and numerical methods, the efficiency of the OWC device could be estimated even at the design stage. The prototype also allows the comparison between theoretical simulated performance with experimental tests. Using the Pico OWC prototype [80], its annual efficiencies in different sea states with different turbine sizes were determined using stochastic modelling approach. The efficiency ranged between 15% and 40% as recorded. The work by [81] also estimated the power performance of an OWC proposed in Spain at 2011. Using data from this site, the monthly average performance of the OWC was estimated using an integrated approach included both the absorbed energy from the waves and the efficiency of the turbines. Turbines with different rated capacities were used. The results showed that the higher rated capacity turbines had lower capacity factor than the lower rated capacity factor turbines. However, the results did not show the efficiency of the device in relative to the total available energy in the region. In other studies [82,83], numerical methods were used in the optimization of OWC that focused on the initiation of damping in the turbine in order to increase energy capture rate.

3.2. Overtopping Devices

The overtopping devices capture water close to the wave crest and introduce it into a chamber that stores water at a higher level than the average sea water level. The potential energy in the stored water is converted to useful energy by a low head hydraulic turbine as shown in Figure 1b. The hydrodynamics of overtopping devices is non-linear, so its power capture is not subjected to the hydrodynamic theories described in Section 4.2. Tapchan [84] is an example of WECs based on this technology, while the more popular ones are Wave Dragon [85] and Seawave Slot Cone Generator [86]. Overtopping devices can be used on the shoreline when they are integrated with breakwater facilities. However, due to the reduction in wave heights as the waves travel onshore from offshore, more power would be collected by the floating types installed offshore since the wave is more powerful, such as Wave Dragon.

Modifications have been made in the Tapchan [84] to the collecting channel, which allowed the approaching wave height to be increased as the collecting channel narrowed from the mouth towards the reservoir end where the propagating waves were concentrated. This increased the potential energy available for the low head hydraulic turbine. The case of the Tapchan exemplifies the different modifications and designs made to the basic structure of a typical overtopping WEC in order to increase the potential energy input into the turbine and provide a steady supply of water into the reservoir. Another example is the Wave Dragon (Figure 3) [85], which has two reflectors in the collecting structure to concentrate the incoming waves towards the reservoir. And the Sea Wave Slot Cone Generator [86] has a slopping wall to introduce the collected water in the reservoir smoothly into the hydraulic turbine.

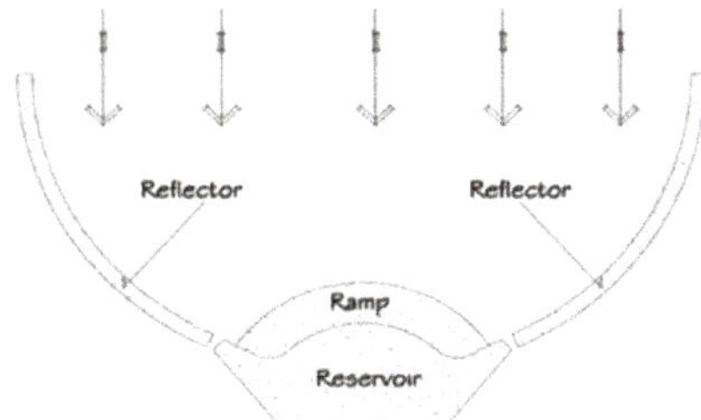

Figure 3. Schematic Representation of Wave Dragon [85].

Basically, the geometry of the reflector and the turbine efficiency determine the overall performance of the overtopping device. The overtopping device has the advantage that its power conversion process is one-step, i.e., the turbine converts energy directly from the captured sea water. Investigating the energy conversion from the captured water is the first step to evaluate the overtopping device. The tests performed on a Wave Dragon prototype by [87] for about 18 month showed overall average energy conversion efficiency around 12%.

3.3. Oscillating Body Systems

Oscillating body system moves with the motion of the ocean waves, which can be translational (mainly heave) or rotational (mainly pitch) as shown in Figure 1c. It can either be floating on or submerged in the ocean. Depending on the wave energy capture concept, oscillating modes in the surge/sway, heave and pitch are possible. This capture method is more suitable for regions with considerable depth because these regions have more powerful waves and the bottom of the floating device will not scour the seabed. This method could capture maximum power when the body is in resonance with the ocean waves [88].

Heaving systems can be single body heaving buoys [89], in which a heaving body moves relatively to a fixed frame of reference. The relative motion between the oscillatory part and the fixed part usually drives a power take-off system or a turbine. Different hydraulic fluids were used in this system, which was discussed in subsequent sections. Most of the heaving systems were connected directly to a linear generator [89] as shown in Figure 4a. A model with 10 kW power output deployed in Oregon (Figure 4b) to undergo series of tests had 3.5 m radius and 6.7 m spar length [90]. Another concept is the multibody system, which was designed to solve the problem that may arise in deep offshore areas as it is difficult to have a body reacting against another structure fixed to the ocean floor due to depth. In a multibody system, the energy was produced from the motion of two connected bodies that oscillates out of sync [30]. Some prototypes based on this concept had been tested to produce useful energy, including AquaBuoy [91], Interproject Service (IPS) buoy (Figure 5a) [92], and Wave Bob (Figure 5b) [93].

Heaving systems can also be fully submerged in which the bottom part is fixed to the seabed and the upper part oscillates vertically according to the wave action. The motion of the buoyant part drives a generator situated in the bottom part of the device. A device can also be hinged to the seabed while its top made of a buoyant material oscillates in response to the action of the waves. A very good example is the Archimedes Wave Swing (AWS) [94] as shown in Figure 5c, which used same concept similar to a single heaving system but fully submerged.

The mechanism of single body or two bodies heaving systems make them unsuitable for shoreline areas. Depending on their sizes, a single heaving body and submerged system can be adapted for nearshore areas. However, the captured energy will be reduced since less energy is captured in the heaving mode. They are much more suitable for offshore areas.

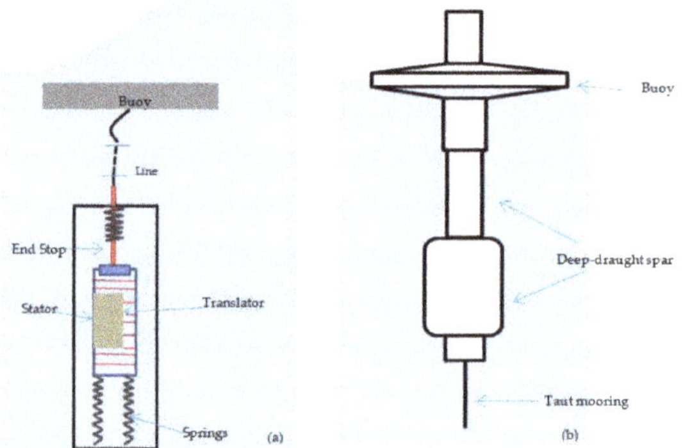

Figure 4. Schematic Representations of (**a**) a single heaving buoy developed in Sweden [89] and (**b**) single heaving body system developed in Oregon University, USA [90].

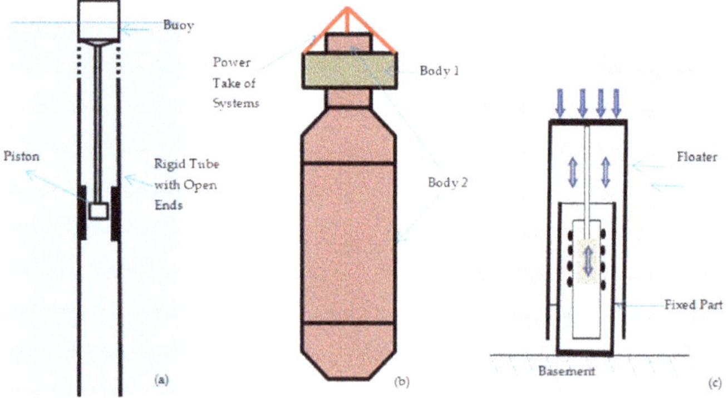

Figure 5. Schematic Representations of the 2-Body systems: (**a**) IPS buoy [92], (**b**) Wavebob [93], and (**c**) Archimedes Wave Swing (Submerged) [94].

Heaving systems interact with waves to produce motion that drives a power take-off (PTO) system to generate useful energy. Optimization of this body motion will lead to increase in captured energy. The hydrodynamics of oscillating bodies leads the fact that the optimum energy is absorbed when the body oscillates at resonance with the incoming waves. The controllable parameters related to the hydrodynamics of an oscillating body are the inertia, geometry, and the PTO damping [95–97]. The optimization approach adopted by [95,96] was to tune the geometry of the WEC to capture most available energy in a particular wave climate. The major limitation of this method is that the optimum capture bandwidth was narrow and also it was location specified. The power capture was improved by varying the mass of device to enable it resonate with the incoming waves [97]. However, practical methods to achieve the improvement were not discussed while only theoretical calculations were presented. The optimization of geometry, shape, and inertia provides a good way of increasing the captured energy of an oscillating body. The main challenge is to practically change these parameters during operation so that the optimum energy capture bandwidth could be increased.

3.4. Rotational (Pitching) Systems

Because of the methods that most WECs were moored, other motion directions (surge/sway and pitching/rolling) are possible. Some WECs have been made to convert energy based on pitch motion, e.g., the Duck [1]. Another successful WEC that converts energy based on the pitch motion is the Pelamis (Figure 6c) [98], which is the first grid connected WEC. The Pelamis is the most studied among all the pitch motion based WECs. It was usually moored loosely with sections hinged together. These hinged sections were aligned with the wave direction. When the joints experience wave action, they were resisted by the hydraulic rams that pressurized the hydraulic fluids through a motor to drive an electric generator. Another example is the Searev (Figure 6b) [99], which was a bottom hinged device suitable for nearshore areas. The Searev was hinged to the seabed with the top part moving in a rotational manner in response to the wave action. Others pitch motion based WECs include McCabe wave pump [30], wave roller [30], Oyster [100], The Mace (Figure 6a) [101], etc. The pitching devices can be adapted for both offshore and nearshore areas since the pitch motion is a quasi-surge motion, which can benefit from the concentration of ocean waves direction in the nearshore region and produce more power from the surge motions.

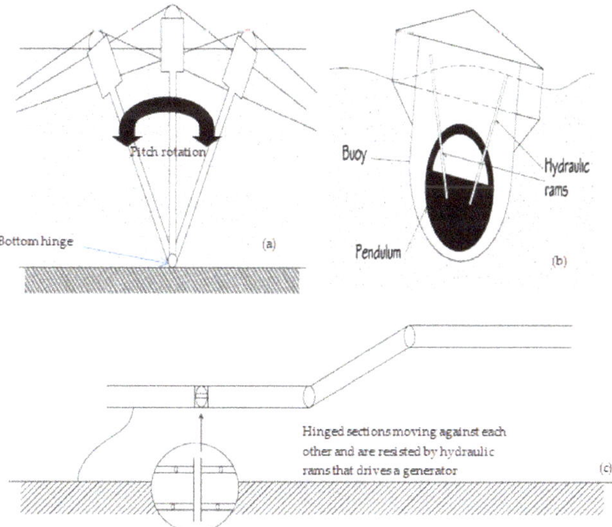

Figure 6. Schematic Representations of the 2-Body systems: (**a**) The Swinging Mace [101], (**b**) Searev [99], and (**c**) Pelamis [102].

As discussed above, Pelamis is one of the most studied WECs due to the fact that it is the first commercially operated WEC. Despite the cessation of operation in 2014 when its parent company went into administration [73], Pelamis had provided researchers a benchmark for the evaluation of wave energy extraction in different locations. Tests of energy extraction based on scaled prototype [98,103] and simulations by numerical methods and computer tools based on the Pelamis concept have been reported. From those studies, it can be seen that the power take-off system could act as a mechanical damper to not only transmit absorbed energy but also to regulate the interaction between the device and the ocean waves in order to optimize the total useful energy captured by Pelamis [104,105].

Table 3 summarizes different WEC concepts with aforementioned examples.

Table 3. Summary of Wave Energy Converters with Examples.

Wave Energy Harvesting Method			Examples
Oscillating Water Column	Fixed Structure		Pico [76], LIMPET [77], Sakata [26], REWEC3 [78], Mutriku [79]
	Floating Structure		Mighty Whale [106]
Oscillating Bodies	Floating	Heave	IPS buoy [92], Wavebob [93], AquaBuoy [91]
		Rotation	Pelamis [98], SEAREV [99]
	Submerged	Heave	AWS [94]
		Rotation	Wave Roller [30], Oyster [100], The Mace [101]
Overtopping Devices	Fixed Structure		TAPCHAN [84], SSG [86]
	Floating Structure		Wave Dragon [85]

3.5. Performance Comparison of Different WECs

The non-convergence of energy conversion technologies has created some challenges in terms of comparing their performances and efficiencies. Many researchers introduced different devices claiming increase in performance through optimizing some parameters. However, the question remains—what is the benchmark of quantifying or qualifying such increase? As discussed in Equation (1) of Section 4.2, maximum theoretical energy absorbed by an oscillating body depends on the direction of its motion, so will it be appropriate to compare the efficiencies of designs with the same or similar parameters under different oscillating directions? It is one of the questions to be answered when comparing the performance of these devices.

Since the aim of different WECs is to reach competitive technical and commercial level, parameters related to this aim should form part of the framework for performance comparison of different WECs. One of such attempt was by [107], who compared the performance of Wavestar (heaving) and Pelamis (pitching) at different locations in Europe. Multiple power ratings of the devices were tested while incorporating variables such as availability and capacity factors. Their results showed that Pelamis performed better in regions with high wave energy potential while Wavestar performed better in regions with low wave energy potential. A more detailed comparison analysis was performed by [108] as shown in Figure 7, in which eight different WECs were investigated including heaving systems, surging systems, and oscillating water columns. Indices used to compare performance included absorbed energy per unit weight of WEC, absorbed energy per surface area, and absorbed energy per power take-off force in five different test sites in Europe. The same methodology was used by [109] in an expanded setting as the WECs' performance were tested in selected regions of high wave energy density in Europe, Africa, Asia, America and Australia. Comparison of efficiency through the power absorbed per weight, surface area or PTO force is a good comparison technique, and can also be expanded to include costs when more technically reliable capture methods become available.

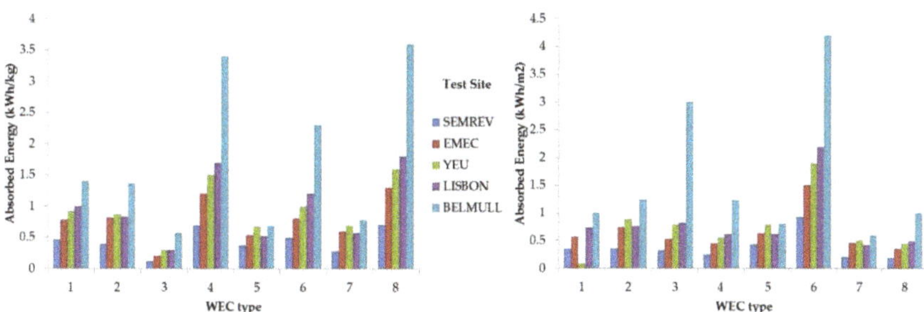

Figure 7. Comparison of Different WECs' performance [108] based on: (**a**) absorbed energy per WEC weight, and (**b**) absorbed energy per contacted surface area. (WEC type: 1 = heaving buoy, 2 = submerged heaving buoy, 3 = 2-body heaving buoy, 4 = fixed buoy array, 5 = floating buoy array, 6 = fixed oscillating flap, 7 = floating 3-body oscillating flap, and 8 = floating OWC).

4. Wave Energy Converter Design

When designing a wave energy converter, the type of harvesting technique should be chosen first. To decide the harvesting technique, one should consider characterization of the ocean wave energy properties and resources, site assessment, and other factors that determine the survivability and successful operation of the WEC in the open sea. For the energy capture of a WEC, numerical and theoretical analysis can be applied to determine the magnitude of the captured energy. Basic theoretical analysis has limitations since real waves are nonlinear and vary greatly both temporally and spatially. In addition, viscous and eddy losses cannot be adequately captured using theoretical analysis. Despite these limitations, basic theoretical analysis forms the first step to be considered when designing a WEC.

Most WEC designers first designed and manufactured laboratory models in wave tanks after the harvesting concept was determined. These scaled models were first tested in the laboratories to analyze their performance. Some after successful tests progressed to testing of prototypes in large wave tanks [30], which were reported in Norway and France. While the testing of laboratory models gave a good insight of a WEC's performance, it was still inadequate to fully predict the WEC's performance in the ocean environment. For example, the OSPREY was destroyed at sea not long after it was deployed [30]. Most WECs did not perform optimally in the open sea compared to the capturing efficiencies reported in the laboratories. Most designs and tests were presently at the model stage, and focused on the optimization and the power take-off of the energy captured by the WECs. Structural reliability studies related to WEC designs were still very few. The work done by [110] investigated the failure mode in a laboratory scaled model of a modified point absorber using the failure mode and effect analysis (FMEA) method. Using a laboratory scaled model, the components of the WEC device were analyzed separately, including buoy, bearings, nuts and bolts, fastenings, etc.

4.1. Model Testing of WECs

The design and testing of the WECs have been carried out by different developers and researchers. However, testing methods are as diverse as the capture technologies since there is still no convergence of the ocean wave capture technologies. While some tests were based on very small-scale laboratory models, some tests were conducted in larger wave testing facilities on bigger scale models that were lower in magnitude up to a factor of three or four of the designed WEC. The performances of WECs can also be estimated based on complex mathematical models through powerful computer simulations. Meanwhile, testing methods are significantly influenced by the wave energy capturing methods. For example, setting up WECs to be installed in breakwater or shoreline areas may be more difficult and expensive to be replicated in the laboratory scale, while a floating system can easily be scaled down for laboratory tests. Some tests were performed considering a WEC as a whole completed device, while some tests focused on different modules of a WEC separately, such as PTO, electrical systems [103,110,111], etc. No matter which testing mode or method is used, it is affected by the aim of the developers/researchers, cost, harvesting technology, and available facilities.

While the testing of small scale model gives a lot of insight about behavior and performance of the device, there are limitations of the method because as the uncertainties in the results greatly increase as the wave properties and size of device increases. With the increase in the number of test sites around the world, larger scale prototypes can be tested now [112]. To reduce the uncertainties associated with small scale tests and cost with large scale tests, it is possible to first used numerical and analytical tools to analyze the survivability, structural reliability, stability, and efficiency of a laboratory scaled model [113]. They compared the results with experimental studies, and then tested a larger scale using real ocean wave conditions in which they also analyzed the maintainability of an oscillating surging WEC [113]. The numerical simulation covered the effect inertial force, drag forces, hydrodynamic radiation and diffraction effects. The complex nonlinear effects such as turbulence, wake impact and flow separation could also be considered. One major advantage of the computer simulations is that changes can be made to the shape of a WEC and wave conditions easily. The results of the simulation of the motion flab angle showed very good convergence between the experimental models and computer simulation model. However, some noticeable differences were seen between the time histories of the sensors embedded in the WEC with similar trend.

The classical approach of WEC design, testing, and deployment was adopted by [87]. They worked on a wave dragon that is an overtopping device. A laboratory model was first tested for about three years. A prototype was then deployed in the open sea with extensive monitoring and measurements on its efficiency, behavior, and survivability for two years. The prototype tested was connected to the grid, and fully equipped with turbines, mooring systems, etc. The prototype is 4.5 times lesser in size than the proposed WEC design. Although they differ in size by a factor of about 12, results from the hydrodynamic performance showed a correlation between the laboratory scale and the prototype model. The whole testing process gave very good feedback to understand the WEC operations, however, the total time spent (about five years) in the testing process is too long.

Another study [114] used numerical analysis to calculate the power absorption under real ocean wave conditions with the equations optimized using MATLAB. A new capturing concept was proposed, which used the pressure differences occurred at various points in the wave field to drive a fluid flow. They compared two different designs of the concept, one having the moving parts at the bottom of the air chambers while the other one having the moving parts at the top. Using purely numerical methods, the device had an estimated yearly capacity factor greater than 50% when it was simulated in a nearshore wave condition of 16.8 kW/m generating 82 kW out of 150 kW rated power.

The characterization, design, and testing of WECs have moved gradually from extensive physical testing of very small scales in the laboratory to using numerical methods and computer simulations with powerful computing tools. This not only saves a lot of time, it also saves costs and is more efficiency in the design and testing process. New computing tools take into consideration complex process in the interactions between WEC and ocean waves that are nonlinear in nature. Large multi-dimensional wave conditions can now be modelled easily. Therefore, real ocean wave conditions can be incorporated into the early stages of design and testing.

Some works used combinations of both numerical and computational analyses. The numerical and computational combination method will not in any event replace the testing of either prototypes or laboratory scale model. Instead, it will act as a compliment to each other. Hence, testing time will be greatly reduced with more accurate understanding of the behavior of a WEC in real ocean conditions.

4.2. Hydrodynamics of Wave Energy Converters

If all the wave energy resource potential at a particular location and time is absorbed and converted, there will be no more waves to continually power the device. In fact, this scenario is theoretically impossible. A lot of theories have been propounded to describe the relationship between the WECs and the sea waves [30]. Apart from the overtopping devices, the rest of the WECs utilized the oscillatory motions to generate energy. The oscillatory motions could be either the oscillatory mode of the body relative to the ocean waves or the oscillatory mode similar with the OWCs in which the water column oscillates in a column chamber. All oscillating bodies employed one or more oscillatory mode. The Oyster and Wave Roller were primarily pitching devices, but they also experienced some quasi surge motions and were usually hinged to the sea bottom (Figure 1c). On the other hand, the Pelamis experienced solely the pitching motion while hinged at the sea surface, and the AWS oscillated only in the heave mode.

The motion of any oscillating body at sea can be described by the Newton second law of motion [30] as expressed in Equation (1).

$$\sum_{j=1}^{6} \left(M_{ij} + A_{ij} \right) \ddot{x}_j + B_{ij} \dot{x}_j + C_{ij} x_j = F_i, \ i = 1, 2, \ldots 6 \tag{1}$$

where i and j are subscripts representing the hydrodynamic properties in the ith mode due to the motion in the jth mode. M_{ij}, A_{ij}, B_{ij}, C_{ij} and F_i are the mass, added mass, radiation damping, hydrostatic force, and external forces matrices, respectively. The external forces include the excitation forces, PTO forces, viscous forces, etc. The value x represents displacement and its derivatives with respect to time. Using a one-dimensional analysis in a linearized frequency domain, Equation (1) can be written as

$$(m + A)\ddot{x} = f_d - B\dot{x} - \rho g S x + f_{PTO} \tag{2}$$

In Equation (2), the viscous effects are neglected. f_d is the excitation force that is zero in calm water. S is the cross-sectional area of the body on the free surface plane while PTO force is represented by f_{PTO}. The PTO force f_{PTO} equals $-C\dot{x} - Kx$. C is the linear damper coefficient, and K is the stiffness of a linear spring. Assuming a regular wave of frequency (ω), $\{x, f_d\} = Re(\{X, F_d\}^{i\omega t})$. X and F_d are complex amplitudes. Based on Equation (2), it can be obtained as

$$X = \frac{F_d}{-\omega^2(m + A) + i\omega(B + C) + \rho g s + K} \tag{3}$$

The absorbed power averaged over a given period of time is given as $\overline{P} = \overline{f_d \dot{x}}$, and the maximum power occurs when

$$\omega = \sqrt{\left(\frac{\rho g s + K}{m + A_w} \right)} \text{ and } C = B(\omega) \tag{4}$$

The condition in Equation (4) is referred as the resonance condition, and ω at that point is the resonant frequency. Some computer applications such as WAMIT developed by the National Renewable Energy Laboratory (NREL) in the U.S. and ANSYS/AQWA [108] can be used to obtain the stiffness, excitation forces, added mass, and the damping coefficient in any body with any geometry and degrees of freedom of the body's motion.

The consequence of the Equations (1) to (4) above is that maximum theoretical power to be extracted from a sea wave has a relationship between the wave energy fluxes per unit length of a wave crest (the resource potential) as described in the following equations [115–117].

$$L_{max} = \frac{P_{max}}{E} = \frac{\lambda n}{2\pi} \tag{5}$$

$n = 1$ if body is in heave motion;

$n = 2$ if body is in surge motion

P_{max} = Maximum Extractable power;

L_{max} = Absorption width at maximum power

λ = ocean wavelength;

$$E = \text{resource potential} = \frac{\rho g^2 T H^2}{64\pi} \tag{6}$$

ρ = seawater density;

g = gravitational acceleration; T = wave period

H = significant wave height

$$\text{Efficiency } \eta = \frac{P}{P_{max}} \tag{7}$$

P is the average power produced at a particular period.

This relation is similar to the Betz limit for the power coefficient in wind energy systems. One of the assumptions of these equations is that the sea wave is regular with small amplitude and oscillates in a single mode so that the power take-off is linear. Suffice to say, the WEC should resonate with the sea waves in order to reach P_{max}. However, it is almost impossible since sea waves are polychromatic in nature. Meanwhile, the extraction technologies have their unique motion modes that may affect the actual captured energy. According to Equation (5), a surging body can theoretically capture twice of what a heaving body does if both devices are operated under similar wave conditions.

From the hydrodynamics analyses of WECs, particularly oscillating bodies, it seems that achieving the optimum energy capture is impossible because resonance of the WEC device will only occur in a narrow wave period bandwidth due to multiple waves and periods in the ocean. A lot of studies [104,118–121] have been performed in order to optimize and control the behavior of a WEC to regulate the ocean energy conversion process in complex ocean conditions. Different control methods were studied. The optimum phase control regulated the reactive power in order to achieve maximum power [122]. Latching [118] is a partial phase control method that was proposed theoretically in the late seventies through to the late eighties [121–128], and is more popular in the design of most proposed WECs. However, the wave conditions of latching method must be predicted in advance in a time frame of about half of the resonant wave period of a WEC, which is very difficult given the high variability of real sea waves conditions.

4.3. Power Take-Off (PTO)

The conversion of wave energy to useful energy is the main goal of the wave energy conversion process. Most of the focus in the design of a WEC is to make power available as electric power to be transferred into the utility grid. There have also been some studies on small to medium scale systems as stand-alone systems, where the captured energy is consumed in-situ or very close to the harvesting area with very little or no need of extensive storage and transmission systems. There are many different existing technologies for wave energy conversion, the choice of power take-off is diverse as well. However, some power take-off systems are standard, and the others are usually modifications and upgrades of the standard systems.

The OWC is one of the oldest wave capture technologies, and the first set of WECs were designed based on this technology. The major challenge the OWC was that the air flow due to the oscillating column of water caused by the wave action was reciprocating in nature. This problem was solved with the invention of the Wells turbine [129,130]. The Wells turbine is an axial flow turbine that can self rectifies. Therefore, the torque of the turbine is not affected by the airflow's direction. Since the first invention and deployment of the Wells turbine, different upgrades and modifications have been performed on it to improve its efficiency. Most OWC preferred using the Wells turbine because the ratio between the airflow and blade velocity was high while the peak efficiency was considerably high and the cost of manufacture was low. Some of the disadvantages of the Wells turbine include (1) the torque can be very low when the airflow is low; (2) their sizes are relatively large when compared to other turbines with similar power ratings; and (3) they are susceptible to aerodynamic noise [131–136]. Other turbines with similar working process of the Wells turbine are the self-rectifying impulse turbine [137–142] and the Dennis-Auld turbine [143]. Compared to conventional turbines, these self-rectifying turbines' performance were relatively acceptable if the performance was averaged over a given period of time considering the fact that they were operating under a highly unsteady state environment.

The hydraulic turbines used in overtopping devices were similar to the ones used in the low head hydroelectric [144,145] because their modes of operation are very similar. The Pelton impulse turbine [146,147] is suitable for use by oscillating bodies as well as overtopping devices because they can use water as an alternative to hydraulic driven motors. Majority of oscillating body conversion systems used the hydraulic oil. They also have gas accumulators that can store energy for some time [105]. This type of system was used in the Pelamis. A comprehensive review of hydraulic systems used in wave energy converters can be found in [104].

Another type of power take-off is the linear electrical generator. This method converts mechanical wave energy directly to electrical energy, so energy loss due to friction from moving parts is greatly reduced. This method was mainly used in oscillating body systems. The downside of this method is that the requirement of using heavy magnets causes increase in mass of the whole system and reduction in efficiency due to the low velocity of wave oscillations. With the increase in popularity of oscillating body systems, the use of linear generator has been proposed for the power take-off of new WEC designs. Extensive reviews on the use of linear generators can be found in [90,91,148–152]. There are ongoing studies to design new systems, improve existing power take-off systems, and develop hybrids of existing methods [153–156].

The energy desired mostly from the wave energy conversion process is the electricity. When connecting into the electricity grid, the requirements and regulations may vary from countries to countries and even vary on sub regional cases. The WEC systems have to be modified to meet these requirements. Like energy from other distributed generation sources, wave energy conversion systems produce energy in an irregular and intermittent way, and needs to go through some conditioning before being fed into the electric grid as regulated in IEEE-1547-2003 [157]. According to [104,158], they have identified three main requirements for a reliable WEC system, including

- a good power take-off system to convert mechanical power to electricity,
- regularization of the unstable electricity to meet grid requirement or meet the electrical load if it is to be supplied to a stand -alone system, and
- the power electronics to ensure the quality of power at the user's end.

Similar to other distributed generation sources, power conversion can be DC/AC, DC/DC, AC/DC, AC/DC/AC and AC/AC. A lot of studies and reviews about wave energy electrical systems can be found in [119,158–168]. Table 4 summarized examples of WECs with their PTO systems.

Table 4. Examples of some Wave Energy Converters with their Power Take-Off (PTO) Systems.

WEC Name	WEC Type	PTO	Remarks
Pico [76]	Oscillating water column	Wells turbine	Horizontal axis turbine used
LIMPET [77]	Oscillating water column	Wells turbine	*
REWEC3 [78]	Oscillating water column	Wells turbine	Multi turbines with 136 chambers
Mutriku [79]	Oscillating water column	Wells turbine	Multi turbines with 16 chambers
Mighty Whale [106]	Oscillating water column	Wells turbine	-
Sakata [26]	Oscillating water column	Wells turbine	Two units installed in breakwater
IPS buoy [92]	Heaving body	Hydraulic oil system & linear generator	Tests performed using the two listed PTO systems
AquaBuoy [91]	Heaving system	Impulse turbine	-
Wavebob [93]	Heaving body	Hydraulic System	Biodegradable hydraulic fluid
SEAREV [99]	Pitching system	Hydraulic System	Hydraulic pumps connected to the pendulum wheel charged the high pressure accumulators, which discharged their energy into hydraulic engines to drive electric generators
Pelamis [98]	Pitching system	Hydraulic System	Hydraulic pumps utilized the bending motion (caused by the passing waves) inside the joints
Oyster [100]	Pitching/quasi surge system	Hydraulic System	Bottom mounted WEC with water-pump serving as PTO
Wave Dragon [85]	Overtopping device	-	Turbines were used to convert the low head of the water into mechanical energy
TAPCHAN [84]	Overtopping device	-	Low head turbines and conventional hydroelectric turbines of appropriate size

5. Challenges of Wave Energy Systems

The challenges of WECs range from the techno-economic problems to issues affecting its operation and maintenance [168] in the harsh ocean environment due to ocean salinity and extreme weather conditions [169]. The environmental impacts of wave energy converters might not be easy to evaluate since very few if at all any device have been deployed long enough to comprehensively determine the environmental impacts both at the coastal areas and offshore locations. Although there are many existing studies focusing on technical aspects of WECs, different capture technologies also face different challenges.

One of the major challenges militating against WEC design and deployment is that the technologies are still at a nascent stage compared with other matured renewable energy technologies such as wind and solar [170–173], despite many prototypes and patents reported in so many literatures. Different technologies are being considered for harvesting wave energy, which also means that there has not been any convergence of wave conversion technology. Therefore, it is difficult to predict the challenges and address them at the design stage of a WEC. Moreover, even the most successful capturing device (Pelamis) to date has not achieved the same levelized cost of energy with wind or solar energy.

5.1. Design, Installation and Operation

The design, operation and installation of any structure or facility in the ocean environment are always a challenge compared to land-based structures. The case of the WEC design is even more pronounced as the WEC is expected to actively interact with the ocean waves in a specific way to harvest energy from it. Apart from the WEC being able to withstand operational loads, its performance and survival under extreme loading conditions during events such as hurricanes and storms are highly important. The corrosive nature of sea water [174–176] can be problems for some parts of a WEC as well. The implication is that a comprehensive operation and maintenance strategy have to be planned for WEC systems at the design stage, which will definitely incur additional cost to the lifecycle cost.

Another important consideration for operation and maintenance planning activity is the accessing of the facility if it is offshore [177–181]. The experience gained from the offshore energy industry including offshore wind and oil and gas industries [182] can help to appreciate the risks and costs associated with maintaining an offshore facility. Because of this, a system in which maintenance activities are well spaced will be a good option, limiting costs associated with mobilizing crews to the facility many times. Compared with WECs in deep offshore locations, issues associated with construction are reduced for WECs to be located in coastal and nearshore areas, and the accessibility is also easier. However, nearshore WEC structures have to compete with other coastal uses such as recreation, docks, etc. If the WEC can be incorporated into existing structures, such as breakwater structures, costs can be shared and minimized.

5.2. Environmental and Other Issues

The impact of WEC structures and its operation on its immediate environment needs to be investigated. Similar with all renewable energy system, its operation produces little or no greenhouse gas that makes WEC environmentally friendly in the first place. However, analysis should be conducted to ensure that the WEC facility have little or no negative effect on the aquatic plant and marine life in the place where WEC is going to be deployed. It should be noted that WECs on shoreline areas will reduce shoreline erosion because energy have been extracted from the waves [15]. There have been concerns that WEC structures can introduce new species into the immediate environment and also become artificial reefs [183] thereby causing a shift in balance of the local ecosystem. For systems that use hydraulic fluid such as oscillating body systems, precaution measurements should be taken to prevent leakages or better make the hydraulic fluid biodegradable. The majority of environmental impacts of WECs may occur during the construction and installation stage. Drilling, dredging and other construction activities can cause pollution or generate unbalance in the natural habitat of the ocean plants and animals. Other adverse effects on marine life could be trapping, collision, etc. [184]. The effect of seaweeds, biofouling on the WEC structure, can cause reduction in performance and also accelerate structural degradation [185–188].

Ocean wave conditions are hard to predict and power production is intermittent. The power density depends on wave height and wave period that increases the challenge in the prediction of expected power production. Other problems that might arise for the WEC system is the transmission of the produced energy because electricity grid facilities may be almost non-existent where the power is produced. Where there is a developed offshore wind energy system, those facilities may be made available for WEC systems. Another issue is the 'aesthetics', especially for shoreline devices since the local coastal community may want to resist the installation of a large structure close to them. In addition, most of the structural designs performed on WECs were completed in isolation of the power capture. WEC designers should to be careful when doing it since changes on some of the structural parameters could affect the power capture. For example, the mooring system of a heaving device goes a long way to affect the damping coefficient, which in turn affects its energy absorbing capability.

6. Conclusions

Many studies are currently taking place around the world, especially in the United States, Europe, and China, on the wave energy systems as it can be seen from the volume of literatures published from these regions. Upon all these studies about different conversion technologies, none have really made it to large commercial stages when compared to wind and solar energy. There has been progress in theoretical studies, experimental and model testing of WEC prototypes in the laboratory, however, translating these successes to actual field deployment have been very difficult. The developments in the methods for wave resource potential characterizations have been very encouraging since there are powerful computer software and applications available now to predict and simulate wave conditions with finer temporal and spatial resolutions. But most of the characterizations were still being done using static methods. Therefore, the complexity of the wave energy temporal and spatial variability was not adequately captured due to the dynamic nature of these properties.

Recent advances in the use of numerical simulations with computational fluid dynamics have seen the use of 'testing' large scale prototypes while incorporating complex interactions between the WEC and the sea waves. This method has been applied to existing technologies and the new ones that are modified from the existing capture methods. The use of both the simulations and laboratory tests will lead to higher efficiency in the nascent wave energy industry. Actually, no 'champion' may not necessarily emerge from the present competing wave energy capture technologies. Instead, all capturing methods may be combined to optimize the wave energy capture in a particular wave energy farm. Future studies will focus on the best placement of WECs in a typical wave energy farm.

When there is a breakthrough in any of the presently wave capture technologies, the challenges that may arise due to construction, operation and maintenance, wave energy farm optimization, lifecycle costs, return on investments, environmental issues, socio-economic issues, and permits will be fully studied. Until then, these challenges that are presently at the level of hypothesis and predictions will remain so.

Author Contributions: Tunde Aderinto wrote the initial draft paper under the supervision of Hua Li. Hua Li made major revision on the initial draft paper, and approved the final version to be published.

Acknowledgments: The authors are thankful to the support from Texas A&M University-Kingsville and National Science Foundation (award # EEC-1359414).

References

1. Clément, A.; McCullen, P.; Falcão, A.; Fiorentino, A.; Gardner, F.; Hammarlund, K.; Pontes, M.T. Wave energy in Europe: Current status and perspectives. *Renew. Sustain. Energy Rev.* **2002**, *6*, 405–431. [CrossRef]
2. Falnes, J. A review of wave-energy extraction. *Mar. Struct.* **2007**, *20*, 185–201. [CrossRef]
3. McCormick, M.E. *Ocean Wave Energy Conversion*; Courier Corporation: North Chelmsford, MA, USA, 1981.
4. Pelc, R.; Fujita, R.M. Renewable energy from the ocean. *Mar. Policy* **2002**, *26*, 471–479. [CrossRef]
5. Masuda, Y. An experience of wave power generator through tests and improvement. In *Hydrodynamics of Ocean Wave-Energy Utilization*; Springer: Berlin/Heidelberg, Germany, 1986; pp. 445–452. [CrossRef]
6. Masters, C.D.; Root, D.H.; Dietzman, W.D. Distribution and quantitative assessment of world crude-oil reserves and resources. In *The Changing Carbon Cycle*; Springer: New York, NY, USA, 1986; pp. 491–507. [CrossRef]
7. McGlade, C.; Ekins, P. The geographical distribution of fossil fuels unused when limiting global warming to 2 [deg] C. *Nature* **2015**, *517*, 187–190. [CrossRef] [PubMed]
8. Grove-Palmer, C.O.J. Wave energy in the United Kingdom: A review of the programme June 1975 to March 1982. In Proceedings of the 2nd International Symposium on Wave Energy Utilization, Trondheim, Norway, 22–24 June 1982; pp. 22–24.
9. Previsic, M.; Moreno, A.; Bedard, R.; Polagye, B.; Collar, C.; Lockard, D.; Rocheleau, R. Hydrokinetic energy in the United States—Resources, challenges and opportunities. In Proceedings of the 8th European Wave Tidal Energy Conference, Uppsala, Sweden, 7–10 September 2009; pp. 76–84.
10. Annual Report 2015. International Energy Agency Implementing Agreement on Ocean Energy Systems. 2016. Available online: https://www.ocean-energy-systems.org/documents/82577_oes_annual_report_2016.pdf/ (accessed on 8 August 2017).
11. Wuebbles, D.J.; Jain, A.K. Concerns about climate change and the role of fossil fuel use. *Fuel Process. Technol.* **2001**, *71*, 99–119. [CrossRef]
12. Thorpe, T.W. *A Brief Review of Wave Energy*; Harwell Laboratory, Energy Technology Support Unit: Didcot, UK, 1999.
13. Ilyas, A.; Kashif, S.A.; Saqib, M.A.; Asad, M.M. Wave electrical energy systems: Implementation, challenges and environmental issues. *Renew. Sustain. Energy Rev.* **2014**, *40*, 260–268. [CrossRef]
14. Cruz, J. *Ocean Wave Energy: Current Status and Future Prespectives*; Springer Science & Business Media: Berlin/Heidelberg, Germany, 2007; e-ISBN 978-3-540-74895-3.
15. Mustapa, M.A.; Yaakob, O.B.; Ahmed, Y.M.; Rheem, C.K.; Koh, K.K.; Adnan, F.A. Wave energy device and breakwater integration: A review. *Renew. Sustain. Energy Rev.* **2017**, *77*, 43–58. [CrossRef]
16. Wahyudie, A.; Jama, M.A.; Susilo, T.B.; Saeed, O.; Nandar, C.S.A.; Harib, K. Simple bottom-up hierarchical control strategy for heaving wave energy converters. *Int. J. Electr. Power Energy Syst.* **2017**, *87*, 211–221. [CrossRef]
17. Mork, G.; Barstow, S.; Kabuth, A.; Pontes, M.T. Assessing the global wave energy potential. In Proceedings of the ASME 2010 29th International Conference on Ocean, Offshore and Arctic Engineering, Shanghai, China, 6–11 June 2010; pp. 447–454.
18. Energy Information Administration. International Energy Outlook. Available online: www.eia.doe.gov/1eo/index.html (accessed on 12 April 2017).
19. IRENA. Wave Energy Technology Brief [Www Document]. 2014. Available online: www.irena.org (accessed on 10 April 2018).
20. Jacobson, P.T.; Hagerman, G.; Scott, G. *Mapping and Assessment of the United States Ocean Wave Energy Resource*; No. DOE/GO/18173-1; Electric Power Research Institute: Palo Alto, CA, USA, 2011.
21. Carbon Trust. *Accelerating Marine Energy CTC797*; Carbon Trust: London, UK, 2011.
22. The Crown Estate. UK Wave and Tidal Key Resource Areas Project—Summary Report [WWW Document]. 2012. Available online: https://www.thecrownestate.co.uk (accessed on 10 April 2018).
23. Younesian, D.; Alam, M.R. Multi-stable mechanisms for high-efficiency and broadband ocean wave energy harvesting. *Appl. Energy* **2017**, *197*, 292–302. [CrossRef]
24. Ocean Energy Systems. Annual Report Ocean Energy Systems 2016. 2017. Available online: https://report2016.ocean-energy-systems.org/ (accessed on 11 March 2018).

25. Bønke, K.; Ambli, N. Prototype wave power stations in Norway. In *Utilization of Ocean Waves—Wave to Energy Conversion*; ASCE: Reston, VA, USA, 1986; pp. 34–45.

26. Ohneda, H.; Igarashi, S.; Shinbo, O.; Sekihara, S.; Suzuki, K.; Kubota, H.; Morita, H. Construction procedure of a wave power extracting caisson breakwater. In Proceedings of the 3rd Symposium on Ocean Energy Utilization, Tokyo, Japan, 22–23 January 1991; pp. 171–179.

27. Falcão, A.F. *Modelling of Wave Energy Conversion*; Instituto Superior Técnico, Universidade Técnica de Lisboa: Lisboa, Portugal, 2014.

28. Sjolte, J. Marine Renewable Energy Conversion: Grid and Off-Grid Modeling, Design and Operation. Doctoral Thesis, Norges Teknisk-Naturvitenskapelige Universitet, Trondheim, Norway, 2014.

29. Whittaker, T.J.T.; McIlwaine, S.J.; Raghunathan, S. A review of the Islay shoreline wave power station. In Proceedings of the First European Wave Energy Symposium, Thorntonhall, UK, 1993; pp. 283–286.

30. Antonio, F.D.O. Wave energy utilization: A review of the technologies. *Renew. Sustain. Energy Rev.* **2010**, *14*, 899–918. [CrossRef]

31. Lehmann, M.; Karimpour, F.; Goudey, C.A.; Jacobson, P.T.; Alam, M.R. Ocean wave energy in the United States: Current status and future perspectives. *Renew. Sustain. Energy Rev.* **2017**. [CrossRef]

32. Filipot, J.F.; Roeber, V.; Boutet, M.; Ody, C.; Lathuiliere, C.; Louazel, S.; Suanez, S. Nearshore wave processes in the Iroise Sea: Field measurements and modelling. In Proceedings of the Coastal Dynamics 2013—7th International Conference on Coastal Dynamics, Arcachon, France, 24–28 June 2013; Available online: http://www.coastaldynamics2013.fr/pdf_files/055_Filipot_Jean_Francois.pdf (accessed on 12 December 2017).

33. Chen, Z.; Yu, H.; Hu, M.; Meng, G.; Wen, C. A review of offshore wave energy extraction system. *Adv. Mech. Eng.* **2013**, *5*, 623020. [CrossRef]

34. Veigas, M.; López, M.; Iglesias, G. Assessing the optimal location for a shoreline wave energy converter. *Appl. Energy* **2014**, *132*, 404–411. [CrossRef]

35. Gunn, K.; Stock-Williams, C. Quantifying the global wave power resource. *Renew. Energy* **2012**, *44*, 296–304. [CrossRef]

36. Izadparast, A.H.; Niedzwecki, J.M. Estimating the potential of ocean wave power resources. *Ocean Eng.* **2011**, *38*, 177–185. [CrossRef]

37. Mollison, D. Wave climate and the wave power resource. In *Hydrodynamics of Ocean Wave-Energy Utilization*; Springer: Berlin/Heidelberg, Germany, 1986; pp. 133–156. [CrossRef]

38. Lenee-Bluhm, P.; Paasch, R.; Özkan-Haller, H.T. Characterizing the wave energy resource of the US Pacific Northwest. *Renew. Energy* **2011**, *36*, 2106–2119. [CrossRef]

39. Iglesias, G.; López, M.; Carballo, R.; Castro, A.; Fraguela, J.A.; Frigaard, P. Wave energy potential in Galicia (NW Spain). *Renew. Energy* **2009**, *34*, 2323–2333. [CrossRef]

40. Citiroglu, H.K.; Okur, A. An approach to wave energy converter applications in Eregli on the western Black Sea coast of Turkey. *Appl. Energy* **2014**, *135*, 738–747. [CrossRef]

41. Haces-Fernandez, F. Investigation on the Possibility of Extracting Wave Energy from the Texas Coast. Master's Thesis, Texas A&M University, Kingsville, TX, USA, 2014.

42. Wang, Z.; Dong, S.; Dong, X.; Zhang, X. Assessment of wind energy and wave energy resources in Weifang sea area. *Int. J. Hydrogen Energy* **2016**, *41*, 15805–15811. [CrossRef]

43. Liang, B.; Shao, Z.; Wu, G.; Shao, M.; Sun, J. New equations of wave energy assessment accounting for the water depth. *Appl. Energy* **2017**, *188*, 130–139. [CrossRef]

44. Liang, B.; Shao, Z.; Wu, Y.; Shi, H.; Liu, Z. Numerical study to estimate the wave energy under Wave-Current Interaction in the Qingdao coast, China. *Renew. Energy* **2017**, *101*, 845–855. [CrossRef]

45. Chen, X.; Wang, K.; Zhang, Z.; Zeng, Y.; Zhang, Y.; O'Driscoll, K. An assessment of wind and wave climate as potential sources of renewable energy in the nearshore Shenzhen coastal zone of the South China Sea. *Energy* **2017**, *134*, 789–801. [CrossRef]

46. Gallagher, S.; Tiron, R.; Whelan, E.; Gleeson, E.; Dias, F.; McGrath, R. The nearshore wind and wave energy potential of Ireland: A high resolution assessment of availability and accessibility. *Renew. Energy* **2016**, *88*, 494–516. [CrossRef]

47. Silva, D.; Bento, A.R.; Martinho, P.; Soares, C.G. High resolution local wave energy modelling in the Iberian Peninsula. *Energy* **2015**, *91*, 1099–1112. [CrossRef]

48. Mentaschi, L.; Besio, G.; Cassola, F.; Mazzino, A. Performance evaluation of Wavewatch III in the Mediterranean Sea. *Ocean Model.* **2015**, *90*, 82–94. [CrossRef]

49. Pelli, D.; Cappietti, L.; Oumeraci, H. Assessing the wave energy potential in the Mediterranean Sea using WAVEWATCH III. In *Progress in Renewable Energies Offshore, Proceedings of the 2nd International Conference on Renewable Energies Offshore, Lisbon, Portugal, 24–26 October 2016*; Taylor & Francis Group: London, UK, 2016; pp. 21–26; ISBN 978-1-138-62627-0.
50. Arena, F.; Laface, V.; Malara, G.; Romolo, A.; Viviano, A.; Fiamma, V.; Carillo, A. Wave climate analysis for the design of wave energy harvesters in the Mediterranean Sea. *Renew. Energy* **2015**, *77*, 125–141. [CrossRef]
51. Jadidoleslam, N.; Özger, M.; Agiralioglu, N. Wave power potential assessment of Aegean Sea with an integrated 15-year data. Renew. *Energy* **2016**, *86*, 1045–1059.
52. Lavidas, G.; Venugopal, V. A 35 year high-resolution wave atlas for nearshore energy production and economics at the Aegean Sea. *Renew. Energy* **2017**, *103*, 401–417. [CrossRef]
53. Seemanth, M.; Bhowmick, S.A.; Kumar, R.; Sharma, R. Sensitivity analysis of dissipation parameterizations in a third-generation spectral wave model, WAVEWATCH III for Indian Ocean. *Ocean Eng.* **2016**, *124*, 252–273. [CrossRef]
54. Fernandez, F.H.; Martinez, A.; Ramirez, D.; Li, H. Characterization of wave energy patterns in Gulf of Mexico. In Proceedings of the 2017 IISE Annual Conference, Pittsburgh, PA, USA, 20–23 May 2017; pp. 1532–1537.
55. Haces-Fernandez, F.; Li, H.; Ramirez, D. Wave energy characterization and assessment in the US Gulf of Mexico, East and West Coasts with Energy Event concept. *Renew. Energy* **2018**, *123*, 312–322. [CrossRef]
56. Mollison, D. Wave energy losses in intermediate depths. *Appl. Ocean Res.* **1983**, *5*, 234–237. [CrossRef]
57. Folley, M.; Whittaker, T.J.T. Analysis of the nearshore wave energy resource. *Renew. Energy* **2009**, *34*, 1709–1715. [CrossRef]
58. Barstow, S.; Mørk, G.; Lønseth, L.; Mathisen, J.P. WorldWaves wave energy resource assessments from the deep ocean to the coast. *J. Energy Power Eng.* **2011**, *5*, 730–742.
59. Zieger, S.; Babanin, A.V.; Rogers, W.E.; Young, I.R. Observation-based source terms in the third-generation wave model WAVEWATCH. *Ocean Model.* **2015**, *96*, 2–25. [CrossRef]
60. Amrutha, M.M.; Kumar, V.S.; Sandhya, K.G.; Nair, T.B.; Rathod, J.L. Wave hindcast studies using SWAN nested in WAVEWATCH III-comparison with measured nearshore buoy data off Karwar, eastern Arabian Sea. *Ocean Eng.* **2016**, *119*, 114–124. [CrossRef]
61. Robertson, B.; Gharabaghi, B.; Hall, K. Prediction of Incipient Breaking Wave-Heights Using Artificial Neural Networks and Empirical Relationships. *Coast. Eng. J.* **2015**, *57*, 1550018. [CrossRef]
62. Castro, A.; Carballo, R.; Iglesias, G.; Rabuñal, J.R. Performance of artificial neural networks in nearshore wave power prediction. *Appl. Soft Comput.* **2014**, *23*, 194–201. [CrossRef]
63. Cornejo-Bueno, L.; Borge, J.N.; Alexandre, E.; Hessner, K.; Salcedo-Sanz, S. Accurate estimation of significant wave height with support vector regression algorithms and marine radar images. *Coast. Eng.* **2016**, *114*, 233–243. [CrossRef]
64. Cornejo-Bueno, L.; Garrido-Merchán, E.C.; Hernández-Lobato, D.; Salcedo-Sanz, S. Bayesian optimization of a hybrid system for robust ocean wave features prediction. *Neurocomputing* **2018**, *275*, 818–828. [CrossRef]
65. Cornejo-Bueno, L.; Nieto-Borge, J.C.; García-Díaz, P.; Rodríguez, G.; Salcedo-Sanz, S. Significant wave height and energy flux prediction for marine energy applications: A grouping genetic algorithm–Extreme Learning Machine approach. *Renew. Energy* **2016**, *97*, 380–389. [CrossRef]
66. Falnes, J.; Lillebekken, P.M. Budal's Latching-Controlled-Buoy Type Wave-Power Plant. In Proceedings of the 5th European Wave Energy Conference, Cork, Ireland, 17–20 September 2003.
67. Wang, L.; Göteman, M.; Engström, J.; Eriksson, M.; Isberg, J. Constrained Optimal Control of Single and Arrays of Point-Absorbing Wave Energy Converters. In Proceedings of the Marine Energy Technology Symposium, Washington, DC, USA, 25–27 April 2016.
68. Wang, L.; Isberg, J. Nonlinear passive control of a wave energy converter subject to constraints in irregular waves. *Energies* **2015**, *8*, 6528–6542. [CrossRef]
69. Valério, D.; Beirão, P.; da Costa, J.S. Optimisation of wave energy extraction with the Archimedes Wave Swing. *Ocean Eng.* **2007**, *34*, 2330–2344. [CrossRef]
70. Wang, L.; Engström, J.; Göteman, M.; Isberg, J. Constrained optimal control of a point absorber wave energy converter with linear generator. *J. Renew. Sustain. Energy* **2015**, *7*, 043127. [CrossRef]
71. Wang, L.; Engström, J.; Leijon, M.; Isberg, J. Coordinated control of wave energy converters subject to motion constraints. *Energies* **2016**, *9*, 475. [CrossRef]
72. Available online: http://aw-energy.com/aboutwaveroller/waveroller-concept (accessed on 20 September 2017).

73. Available online: http://www.emec.org.uk/about-us/wave-clients/pelamis-wave-power/ (accessed on 20 September 2017).
74. Ogata, T.; Washio, Y.; Osawa, H.; Tsuritani, Y.; Yamashita, S.; Nagata, Y. The open sea tests of the offshore floating type wave power device "Mighty Whale": Performance of the prototype. In Proceedings of the ASME 2002 21st International Conference on Offshore Mechanics and Arctic Engineering, Oslo, Norway, 23–28 June 2002; pp. 517–524. [CrossRef]
75. Masuda, Y.; McCormick, M.E. Experiences in pneumatic wave energy conversion in Japan. In *Utilization of Ocean Waves—Wave to Energy Conversion*; ASCE: Reston, VA, USA, 1986; pp. 1–33.
76. Falcão, A.D.O. The shoreline OWC wave power plant at the Azores. In Proceedings of the Fourth European Wave Energy Conference, Aalborg, Denmark, 4–6 December 2000; pp. 4–6.
77. Heath, T.; Whittaker, T.J.T.; Boake, C.B. The design, construction and operation of the LIMPET wave energy converter (Islay, Scotland). In Proceedings of the 4th European Wave Energy Conference, Aalborg, Denmark, 4–6 December 2000; pp. 49–55.
78. Arena, F.; Ascanelli, A.; Romolo, A. On design of the first prototype of a REWEC3 caisson breakwater to produce electrical power from wave energy. In Proceedings of the ASME 2013 32nd International Conference on Ocean, Offshore and Arctic Engineering, Nantes, France, 9–14 June 2013; p. V008T09A102.
79. Torre-Enciso, Y.; Ortubia, I.; de Aguileta, L.L.; Marqués, J. Mutriku wave power plant: From the thinking out to the reality. In Proceedings of the 8th European Wave and Tidal Energy Conference, Uppsala, Sweden, 7–10 September 2009; Volume 710.
80. De O Falcão, A.F.; Rodrigues, R.J.A. Stochastic modelling of OWC wave power plant performance. *Appl. Ocean Res.* **2002**, *24*, 59–71. [CrossRef]
81. Carballo, R.; Iglesias, G. A methodology to determine the power performance of wave energy converters at a particular coastal location. *Energy Convers. Manag.* **2012**, *61*, 8–18. [CrossRef]
82. Lopez, I.; Pereiras, B.; Castro, F.; Iglesias, G. Optimization of turbine-induced damping for an OWC wave energy converter using a RANS-VOF numerical model. *Appl. Energy* **2014**, *127*, 105–114. [CrossRef]
83. Simonetti, I.; Cappietti, L.; Elsafti, H.; Oumeraci, H. Optimization of the geometry and the turbine induced damping for fixed detached and asymmetric OWC devices: A numerical study. *Energy* **2017**, *139*, 1197–1209. [CrossRef]
84. Mehlum, E. Tapchan. In *Hydrodynamics of Ocean Wave-Energy Utilization*; Springer: Berlin/Heidelberg, Germany, 1986; pp. 51–55. [CrossRef]
85. Tedd, J.; Kofoed, J.P. Measurements of overtopping flow time series on the Wave Dragon, wave energy converter. *Renew. Energy* **2009**, *34*, 711–717. [CrossRef]
86. Vicinanza, D.; Frigaard, P. Wave pressure acting on a seawave slot-cone generator. *Coast. Eng.* **2008**, *55*, 553–568. [CrossRef]
87. Kofoed, J.P.; Frigaard, P.; Friis-Madsen, E.; Sørensen, H.C. Prototype testing of the wave energy converter wave dragon. *Renew. Energy* **2006**, *31*, 181–189. [CrossRef]
88. Falnes, J. *Principles for Capture of Energy from Ocean Waves. Phase Control and Optimum Oscillation*; Department of Physics, NTNU: Trondheim, Norway, 1997.
89. Hirohisa, T. Sea trial of a heaving buoy wave power absorber. In Proceedings of the 2nd International Symposium on Wave Energy Utilization, Trondheim, Norway, 22–24 June 1982; pp. 403–417.
90. Waters, R.; Stålberg, M.; Danielsson, O.; Svensson, O.; Gustafsson, S.; Strömstedt, E.; Leijon, M. Experimental results from sea trials of an offshore wave energy system. *Appl. Phys. Lett.* **2007**, *90*, 034105. [CrossRef]
91. Weinstein, A.; Fredrikson, G.; Parks, M.J.; Nielsen, K. AquaBuOY-the offshore wave energy converter numerical modeling and optimization. In Proceedings of the OCEANS'04- MTTS/IEEE TECHNO-OCEAN'04, Kobe, Japan, 9–12 November 2004; Volume 4, pp. 1854–1859. [CrossRef]
92. Salter, S.H.; Lin, C.P. Wide tank efficiency measurements on a model of the sloped IPS buoy. In Proceedings of the 3rd European Wave Energy Conference, Patras, Greece, 30 September–2 October 1998; pp. 200–206.
93. Weber, J.; Mouwen, F.; Parish, A.; Robertson, D. Wavebob—Research & development network and tools in the context of systems engineering. In Proceedings of the Eighth European Wave and Tidal Energy Conference, Uppsala, Sweden, 7–10 September 2009.
94. Gardner, F.E. Learning experience of AWS pilot plant test offshore Portugal. In Proceedings of the 6th European Wave Energy Conference, Glasgow, UK, 29 August–2 September 2005; pp. 149–154.

95. Shadman, M.; Estefen, S.F.; Rodriguez, C.A.; Nogueira, I.C. A geometrical optimization method applied to a heaving point absorber wave energy converter. *Renew. Energy* **2018**, *115*, 533–546. [CrossRef]

96. Goggins, J.; Finnegan, W. Shape optimisation of floating wave energy converters for a specified wave energy spectrum. *Renew. Energy* **2014**, *71*, 208–220. [CrossRef]

97. Flocard, F.; Finnigan, T.D. Increasing power capture of a wave energy device by inertia adjustment. *Appl. Ocean Res.* **2012**, *34*, 126–134. [CrossRef]

98. Yemm, R.; Pizer, D.; Retzler, C.; Henderson, R. Pelamis: Experience from concept to connection. *Philos. Trans. R. Soc. A* **2012**, *370*, 365–380. [CrossRef] [PubMed]

99. Ruellan, M.; BenAhmed, H.; Multon, B.; Josset, C.; Babarit, A.; Clement, A. Design methodology for a SEAREV wave energy converter. *IEEE Trans. Energy Convers.* **2010**, *25*, 760–767. [CrossRef]

100. Whittaker, T.; Collier, D.; Folley, M.; Osterried, M.; Henry, A.; Crowley, M. The development of Oyster—A shallow water surging wave energy converter. In Proceedings of the 7th European Wave and Tidal Energy Conference, Porto, Portugal, 11–13 September 2007; pp. 11–14.

101. Salter, S.H. The swinging mace. In Proceedings of the Workshop Wave Energy R&D, Cork, Ireland, 1–2 October 1992; pp. 197–206.

102. Available online: http://tpe-pelamis.e-monsite.com/pages/mes-pages/i-b-presentation-du-pelamis-schemas-et-videos.html (accessed on 6 December 2017).

103. Henderson, R. Design, simulation, and testing of a novel hydraulic power take-off system for the Pelamis wave energy converter. *Renew. Energy* **2006**, *31*, 271–283. [CrossRef]

104. Liu, Z.; Qu, N.; Han, Z.; Zhang, J.; Zhang, S.; Li, M.; Shi, H. Study on energy conversion and storage system for a prototype buoys-array wave energy converter. *Energy Sustain. Dev.* **2016**, *34*, 100–110. [CrossRef]

105. Antonio, F.D.O. Modelling and control of oscillating-body wave energy converters with hydraulic power take-off and gas accumulator. *Ocean Eng.* **2007**, *34*, 2021–2032. [CrossRef]

106. Washio, Y.; Osawa, H.; Nagata, Y.; Fujii, F.; Furuyama, H.; Fujita, T. The offshore floating type wave power device "Mighty Whale": Open sea tests. In Proceedings of the Tenth International Offshore and Polar Engineering Conference, Seattle, WA, USA, 28 May–2 June 2000.

107. O'Connor, M.; Lewis, T.; Dalton, G. Techno-economic performance of the Pelamis P1 and Wavestar at different ratings and various locations in Europe. *Renew. Energy* **2013**, *50*, 889–900. [CrossRef]

108. Babarit, A.; Hals, J.; Muliawan, M.J.; Kurniawan, A.; Moan, T.; Krokstad, J. Numerical benchmarking study of a selection of wave energy converters. *Renew. Energy* **2012**, *41*, 44–63. [CrossRef]

109. Rusu, L.; Onea, F. The performance of some state-of-the-art wave energy converters in locations with the worldwide highest wave power. *Renew. Sustain. Energy Rev.* **2017**, *75*, 1348–1362. [CrossRef]

110. Chandrasekaran, S.; Raghavi, B. Design, Development and Experimentation of Deep Ocean Wave Energy Converter System. *Energy Procedia* **2015**, *79*, 634–640. [CrossRef]

111. Eriksson, M.; Isberg, J.; Leijon, M. Hydrodynamic modelling of a direct drive wave energy converter. *Int. J. Eng. Sci.* **2005**, *43*, 1377–1387. [CrossRef]

112. Viviano, A.; Naty, S.; Foti, E.; Bruce, T.; Allsop, W.; Vicinanza, D. Large-scale experiments on the behaviour of a generalised Oscillating Water Column under random waves. *Renew. Energy* **2016**, *99*, 875–887. [CrossRef]

113. Dias, F.; Renzi, E.; Gallagher, S.; Sarkar, D.; Wei, Y.; Abadie, T.; Rafiee, A. Analytical and computational modelling for wave energy systems: The example of oscillating wave surge converters. *Acta Mech. Sin.* **2017**, 1–16. [CrossRef] [PubMed]

114. Babarit, A.; Wendt, F.; Yu, Y.H.; Weber, J. Investigation on the energy absorption performance of a fixed-bottom pressure-differential wave energy converter. *Appl. Ocean Res.* **2017**, *65*, 90–101. [CrossRef]

115. Aderinto, T.O.; Haces-Fernandez, F.; Li, H. Design and Potential Application of Small Scale Wave Energy Converter. In Proceedings of the ASME 2017 International Mechanical Engineering Congress and Exposition, Tampa, FL, USA, 3–9 November 2017; p. V006T08A084. [CrossRef]

116. Evans, D.V. A theory for wave-power absorption by oscillating bodies. *J. Fluid Mech.* **1976**, *77*, 1–25. [CrossRef]

117. Budar, K.; Falnes, J. A resonant point absorber of ocean-wave power. *Nature* **1975**, *256*, 478–479. [CrossRef]

118. Mei, C.C. Power extraction from water waves. *J. Ship Res.* **1976**, *20*, 63–66.

119. Falnes, J. Optimum control of oscillation of wave-energy converters. *Int. J. Offshore Polar Eng.* **2002**, *12*, 147–155.

120. Wang, L.; Isberg, J.; Tedeschi, E. Review of control strategies for wave energy conversion systems and their validation: The wave-to-wire approach. *Renew. Sustain. Energy Rev.* **2018**, *81*, 366–379. [CrossRef]
121. Penalba, M.; Cortajarena, J.A.; Ringwood, J.V. Validating a wave-to-wire model for a wave energy converter—Part II: The electrical system. *Energies* **2017**, *10*, 1002. [CrossRef]
122. Lin, Y.; Bao, J.; Liu, H.; Li, W.; Tu, L.; Zhang, D. Review of hydraulic transmission technologies for wave power generation. *Renew. Sustain. Energy Rev.* **2015**, *50*, 194–203. [CrossRef]
123. Hals, J.; Bjarte-Larsson, T.; Falnes, J. Optimum reactive control and control by latching of a wave-absorbing semisubmerged heaving sphere. In Proceedings of the 21st International Conference on Offshore Mechanics and Arctic Engineering, Oslo, Norway, 23–28 June 2002; Volume 4, pp. 415–423. [CrossRef]
124. Budal, K.; Falnes, J. Wave power conversion by point absorbers: A Norwegian project. *Int. J. Ambient Energy* **1982**, *3*, 59–67. [CrossRef]
125. Budal, K.; Falnes, J. Interacting point absorbers with controlled motion. In *Power from Sea Waves*; Academic Press: London, UK, 1980; pp. 381–399.
126. Naito, S.; Nakamura, S. Wave energy absorption in irregular waves by feedforward control system. In *Hydrodynamics of Ocean Wave-Energy Utilization*; Springer: Berlin/Heidelberg, Germany, 1986; pp. 269–280. [CrossRef]
127. Hoskin, R.E.; Count, B.M.; Nichols, N.K.; Nicol, D.A.C. Phase control for the oscillating water column. In *Hydrodynamics of Ocean Wave-Energy Utilization*; Springer: Berlin/Heidelberg, Germany, 1986; pp. 257–268.
128. Hoskin, R.E.; Nichols, N. Optimal strategies for phase control of wave energy devices. In *Utilization of Ocean Waves—Wave to Energy Conversion*; ASCE: New York, NY, USA, 1987. [CrossRef]
129. Korde, U.A. Phase control of floating bodies from an on-board reference. *Appl. Ocean Res.* **2001**, *23*, 251–262. [CrossRef]
130. Wells, A.A. Fluid driven rotary transducer. *Br. Patent Spec.* **1976**, *1*, 595–700.
131. Whittaker, T.J.T.; Wells, A.A. Experiences with a hydropneumatic wave power device. In Proceedings of the International Symposium on Wave and Tidal Energy, Canterbury, UK, 27–29 September 1978; Volume 1, pp. B4–B57.
132. Raghunathan, S. The Wells air turbine for wave energy conversion. *Prog. Aerosp. Sci.* **1995**, *31*, 335–386. [CrossRef]
133. Curran, R.; Gato, L.M.C. The energy conversion performance of several types of Wells turbine designs. *Proc. Inst. Mech. Eng. Part A J. Power Energy* **1997**, *211*, 133–145. [CrossRef]
134. De Falcão, A.F. Control of an oscillating-water-column wave power plant for maximum energy production. *Appl. Ocean Res.* **2002**, *24*, 73–82. [CrossRef]
135. Gato, L.M.C.; Falcao, A.D.O. Aerodynamics of the wells turbine: Control by swinging rotor-blades. *Int. J. Mech. Sci.* **1989**, *31*, 425–434. [CrossRef]
136. Gato, L.M.C.; Eça, L.R.C.; Falcao, A.D.O. Performance of the Wells turbine with variable pitch rotor blades. *J. Energy Resour. Technol.* **1991**, *113*, 141–146. [CrossRef]
137. Taylor, J.R.M.; Caldwell, N.J. Design and construction of the variable-pitch air turbine for the Azores wave energy plant. In Proceedings of the Third European Wave Power Conference, Patras, Greece, 30 September–2 October 1998; Volume 30.
138. Babintsev, I.A. Apparatus for Converting Sea Wave Energy into Electrical Energy. U.S. Patent No. 3,922,739, 2 December 1975.
139. Liu, Z.; Cui, Y.; Kim, K.W.; Shi, H.D. Numerical study on a modified impulse turbine for OWC wave energy conversion. *Ocean Eng.* **2016**, *111*, 533–542. [CrossRef]
140. Pereiras, B.; Valdez, P.; Castro, F. Numerical analysis of a unidirectional axial turbine for twin turbine configuration. *Appl. Ocean Res.* **2014**, *47*, 1–8. [CrossRef]
141. Pereiras, B.; López, I.; Castro, F.; Iglesias, G. Non-dimensional analysis for matching an impulse turbine to an OWC (oscillating water column) with an optimum energy transfer. *Energy* **2015**, *87*, 481–489. [CrossRef]
142. Setoguchi, T.; Santhakumar, S.; Maeda, H.; Takao, M.; Kaneko, K. A review of impulse turbines for wave energy conversion. *Renew. Energy* **2001**, *23*, 261–292. [CrossRef]
143. Scuotto, M.; Falcão, A.D.O. Wells and impulse turbines in an OWC wave power plant: A comparison. In Proceedings of the 6th European Wave Tidal Energy Conference, Glasgow, UK, 29 August–2 September 2005.

144. Curran, R.; Denniss, T.; Boake, C. Multidisciplinary design for performance: Ocean wave energy conversion. In Proceedings of the Tenth International Offshore and Polar Engineering Conference, Seattle, WA, USA, 28 May–2 June 2000.

145. Paish, O. Small hydro power: Technology and current status. *Renew. Sustain. Energy Rev.* **2002**, *6*, 537–556. [CrossRef]

146. Raabe, J. *Hydro Power*; VDI Verlag: Duseeldorf, Germany, 1985.

147. Estefen, S.F.; Esperança, P.T.T.; Ricarte, E.; Costa, P.R.; Pinheiro, M.M.; Clemente, C.H.; Souza, J.A. Experimental and numerical studies of the wave energy hyperbaric device for electricity production. In Proceedings of the 27th International Conference Offshore Mechanics Arctic Engineering, Estoril, Portugal, 15–20 June 2008. [CrossRef]

148. Boström, C.; Lejerskog, E.; Stålberg, M.; Thorburn, K.; Leijon, M. Experimental results of rectification and filtration from an offshore wave energy system. *Renew. Energy* **2009**, *34*, 1381–1387. [CrossRef]

149. Leijon, M.; Boström, C.; Danielsson, O.; Gustafsson, S.; Haikonen, K.; Langhamer, O.; Tyrberg, S. Wave energy from the North Sea: Experiences from the Lysekil research site. *Surv. Geophys.* **2008**, *29*, 221–240. [CrossRef]

150. Lejerskog, E.; Gravråkmo, H.; Savin, A.; Strömstedt, E.; Tyrberg, S.; Haikonen, K.; Svensson, O. Lysekil research site, sweden: A status update. In Proceedings of the 9th European Wave and Tidal Energy Conference, Southampton, UK, 5–9 September 2011.

151. Elwood, D.; Yim, S.C.; Prudell, J.; Stillinger, C.; von Jouanne, A.; Brekken, T.; Paasch, R. Design, construction, and ocean testing of a taut-moored dual-body wave energy converter with a linear generator power take-off. *Renew. Energy* **2010**, *35*, 348–354. [CrossRef]

152. Scruggs, J.; Jacob, P. Harvesting Ocean Wave Energy. *Science* **2009**, *323*, 1176–1178. [CrossRef] [PubMed]

153. Brekken, T.K.; Von Jouanne, A.; Han, H.Y. Ocean wave energy overview and research at Oregon State University. In Proceedings of the Power Electronics and Machines in Wind Applications, Lincoln, NE, USA, 24–26 June 2009; pp. 1–7. [CrossRef]

154. Papini, G.P.; Fontana, M. Reduced Model and Application of Inflating Circular Diaphragm Dielectric Elastomer Generators for Wave Energy Harvesting. *J. Vibr. Acoust.* **2015**, *137*, 011004. [CrossRef]

155. Moretti, G.; Fontana, M.; Vertechy, R. Parallelogram-shaped dielectric elastomer generators: Analytical model and experimental validation. *J. Intell. Mater. Syst. Struct.* **2015**, *26*, 740–751. [CrossRef]

156. Scherber, B.; Grauer, M.; Köllnberger, A. Electroactive polymers for gaining sea power. *Proc. SPIE* **2013**, *8687*, 86870K. [CrossRef]

157. Jean, P.; Wattez, A.; Ardoise, G.; Melis, C.; Van Kessel, R.; Fourmon, A.; Queau, J.P. Standing wave tube electro active polymer wave energy converter. *Proc. SPIE* **2012**, *8340*, 83400C. [CrossRef]

158. Basso, T.S.; DeBlasio, R. IEEE 1547 series of standards: Interconnection issues. *IEEE Trans. Power Electron.* **2004**, *19*, 1159–1162. [CrossRef]

159. O'Sullivan, D.; Griffiths, J.; Egan, M.G.; Lewis, A.W. Development of an electrical power take off system for a sea-test scaled offshore wave energy device. *Renew. Energy* **2011**, *36*, 1236–1244. [CrossRef]

160. Blanco, M.; Navarro, G.; Lafoz, M. Control of power electronics driving a switched reluctance linear generator in wave energy applications. In Proceedings of the 13th European Conference on Power Electronics and Applications, Barcelona, Spain, 8–10 September 2009; pp. 1–9.

161. Boström, C.; Leijon, M. Operation analysis of a wave energy converter under different load conditions. *IET Renew. Power Gener.* **2011**, *5*, 245–250. [CrossRef]

162. Tedeschi, E.; Carraro, M.; Molinas, M.; Mattavelli, P. Effect of control strategies and power take-off efficiency on the power capture from sea waves. *IEEE Trans. Energy Convers.* **2011**, *26*, 1088–1098. [CrossRef]

163. Luan, H.; Onar, O.C.; Khaligh, A. Dynamic modeling and optimum load control of a PM linear generator for ocean wave energy harvesting application. In Proceedings of the Twenty-Fourth Annual IEEE Applied Power Electronics Conference and Exposition, Washington, DC, USA, 15–19 February 2009; pp. 739–743. [CrossRef]

164. Igic, P.; Zhou, Z.; Knapp, W.; MacEnri, J.; Sørensen, H.C.; Friis-Madsen, E. Multi-megawatt offshore wave energy converters–electrical system configuration and generator control strategy. *IET Renew. Power Gener.* **2011**, *5*, 10–17. [CrossRef]

165. Barnes, M.; El-Feres, R.; Kromlides, S.; Arulampalam, A. Power quality improvement for wave energy converters using a D-STATCOM with real energy storage. In Proceedings of the 2004 First International Conference on Power Electronics Systems and Applications, Hong Kong, China, 9–11 November 2004; pp. 72–77. [CrossRef]

166. Kazmierkowski, M.P.; Jasinski, M. Power electronics for renewable sea wave energy. In Proceedings of the 2010 12th International Conference on Optimization of Electrical and Electronic Equipment (OPTIM), Brasov, Romania, 20–22 May 2010; pp. 4–9. [CrossRef]

167. Wang, L.; Lee, D.J.; Lee, W.J.; Chen, Z. Analysis of a novel autonomous marine hybrid power generation/energy storage system with a high-voltage direct current link. *J. Power Sources* **2008**, *185*, 1284–1292. [CrossRef]

168. Shek, J.K.H.; Macpherson, D.E.; Mueller, M.A. Experimental verification of linear generator control for direct drive wave energy conversion. *IET Renew. Power Gener.* **2010**, *4*, 395–403. [CrossRef]

169. Veritas, D.N. *Guidelines on Design and Operation of Wave Energy Converters*; Carbon Trust: London, UK, 2005.

170. Mueller, M.; Jeffrey, H. *UKERC Marine (Wave and Tidal Current) Renewable Energy Technology Roadmap: Summary Report*; UK Energy Research Centre, University of Edinburgh: Edinburgh, UK, 2008.

171. McDonald, A.; Schrattenholzer, L. Learning rates for energy technologies. *Energy Policy* **2001**, *29*, 255–261. [CrossRef]

172. Guanche, R.; De Andres, A.D.; Simal, P.D.; Vidal, C.; Losada, I.J. Uncertainty analysis of wave energy farms financial indicators. *Renew. Energy* **2014**, *68*, 570–580. [CrossRef]

173. Krohn, D.; Woods, M.; Adams, J.; Valpy, B.; Jones, F.; Gardner, P. *Wave and Tidal Energy in the UK: Conquering Challenges, Generating Growth*; document Version; Renewable UK: London, UK, 2013.

174. Starling, M. *Guidelines for Reliability, Maintainability and Survivability of Marine Energy Conversion Systems: Marine Renewable Energy Guides*; European Marine Energy Centre: Orkney, UK, 2009.

175. Faulkner, D. Rogue waves-defining their characteristics for marine design. *ACTES DE COLLOQUES-IFREMER* **2012**, 3–18.

176. Roth, I.F.; Ambs, L.L. Incorporating externalities into a full cost approach to electric power generation life-cycle costing. *Energy* **2004**, *29*, 2125–2144. [CrossRef]

177. O'Connor, M.; Lewis, T.; Dalton, G. Operational expenditure costs for wave energy projects and impacts on financial returns. *Renew. Energy* **2013**, *50*, 1119–1131. [CrossRef]

178. Gallagher, S.; Tiron, R.; Dias, F. A detailed investigation of the nearshore wave climate and the nearshore wave energy resource on the west coast of Ireland. In Proceedings of the ASME 2013 32nd International Conference on Ocean, Offshore and Arctic Engineering, Nantes, France, 9–14 June 2013.

179. O'Connor, M.; Lewis, T.; Dalton, G. Weather window analysis of Irish west coast wave data with relevance to operations & maintenance of marine renewables. *Renew. Energy* **2013**, *52*, 57–66. [CrossRef]

180. Abdulla, K.; Skelton, J.; Doherty, K.; O'Kane, P.; Doherty, R.; Bryans, G. Statistical availability analysis of wave energy converters. In Proceedings of the Twenty-first International Offshore and Polar Engineering Conference, Maui, HI, USA, 19–24 June 2011.

181. Gray, A.; Dickens, B.; Bruce, T.; Ashton, I.; Johanning, L. Reliability and O&M sensitivity analysis as a consequence of site specific characteristics for wave energy converters. *Ocean Eng.* **2017**, *141*, 493–511. [CrossRef]

182. Maples, B.; Saur, G.; Hand, M.; van de Pietermen, R.; Obdam, T. *Installation, Operation, and Maintenance Strategies to Reduce the Cost of Offshore Wind Energy*; NREL: Denver, CO, USA, 2013. [CrossRef]

183. Witt, M.J.; Sheehan, E.V.; Bearhop, S.; Broderick, A.C.; Conley, D.C.; Cotterell, S.P.; Hosegood, P. Assessing wave energy effects on biodiversity: The Wave Hub experience. *Philos. Trans. R. Soc. A* **2012**, *370*, 502–529. [CrossRef] [PubMed]

184. Inger, R.; Attrill, M.J.; Bearhop, S.; Broderick, A.C.; James Grecian, W.; Hodgson, D.J.; Godley, B.J. Marine renewable energy: Potential benefits to biodiversity? An urgent call for research. *J. Appl. Ecol.* **2009**, *46*, 1145–1153. [CrossRef]

185. Tiron, R.; Pinck, C.; Reynaud, E.G.; Dias, F. Is biofouling a critical issue for wave energy converters? In Proceedings of the Twenty-Second International Offshore and Polar Engineering Conference, Rhodes, Greece, 17–22 June 2012.

186. Tiron, R.; Mallon, F.; Dias, F.; Reynaud, E.G. The challenging life of wave energy devices at sea: A few points to consider. *Renew. Sustain. Energy Rev.* **2015**, *43*, 1263–1272. [CrossRef]

187. Tiron, R.; Gallagher, S.; Doherty, K.; Reynaud, E.G.; Dias, F.; Mallon, F.; Whittaker, T. An Experimental Study of the Hydrodynamic Effects of Marine Growth on Wave Energy Converters. In Proceedings of the International Conference on Offshore Mechanics and Arctic Engineering, Nantes, France, 9–14 June 2013; Volume 8. [CrossRef]
188. Shields, M.A.; Woolf, D.K.; Grist, E.P.; Kerr, S.A.; Jackson, A.C.; Harris, R.E.; Gibb, S.W. Marine renewable energy: The ecological implications of altering the hydrodynamics of the marine environment. *Ocean Coast. Manag.* **2011**, *54*, 2–9. [CrossRef]

Review

A Survey of WEC Reliability, Survival and Design Practices

Ryan G. Coe [1],*, Yi-Hsiang Yu [2] and Jennifer van Rij [2]

[1] Sandia National Laboratories, Albuquerque, NM 87185, USA
[2] National Renewable Energy Laboratory, Golden, CO 80303, USA; Yi-Hsiang.Yu@nrel.gov (Y.-H.Y.); Jennifer.VanRij@nrel.gov (J.v.R.)
* Correspondence: rcoe@sandia.gov; Tel.: +1-505-845-9064

Received: 30 November 2017; Accepted: 15 December 2017; Published: 21 December 2017

Abstract: A wave energy converter must be designed to survive and function efficiently, often in highly energetic ocean environments. This represents a challenging engineering problem, comprising systematic failure mode analysis, environmental characterization, modeling, experimental testing, fatigue and extreme response analysis. While, when compared with other ocean systems such as ships and offshore platforms, there is relatively little experience in wave energy converter design, a great deal of recent work has been done within these various areas. This paper summarizes the general stages and workflow for wave energy converter design, relying on supporting articles to provide insight. By surveying published work on wave energy converter survival and design response analyses, this paper seeks to provide the reader with an understanding of the different components of this process and the range of methodologies that can be brought to bear. In this way, the reader is provided with a large set of tools to perform design response analyses on wave energy converters.

Keywords: wave energy converter (WEC); design; survival; extreme conditions

---

1. Introduction

As the development of wave energy converters (WECs) has increased in detail and scope, increasing attention is now being given to the prediction of design responses and loads for devices to allow for thorough and efficient WEC design. In the past, failures of WECs have lead to closures of private developers. While less immediately obvious, a lack of confidence in design methods has also likely resulted in over-inflated fabrication and maintenance costs. The need to both ensure device survival and minimize safety factors (and therefore unnecessary structural costs) has thus created an increased incentive to better understand design responses and loads for WECs. At the same time, thanks to increased research and development work on WECs, there is now rapidly growing knowledge about methods that can be applied to the design of WECs.

In 2014, a workshop was held to gather WEC researchers and developers with the goal of better understanding the state of the art in WEC design response analysis [1]. The workshop attendees highlighted that while design response analysis is widely understood as a critical step in the design process for WECs, additional development in this area would be crucial for the success of the industry. Both device modeling tools and methods for determining environmental loads were identified as key areas in need of further development. Additionally, a need for more publicly available experimental datasets for validation of design processes and numerical models was noted. The workshop attendees also agreed that there was a need for additional guidelines, best-practices documents and standards to normalize the design process of the WECs and increase investor confidence. The recommendations from this workshop highlight the need for improvements to components within the WEC design process, as well as increased understanding and confidence in the WEC design process as a whole.

A number of standards/technical specifications (TS) can be considered for WEC design. An even larger number of informal guidelines and best-practices documents are available on the subject, with yet more standards, guidelines and other documents available on related systems, such as ships and offshore structures. However, as of the writing this article, IEC TS 62600-2 is the only standard/TS dedicated specifically to WEC design [2]. This TS gives design guidance for WECs, as well as current energy converters (CECs) and tidal energy converters (TECs). In addition to IEC TS 62600-2, a number of documents for offshore structures can also provide useful guidance. N-003 [3] is the NORSOK offshore structure design standard; DNV-RP-C205 is a Det Norkse Veritas (DNV) recommended practices document targeted at offshore structures [4]; ABS (American Bureau of Shipping) [5], DNV [6] and IEC [7] all have design documents for offshore wind turbines.

This work was completed in the interest of two related goals. Firstly, this survey should provide researchers and practitioners with some general background on available methodologies for WEC design. Secondly, considering that there are currently efforts underway to create, revise and improve design standards for WECs, this paper should serve as an overview both of the WEC design process as a whole and also specific practices. Due to the short history and limited experience in WEC design, it is essential that all of the available knowledge and understanding of best-practices be used to define standards. An initial review of extreme conditions modeling and design response analysis practices was presented by [8]. Extending the previous work, this paper provides a review and general survey of design practices for WECs, by examining the key components of the WEC design process. A generalized workflow diagram is shown in Figure 1. Working from left to right, this diagram illustrates a major steps in determining characteristic loads for a WEC. This characteristic load, which may be the maximum mooring tension or equivalent fatigue load in a structural member, can be evaluated against structural capacities. The workflow blocks in Figure 1 (design framework selection, device configurations, failure modes analysis, environmental characterization, modeling, experimental testing, extreme response statistics and fatigue analysis) are considered in turn. For each of these areas/blocks, some general discussion of relevant methods is provided, and when possible, studies applying these methods for WECs and related systems are considered.

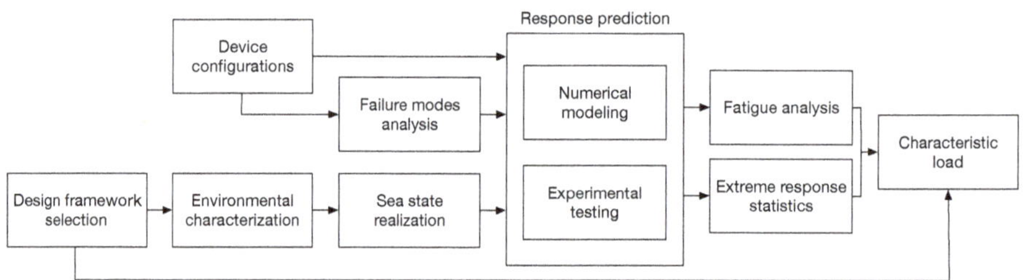

Figure 1. General WEC design workflow with component processes.

2. Device Configuration

In defining a design case, device configuration must be considered in addition to environmental conditions. While operational loads are often considered the most crucial, it is also important to consider the other scenarios/conditions, such as fault conditions and transportation. IEC TS 62600-2 considers transportation, normal operation, faults and emergency shutdown, survival, loss of stability, mooring failure and maintenance scenarios. In addition to defining modeling and experimental design, these configurations can also drive the identification of failure modes.

A variety of survival modes are currently being considered for WECs. These special device configurations are intended to reduce both energy absorption and loading during the most dangerous sea states for a device. In a sense, a WEC survival strategy is intended to be analogous to the way a wind turbine feathers its blades to reduce loading above a certain wind speed. In some cases, the PTO can be "locked" to prevent motion and problems with component end stops [9–11]. Some research has also been focused on applying the concept of life extending control (LEC), in which control algorithms are used to alter a WEC's PTO performance to avoid damage [12,13]. The inclusion of constraints in WEC control design can also be an important factor when considering design loads [14,15]. Such efforts have the potential to interface well with ongoing work to improve WEC energy absorption through control (see, e.g., [14,16]). In some cases, a passive load shedding mechanism can be incorporated into a WEC design. Retzler and Pizer [17] tested a 1/20-scale model of the Pelamis device in regular and irregular waves. This study reported that the Pelamis device's design allowed it to effectively detune in waves of a height larger than the diameter of its cylinders.

Relatively small changes in geometry can also reduce loads. Stallard et al. [18] performed a study to investigate tailoring the above-water geometry of a point absorber in a way that allows for small changes in draft to result in large changes in device response. The study achieved a 50% reduction in device response with a need for only a 10% change in mass. Terminator devices are often designed to change their angle to incoming waves or increase their draft to reduce energy absorption, and therefore loading, in large waves [19–22]. Pecher et al. [22] detailed 1/15-scale testing of the Weptos WEC to assess mooring force and structural bending moments in five extreme wave conditions. The Weptos, which is a terminator-style device, is designed to reduce its opening angle to incoming waves in order to limit loading during extreme wave conditions. This survival strategy was shown to be effective in reducing the structural bending moments within the device down to levels on the order of those in operational conditions. This strategy was also shown to be partially effective for reducing mooring loads, however not to the same degree.

Submergence is an attractive survival strategy. Wan et al. [23] performed numerical modeling and wave tank testing to assess survival modes of a combined floating wind turbine and spar-torus-type point absorber. These tests elicited slamming, green water and Mathieu instabilities. Wan et al. found that when the torus WEC was submerged for survival mode, reducing the natural period of the system in heave, loads were significantly reduced. While not feasible for many devices, the Wavestar device was capable of simply lifting the reacting bodies out of the water in survival mode [24].

3. Failure Modes Analysis

A failure modes and effects analysis (FMEA; also referred to as failure modes, effects and criticality analysis (FMECA)) is a process used to identify possible failure modes in an engineering system. In addition to estimations of maintenance schedules and levelized cost of energy (LCOE), an FMEA process is critical within the design workflow illustrated in Figure 1. As shown in Figure 1, the task of developing an experimental or numerical model is influenced by the FMEA. Since these models will, in many ways, be simplified representations of a full-scale device, it is important that they capture the necessary physics that are needed to assess failure modes.

With limited historical experience and a wide variety of WEC archetypes, the FMEA process and common failure modes for WECs are not as developed as in many other fields. Kenny et al. [25] defined a set of failure modes for a multi-body WEC so that a conditional monitoring system (CMS) could be designed to schedule preemptive maintenance. Here, the WEC was considered as a conglomerate of subsystems with varying groupings of priority. This study defined the high-priority subsystems to include hydraulic, high-voltage electrical and instrumentation systems. In a subsequent study, Kenny et al. [26] assessed the failure log of the deployed WEC against the original FMEA. For this novel WEC, which underwent design updates after deployment, numerous failure modes, often linked to human factors, not originally identified in the FEMEA occurred. Okoro et al. [27] created a system for ranking (giving relative priority to) failure modes of a WEC, which may provide some further insight into prioritizing maintenance and inspections.

4. Design Frameworks

A key area of interest for WEC design is determining which cases should be evaluated to determine design loads and, consequently, how the responses predicted in those case should be interpreted to produce a characteristic load (see Figure 1). For a WEC, this requires two major steps: selection of the design case (e.g., 50-year waves combined with 25-year wind during an operational condition) and definition (i.e., how will the precise conditions, which will be evaluated through experimental testing and/or numerical modeling, be determined). Selection of design load cases depends foremost on the intended usage and lifetime of the system. A demonstration system intended to collect data over the course of three years may not be evaluated using as harsh of conditions as a commercial device expected to last 25 years. Additionally, the consequences of failure can be considered at this stage to influence the return periods of conditions selected. Offshore oil and gas platforms are typically designed using 100-year return conditions [3,4] as a way of including a safety factor since loss of life is a potential consequence of failure. Once a set of conditions with which to compose design cases has been selected, some work is required to define (in terms of actual waves, currents, etc.) the physical condition that must be considered.

Figure 2 illustrates the three main methodologies for selecting the environmental conditions of survival/extreme response design cases. This figure focuses on wave conditions, but similar contours (and at a higher dimension, surfaces) can be used to incorporate additional environmental conditions. The all sea states approach, shown in Figure 2 with red squares, is both the most rigorous and most expensive approach. Additionally, this approach is often also referred to as the full long-term approach. Descriptions for such an approach can be found in a number of guidelines and standards, including N-003 [3] and DNV-RP-C205 [4]. Additionally, this approach has been considered for WECs in a number of studies [28,29]. The short-term responses (i.e., the sea state-specific responses), $\bar{F}_{X_{3hr}\|(t,h)}$, are integrated, along with a weighting for occurrence likelihood of each sea state, $f_{(T_e,H_s)}(t,h)$.

$$\bar{F}_{X_{3hr}}(x) = \int_h \int_t \bar{F}_{X_{3hr}\|(t,h)}(x\|(T_e, H_s))f_{(T_e,H_s)}(t,h)\, dt\, dh. \tag{1}$$

Here, the sea state is defined by the energy period, T_e, and significant wave height, H_s, which are integrated over the relevant ranges via t and h, respectively. Following the common convention, Equation (1) uses 3-h storms to define the short-term responses. The survival function on the left-hand side of Equation (1), also known as a complementary cumulative distribution function (CCDF), can be written as:

$$\bar{F}_X(x) = f(X > x) = 1 - F(x), \tag{2}$$

where $F(x)$ is the cumulative distribution function. In other words, $\bar{F}_X(x)$ represents the probability that some response of interest (e.g., mooring tension, PTO force), X, will exceed some arbitrary limit of x. Thus, Equation (1) gives the probability that the largest value of X in a 3-h sea state will exceed a limit of x. Based on return periods of interest, which can be defined in terms of probability levels, expected limits on the response/load X can be predicted.

The blue dots in Figure 2 illustrate a contour approach for load case selection. This approach, also commonly referred to as the inverse first-order method (I-FORM), was popularized for ocean systems by Winterstein and Haver [30,31]. Contour analyses have been applied for WECs in multiple studies [29,32] and are suggested for offshore structures in N-003 and DNV-RP-C205. For a desired return period, e.g., 50 years, an environmental contour is determined. Figure 2 shows such a contour in the H_s-T_e (significant wave height and energy period) space. Working along the defined contour, device responses are predicted for a series of environmental conditions. From each of the predictions along the contour, the largest response is taken to be the design response. Correction factors are generally applied to account for short-term variation. This can be accomplished by selecting a high percentile from the short-term response on the contour to represent design load. Within the marine industry, percentiles of 75–99% are often applied [4,29,33–35]. An alternative approach is to apply a correction factor to the expected (mean) value from the extreme distribution (Ren et al. [36] suggests a factor of 1.3). This latter approach is attractive since predictions of extreme response distributions often exhibit a large degree of variability in the tail region (see, e.g., [33,37]).

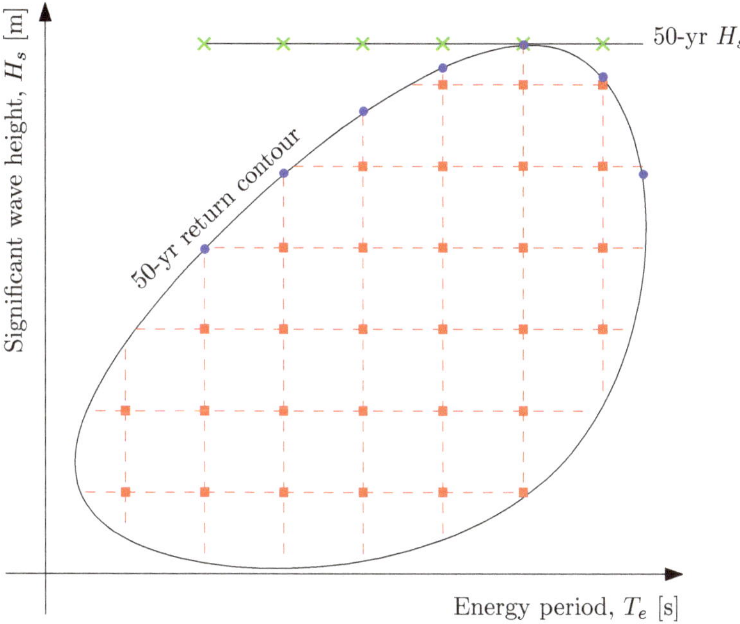

Figure 2. Frameworks for survival/extreme response WEC design cases: all sea state approach shown in red squares; contour approach shown in blue disks; 1D approach shown in green "X" marks.

The green "X" marks in Figure 2 illustrate a one-dimensional design load selection. This approach is suggested in IEC 62600-2 [2] and NORSOK N-003 [3]. The n year significant wave height, $H_{s,n}$, is predicted through an extreme value analysis and a range of sea states with relevant wave periods are evaluated.

$$a\sqrt{H_{s,n}} \leq T \leq b\sqrt{H_{s,n}} \tag{3}$$

Here, a and b are based on empirical data. IEC 62600-2 currently suggests $a = 3.54$ and $b = 4.56$. The relevant wave periods vary depending on location. Thus, when following this approach, it is best to use location-specific information to determine a range of relevant wave periods. The sea state with the largest response is then used as the design condition. Alternatively, for responses driven by excitation period, a wave period can be determined first, then used, in turn, to obtain a wave height. This may be a useful approach for many WECs with strong resonant dynamics. Again, as with the contour approach, it may be necessary to apply a correction factor. Fonseca et al. [38] applied a variation on this 1D analysis of significant wave height buoy data to define design cases for use in the mooring design of a WEC.

Various researchers have compared predictions from contour and all sea states approaches. Muliawan et al. [29] presented a comparison of the contour line and full long-term methods for loading in the mooring lines of a two-body floating WEC using a Morison's equation model. This study showed the contour line method to be capable of predicting the characteristic load in the system's mooring lines using correction factors similar to those employed by the oil and gas industry. A similar study was conducted by Ren et al. [36] on a combined spar offshore wind turbine and torus WEC. Rendon and Manuel [35] implemented a contour analysis in which characteristic extreme values were chosen as a function of the variance of each short-term distribution (that is, higher variance in the short-term distribution requires the usage of a higher percentile to represent that distribution's characteristic extreme). Agarwal [39] found the 2D contour method to potentially be both inaccurate and nonconservative for offshore wind turbines. Baarholm et al. [33] analyzed offshore platforms using both a contour method and a full long-term approach. The percentile level from the contour approach that matched the full long-term approach ranged from 75% to 97%.

5. Environmental Characterization

As discussed in Section 4, characterizing a WEC device's deployment environment represents a crucial step in the design load analysis process. A number of factors combine to make the ocean a challenging environment to characterize for engineering design. The remote nature of many prospective ocean deployment sites makes long-term measurement of conditions challenging and expensive. With IEC TS 62600-2 suggesting a 50-year return period in many design cases, this generally means that the period of record will be less than or equal to the return period for the design case. This makes for a challenging extreme value problem, as is further discussed in Section 8.2.

Additionally, seasonal and longer scale weather variations (e.g., El Niño/La Niña) necessitate longer sampling periods. Climate change may also be an important factor. The use of historical records to predict extreme sea states of a given return period has recently come into question with growing evidence of climate change. Ruggiero et al. [40] investigated the changing wave climate in the U.S. Pacific Northwest via two buoys with 40 years of data. Their results showed that the average yearly significant wave height has increased steadily over that period (at a rate of 0.015 m/year), while the average significant wave height of the five largest storms per year has grown at a rate of 0.071 m/year. Young et al. [41] looked at wind and wave data across the entire globe for roughly 20 years and saw a similar trend, with the extreme waves increasing faster (on the order of 1% annually throughout most U.S. coastal waters) than the mean. Modeling into the future, climate scientists predict that significant wave heights will increase. This increase in extreme conditions is expected to similarly increase downtime in WEC testing at the Wave Hub site [42].

Given these challenges, it is clear that characterizing extreme conditions for a specific ocean environment is a nontrivial task. In general for WEC design, as shown in Figure 2, it is desirable to characterize the environment in terms of significant wave height and energy period. While empirical distributions for commonly-occurring sea states can be obtained directly from buoy data or hindcasts, characterizing extreme waves requires extrapolation into the tails of the joint distribution and thus some statistical framework. In their application of the I-FORM, Haver and Winterstein utilized a Ronsblatt transformation so that a desired return period can be used to define a fixed radius sphere within the new space [30,43]. Alternative transformations, such as a Nataf transformation, have also been utilized [44]. Huseby et al. [45] introduced a second class of methods for extrapolating extreme sea states, which rely on Monte Carlo simulation of sea states based on a joint probability model. Vanem and Bitner-Gregersen [46] provided a useful comparison of these two approaches, both in terms of methodology and performance. An improved version of the I-FORM process is presented in [47]. Here, a principle component analysis (PCA) approach is applied to better utilize the physical relationship between wave period and height. Many of these contour methods have been implemented in the open-source WEC Design Response Toolbox (WDRT) [48]. Figure 3 shows an example where the WDRT was used to produce a set of 50-year contours using a variety of methods. From Figure 3, it is clear that these various processes of determining extreme environmental conditions return a wide variety of results.

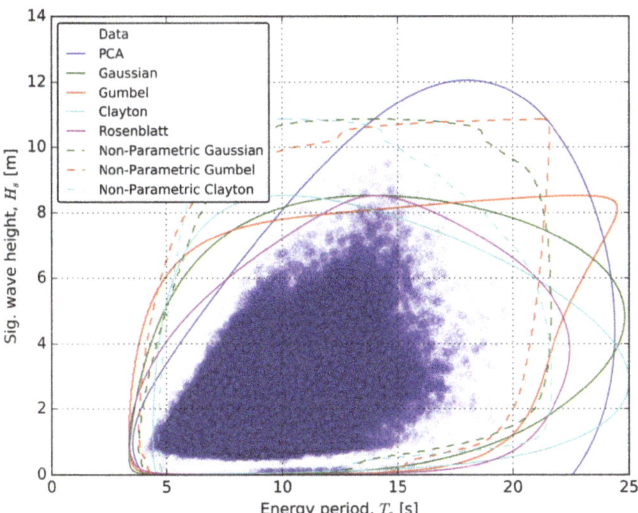

Figure 3. Data and fifty-year contours for NDBC (National Data Buoy Center) 46022 (off the coast of Eureka, CA, USA).

6. Sea State Realization

Once design cases have been defined, the specifics of realizing the design case, in order to allow for response prediction through experimental testing and modeling (see Figure 1), must be determined. While, as previously discussed, other metocean factors may play an important role for certain devices, we focus here on the definition of a design-wave. The three most common general paths for defining a design-wave are illustrated in Figure 4. The most obvious choice for a design-wave is to use either a representative or idealized spectrum to define an irregular sea state. Alternative approaches, which are generally meant to be more efficient than the "brute force" approach of simulating long irregular wave-trains, are to define representative regular waves or even focused waves based on sea state parameters. This can be helpful, or even essential, when faced with the challenges and time-restraints of high-fidelity modeling and experimental testing.

When utilizing an irregular design-wave, three-hour realizations of the sea state are generally recommended. This derives from the well-supported concept that wave amplitudes in a sea state follow a Rayleigh distribution (see, e.g., [49,50]). However, multiple realizations of each sea state should be considered. IEC 62600-2 requires that at least six three-hour realizations (18 h total) be considered for each sea state. If sufficient information is available, a representative spectral density and spreading function should be used to create sea state realizations. When more detailed information is not available, idealized spectral spectral shapes can be utilized. However, studies of offshore platforms have shown that spectral shape can play an important role in determining extreme responses [51], so care should be taken to consider the spectral shape at the deployment location whenever possible. Additional discussion of extreme value analyses in irregular sea states will be presented in Section 8.2.

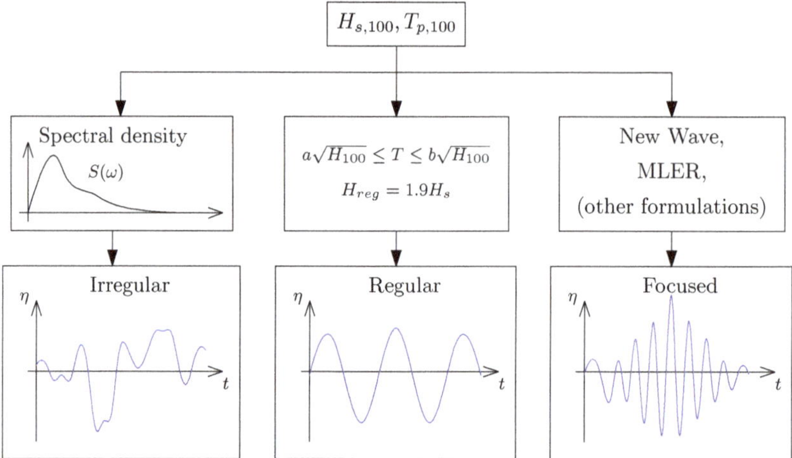

Figure 4. Design-wave paths.

In one variant of the design-wave approach, the irregular ocean sea state is represented by a regular wave. The design-wave height (H_n) is generally chosen as 1.9-times the n year return significant wave height ($H_{s,n}$), which follows from the Raleigh distribution of wave amplitudes discussed previously (see, e.g., [3,49]). While seemingly somewhat simplistic, a regular design wave can be thought of as a limit for extremely narrow-banded spectra. When utilizing a regular design-wave, it is often recommended that, as with the one-dimensional design case approach discussed in Section 4, a range of wave periods be considered. For this, we can simply replace H_s with H_{100} in (3).

$$a\sqrt{H_{100}} \leq T \leq b\sqrt{H_{100}} \tag{4}$$

The bounding constants are fairly similar as in (3): DNV-RP-C205 recommends $a = 2.55$, $b = 3.22$; NORSOK N-003 recommends $a = 2.55$, $b = 3.31$. The wave eliciting the largest response is then used as the design-wave, and the response becomes the characteristic load.

An alternative design-wave approach is to define a focused wave based on the spectral sea state. A number of formulations have been developed to design focused waves. The New Wave formulation [52] uses broad-banded wave theory to define amplitudes and phases of individual waves such that the most probable maximum wave occurs. A number of studies have applied the New Wave approach in wave tank testing of WECs [53–57]. The most-likely-extreme-response (MLER) method was used for a WEC by Quon et al. [58]. This approach utilizes both the desired spectrum and an estimate for the linear response of the system to define a design-wave.

One of the known drawbacks of focused waves is that they are unable to properly incorporate memory/dynamic state effects, which can play a dominant role in system dynamics. A hybrid approach, which utilizes irregular wave series with extremes inserted based on a Gaussian process, has been used with some success to alleviate this issue [59]. This approach creates wave trains on the $O(\text{minutes})$ and thus falls somewhere between the $O(\text{hours})$ irregular wave simulations and $O(\text{seconds})$ focused waves. Göteman et al. [60] employed embedded extreme waves during wave tank tests of a point absorber.

7. Response Prediction

With a design case realized by a design-wave, the next step in the analysis process is to predict the device response. Two main approaches can be used to determine the response of a WEC in a set of conditions. Numerical modeling and physical modeling can both be utilized at this stage and are often used in conjunction with each other to improve efficiency and increase confidence in the results.

7.1. Numerical Modeling

A wide range of numerical modeling approaches is available for prediction of WEC responses. Fortunately, a number of guidance documents and review papers are available on this topic. Topper [61] provides a general discussion of numerical modeling for WECs. This guide covers many subjects that can help a designer assess the relative importance of different physical phenomena for their device and select appropriate modeling approaches. Penalba et al. [62], Folley et al. [63] and Li and Yu [64] each provide useful surveys and reviews of numerical modeling approaches for WECs.

Frequency-domain models represent the most efficient formulation for predicting WEC dynamics. Falnes [65] provides an excellent overview of this style of model. In general, frequency-domain models assume a linear decomposition of the dynamics of a floating body. Assuming the water to be an inviscid and irrotational fluid, a potential flow formulation can be applied along with boundary conditions for the free surface, sea floor and other solid bodies (see, e.g., [66]). By assuming small motions, the radiation and diffraction phenomena can be decoupled. In practice, a frequency-domain model for a WEC can be obtained either numerically, using boundary element model tools such as WAMIT [67] and Nemoh [68], or through experimental system identification (see, e.g., [69]). Note that some nonlinear frequency-domain approaches, such as Volterra models (see, e.g., [70]), can also be applied. One example was shown by Gkikas and Athanassoulis [71], in which the pressure of an oscillating water column WEC was modeled using a Volterra formulation. Another option is to employ local linear models, which may be tuned to provide a suitable representation of WEC dynamics within a regime [69].

Cummins [72] provided the best-known translation of the frequency-domain model for a floating body into the time-domain. The main advantages of working with this type of formulation is the more straightforward ability to include arbitrary nonlinear dynamics as some perturbation from the linear solution. This may include nonlinear hydrodynamics (see, e.g., [73–77]), nonlinear mechanical dynamics, such as due to a mooring system (see, e.g., [78]) or some arbitrary control input (see, e.g., [14,16]). Giorgi and Ringwood [79] performed an assessment of the relative importance of various nonlinear perturbations within a Cummins-style time-domain model and compared computational expenses. In this study, Giorgi and Ringwood found that viscous drag nonlinearities can be significant, even for the simple spherical WEC considered in their study, when motions are large. Nonlinear Froude–Krylov phenomenon were found to have a smaller impact. A similarly focused study on nonlinear hydrodynamics was completed by Bharath et al. [80] using the commercial CFD code STAR-CCM+.

By solving the potential flow problem in a time-dependent fashion, a nonlinear potential flow model can also be created. There are a few well-known nonlinear potential flow solvers from the naval architecture field (AEGIR [81] and LAMP (Large Amplitude Motion Program) [82]). Letournel et al. [83] performed a comparison of such nonlinear potential flow models with those that rely on superposition. Greenhow et al. [84] presented an analysis of Salter's "duck" in extreme waves using a two-dimensional nonlinear potential flow code and experimental testing. Guerber et al. [85] have investigated a nonlinear potential flow model for analyzing a simple submerged WEC.

CFD models solve spatially- and temporally-discretized form of Navier–Stokes equations. In practice for most real engineering systems, these models rely on a Reynolds-averaged turbulence closure to perform simulations. ITTC 7.5-03-02-03 [86] provides useful guidelines for conducting CFD simulations of ships. While some considerations for ships focus on forward speed and resistance issues, which are of lesser interest for most WECs, other issues, such as free surface modeling, mesh density to capture waves and temporal discretization, remain of prime importance. Ransley et al. [87] used OpenFOAM to model the response of simple moored buoy. Using a New Wave focused wave, the response of the buoy predicted by CFD matched the experimental results quite closely. Yu and Li [88] showed a similar approach using the commercial code STAR-CCM+ for analysis of the Reference Model 3 (RM3) [89]. The RM3 was studied in both "locked" and "operational" PTO configurations. When considering the response amplitude operator (RAO) defined as regular, these CFD simulations showed fair agreement with experimental results; in general, the device response predicted in CFD tended to be smaller than predicted by linear models. Sjökvist and Göteman [90] studied the response of a tethered point absorber using regular waves and tsunami waves in OpenFOAM. Westphalen et al. [91] performed CFD simulations for a simple floating body (the "Manchester Bobber") and performed comparisons with experimental data collected by Weller et al. [54]. In Wesphalen's study, both a finite volume Navier–Stokes and smooth particle hydrodynamics (SPH) model generally showed good agreement with the near-focused wave trains considered. Madhi and Yeung [92] applied SPH modeling to analyze the effect of breaking waves on a WEC. The SPH model was able to accurately capture the breaking waves and showed good agreement with experimental results.

7.2. Experimental Testing

Experimental testing of WECs is mostly defined by two components: model design/fabrication and test design. Both components are of key importance and must be considered in conjunction with each other. However, before detailed work may begin on model and test design, a key assessment should be made about the stage of technology development. This topic has been discussed by a number of authors. Both Holmes [93] and ITTC 7.5-02-07-03.7 [94] suggest that WEC survival testing be undertaken during the design exploration phase at a small scale ($1/100 < \lambda < 1/25$) and during the design validation stage at a medium scale ($1/25 < \lambda < 1/10$). Note that working at a smaller scale can allow for extreme sea states to produced in a wider range of wave tanks. ITTC 7.5-02-07-02.3 [95], which focuses experiments on rarely occurring events for ships, recommends that extreme conditions be evaluated at as large a scale as feasible. This opposing recommendation is driven by the desire to limit effects of scaling.

As with other types of model-scale testing, dynamic similitude is desired for testing of a WEC, but not fully obtainable. Froude scaling is generally preferred for most WECs; however, as noted in Section 7.1, when motions are large, viscous damping can begin to play an important role in WEC dynamics. Therefore, Reynolds scaling should at least be calculated in order to understand potential implications [93,96]. The presence of electro-mechanical systems in a WEC adds additional complicating factors. While the use of a survival mode (see discussion in Section 2) may reduce their importance for tests focused on extreme conditions, a number of WEC testing guidelines call out the importance of designing a model's PTO system [94,96]. Mooring systems must also be represented in model scale. These mechanical subsystems may prove challenging to scale with appropriate frictional rates and/or fluid compressibility (e.g., for an oscillating water column device).

Considering the design of a wave tank testing experiment to assess WEC survival, one is primarily concerned with combinations of wave conditions and WEC configurations. It is generally recommended that sea states with a range of angles of incidence to the WEC be considered [93,94], as these can create unexpected responses in the device. ITTC 7.5-02-07-03.7 also stresses the importance of considering short-crested waves [94]. Available guidelines on experimental WEC testing generally recommend performing survival tests corresponding to three hours at full scale (on the order of 30 to 45 min at model scale) [93,94]. This is relevant when applying an irregular sea state approach as discussed in Section 6.

In addition to typical sensors, such as strain gauges, force/torque transducers, position sensors and accelerometers, ITTC 7.5-02-07-02.3 also recommends that the relative motion between the model and free surface be measured when considering extreme conditions. This can be achieved by mounting wave probes to the model itself. Slamming may also be important and thus necessitate surface pressure measurements. This has been investigated for the near-shore flapping Oyster device by Henry et al. [97]. Dedicated slamming tests have also been conducted on a composite point absorber at Ghent University, which found a roughly 50% reduction in slamming loads for the composite structure when compared to steel [98,99].

8. Determining a Characteristic Load

Through experimental testing and numerical modeling of specific design cases, it is possible to generate large amounts of data for responses of interest identified through an FMEA. To utilize these results within the WEC design process, we must distill and/or extrapolate these results to estimate a single design values in either fatigue and/or ultimate loading. Depending on the design case and failure mode, fatigue or extreme responses can be considered.

8.1. Fatigue

In addition to extreme loads, a WEC must also be structurally able to withstand fatigue loading for its design lifetime. Fatigue loads are time-varying loads, which cause cumulative damage to structural components and eventually lead to structural failure. Usually, a component's fatigue life/strength is reported in terms of an *S-N* curve (see, e.g., [100,101]). The *S-N* curve, which is typically obtained empirically, gives the number of load cycles, N, to failure at constant load amplitude, S, as illustrated in Figure 5. Mathematically, the behavior in Figure 5 is given by, $\log(N) = \log(K) - m \log(S)$. Here, S_{ult} is the ultimate strength, S_{end} is the endurance limit below which no failure occurs with constant amplitude loading, m is the slope of the *S-N* curve and K is an empirical material constant determining the level of the *S-N* curve.

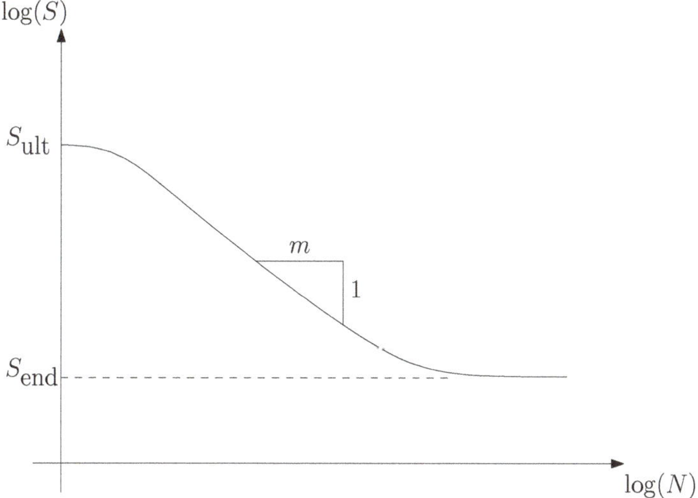

Figure 5. Typical *S-N* curve.

WEC loads, however, are typically highly variable and by no means of constant amplitude. The most common method used to predict the cumulative damage due to variable loading is the Palmgren–Miner rule. The Palmgren–Miner rule is based on the assumption that the cumulative damage of each load cycle is sequence independent, and thus, the total damage equivalent load, S_N, is obtained with a linear summation of the distributed load ranges [102].

$$S_N = \sqrt[m]{\sum_i S_i^m n_i / N} \tag{5}$$

Here, S_i is the load range for bin i and n_i is the number of cycles for load range bin i. The discretized load distribution used in Equation (5) is usually obtained via the rain-flow counting method [103].

Fatigue analyses have been used increasingly in studies of WECs [104–106]. These have begun to provide some guidance on the calibration of design factors when considering fatigue in a WEC and have also allowed for some more refined estimates of reliability and maintenance [107]. Thies et al. [108] investigated mooring fatigue in WECs, which is known to be an important consideration for some WEC designs. Thies et al. found large uncertainties in the spread mooring system's line fatigue loads. A number of studies have also considered fatigue in WEC design in conjunction with the assessment of control strategies [58,109].

8.2. Extreme Response Statistics

As illustrated in Figure 1, extreme response statistics are employed to obtain long-term characteristic design loads for a specific design case. As an intermediary step, a "short-term" extreme response (i.e., the sea state-specific extreme response) is also generally computed. However, as discussed in Section 2, the process to obtain a characteristic load ("long-term" response) depends on the design frame work utilized (i.e., all sea states, contour, or 1D approaches shown in Figure 2). Coles [110] provides an excellent general introduction to extreme response statistics. Naess and Moan [111] cover the application of extreme response statistics to marine systems.

A broad comparison of different short-term extreme response methods applied to WECs is presented by Michelen and Coe [37] (many of these methods have been implemented in an open-source toolbox [48]). In general, short-term extreme response methods attempt to balance the amount of data utilized with the importance of that data to predicting extremes. While more data are typically beneficial, the amount of data required to perform an analysis is highly-dependent on both the method being used and the response being considered. The block-maxima method requires a large amount of data and is therefore often seen as more robust. However, in order to be effective, the block-maxima method (as with other extreme response methods) requires a sufficient amount of data. In the interest of better understanding the amount of required data, Agarwal [39] developed a convergence criterion for short-term load statistics of offshore wind turbines based on bootstrapping. Global and block maxima methods were used to produce extreme response distributions; however, it was shown that the block maxima method offered no advantage over the global maxima method for the system studied.

Other methods, such as all-peaks Weibull fitting, peaks-over-threshold methods and Weibull tail-fitting, attempt to use less data by focusing more on the tail of the distribution (see, e.g., [110,111]). Naess [112] developed a method that exploits an envelope process of some narrow-banded response. Since the extremes of the envelope process are very similar to those of the underlying process, the envelope process can be analyzed instead. Another alternative is the average conditional exceedance rate (ACER) method, which utilizes the conditional exceedance rate of a process to create an extreme response distribution [113].

The frequency response of the load being considered also plays a role in the amount of data required to perform a short-term extreme response analysis. Responses with lower frequency content (e.g., mooring loads) will require longer simulations/experiments than those with higher frequency content. Vazquez-Hernandez et al. [114] used Weibull-tail fitting and a number of other methods to find design loads of mooring lines in floating platforms. This study observed a large degree of variability in extreme distribution extrapolation for short-term extremes. Very rare events, such a slamming and overtopping, carry their own additional challenge for characterization of survival loads. Lian and Haver [115] examined the process to obtain the response statistics due to a slamming on a semisubmersible offshore oil and gas structure. For slamming loads, this study observed large variability in loading due to subtle differences in wave trains. As such, roughly ten-times the number of sea state realizations were needed to accurately characterize the load when compared with other responses.

9. Conclusions

A survey of practices for survival and design response analysis of WECs has been presented. Considering a general workflow for this process, subprocesses, including design case specification, environmental characterization, modeling, experimental testing, fatigue and extreme response statistics, were examined in detail. For each of these component steps in the workflow of considering the design response of WEC, we have presented a survey of available methods and common practices. Together, these tools comprise a set that can be applied to assess WEC survival in a variety of ways.

Further research and application of these tools will lead to refinements and, likely, a down-selection to a smaller subset of tools, which can be considered definitive best-practices. Thus, to achieve this goal, future studies should be conducted to better explore the WEC design process. While, as shown in this survey, there has been a large number of studies that report on the application of some subset of steps within the design process, there is a need for sharing experiences on conducting more complete WEC design analyses. Such case-studies would likely prove very useful to individual designers and researchers, as well as those looking to define technical specifications and standards.

Acknowledgments: This work was funded by the U.S. Department of Energy's Water Power Technologies Office. Sandia National Laboratories is a multi-mission laboratory managed and operated by National Technology and Engineering Solutions of Sandia, LLC., a wholly-owned subsidiary of Honeywell International, Inc., for the U.S. Department of Energy's National Nuclear Security Administration under Contract DE-NA0003525. This work was also supported by the U.S. Department of Energy under Contract No. DE-AC36-08GO28308 with the National Renewable Energy Laboratory.

Author Contributions: R.G.C., Y.-H.Y. and J.v.R. wrote the paper.

Conflicts of Interest: R.G.C. serves as the subject-matter expert for the IEC 62600 MT-2, which is charged with revising the IEC TS 62600-2 ("Design requirements for marine energy systems"). The funding sponsors had no role in the design of the study; in the collection, analyses or interpretation of data; in the writing of the manuscript; nor in the decision to publish the results.

Abbreviations

The following abbreviations are used in this manuscript:

ABS	American Bureau of Shipping
ACER	average conditional exceedance rate
CEC	current energy converter
CFD	computational fluid dynamics
CMS	conditional monitoring system
DNV	Det Norske Veritas
FEA	finite element analysis
FMEA	failure modes and effects analysis
FMECA	failure mode, effects and criticality analysis
IEC	International Electrotechnical Commission
I-FORM	inverse first-order reliability method
ITTC	International Towing Tank Conference
LCOE	levelized cost of energy
LEC	life extending control
NDBC	National Data Buoy Center
PCA	principle component analysis
PTO	power take-off
RM3	Reference Model 3
TEC	tidal energy converter
TS	technical specification
WEC	wave energy converter

References

1. Coe, R.G.; Neary, V.S.; Lawson, M.J.; Yu, Y.H.; Weber, J. *Extreme Conditions Modeling Workshop Report*; Technical Report NREL/TP-5000-62305 SNL/SAND2014-16384R; Sandia National Laboratories: Albuquerque, NM, USA, 2014.
2. International Electrotechnical Commission. *Marine Energy—Wave, Tidal and Other Water Current Converters—Part 2: Design Requirements for Marine Energy Systems*; IEC TS 62600-2:2016; IEC: Geneva, Switzerland, 2016.

3. NORSOK. *Actions and Action Effects (N-003)*; N-003; Standards Norway: Oslo, Norway, 2007.
4. Det Norske Veritas. *Environmental Conditions and Environmental Loads*; DNV-RP-C205; Det Norske Veritas: Oslo, Norway, 2007.
5. American Bureau of Shipping. *Design Standards for Offshore Wind Farms*; M10PC00105; American Bureau of Shipping: Houston, TX, USA, 2011.
6. Det Norske Veritas. *Design of Offshore Wind Turbine Structures*; DNV-OS-J101; Det Norske Veritas: Oslo, Norway, 2014.
7. International Electrotechnical Commission. *Design Requirements for Offshore Wind Turbines*; IEC 61400-3:2009; International Electrotechnical Commission: Geneva, Switzerland, 2014.
8. Coe, R.G.; Neary, V.S. Review of Methods for Modeling Wave Energy Converter Survival in Extreme Sea States. In Proceedings of the 2nd Marine Energy Technology Symposium, Seattle, WA, USA, 15–17 April 2014.
9. Whittaker, T.; Collier, D.; Folley, M.; Osterried, M.; Henry, A.; Crowley, M. The development of Oyster—A shallow water surging wave energy converter. In Proceedings of the 7th European Wave and Tidal Energy Conference (EWTEC2007), Porto, Portugal, 11–14 September 2007.
10. Beatty, S.J.; Hiles, C.E.; Nicoli, R.S.; Adamson, J.E.; Buckham, B.J. Design Synthesis of a Wave Energy Converter. In Proceedings of the 28th International Conference on Ocean, Offshore and Arctic Engineering (OMAE2009), Honolulu, HI, USA, 31 May–5 June 2009.
11. Yu, Y.H.; Li, Y. Preliminary results of a RANS simulation for a floating point absorber wave energy system under extreme wave conditions. In Proceedings of the 30th International Conference on Ocean, Offshore and Arctic Engineering (OMAE2011), Rotterdam, The Netherlands, 19–24 June 2011.
12. Stillinger, C.; Brekken, T.; von Jouanne, A. Furthering the study of real-time life extending control for ocean energy conversion. In Proceedings of the IEEE Power and Energy Society General Meeting, San Diego, CA, USA, 22–26 July 2012.
13. Tom, N.M.; Yu, Y.H.; Wright, A.D.; Lawson, M.J. Pseudo-spectral control of a novel oscillating surge wave energy converter in regular waves for power optimization including load reduction. *Ocean Eng.* **2017**, *137*, 352–366.
14. Hals, J.; Falnes, J.; Moan, T. A Comparison of Selected Strategies for Adaptive Control of Wave Energy Converters. *J. Offshore Mech. Arct. Eng.* **2011**, *133*, 031101.
15. Wang, L.; Engström, J.; Göteman, M.; Isberg, J. Constrained optimal control of a point absorber wave energy converter with linear generator. *J. Renew. Sustain. Energy* **2015**, *7*, 043127.
16. Coe, R.G.; Bacelli, G.; Wilson, D.; Abdelkhalik, O.; Korde, U.A.; Robinett, R.D., III. A comparison of control strategies for wave energy converters. *Int. J. Mar. Energy* **2017**, *20*, 45–63.
17. Retzler, C.H.; Pizer, D.J. The hydrodynamics of the PELAMIS wave energy device: experimental and numerical results. In Proceedings of the 20th International Conference on Offshore Mechanics and Arctic Engineering (OMAE2001), Rio de Janeiro, Brazil, 3–8 June 2001.
18. Stallard, T.J.; Weller, S.D.; Stansby, P.K. Limiting heave response of a wave energy device by draft adjustment with upper surface immersion. *Appl. Ocean Res.* **2009**, *31*, 282–289.
19. Palm, J.; Eskilsson, C.; Moura Paredes, G.; Bergdahl, L. CFD Simulation of a Moored Floating Wave Energy Converter. In Proceedings of the 10th European Wave and Tidal Energy Conference (EWTEC), Aalborg, Denmark, 2–6 September 2013.
20. Parmeggiani, S.; Kofoed, J.P.; Friis-Madsen, E. Extreme Loads on the Mooring Lines and Survivability Mode for the Wave Dragon Wave Energy Converter. In Proceedings of the World Renewable Energy Congress (WREC), Linköping, Sweden, 8–13 May 2011.
21. Parmeggiani, S.; Muliawan, M.J.; Gao, Z.; Moan, T.; Friis-Madsen, E. Comparison of Mooring Loads in Survivability Mode on the Wave Dragon Wave Energy Converter Obtained by a Numerical Model and Experimental Data. In Proceedings of the 31st International Conference on Ocean, Offshore and Arctic Engineering (OMAE2012), Rio de Janeiro, Brazil, 1–6 July 2012; pp. 341–350.
22. Pecher, A.; Kofoed, J.P.; Larsen, T. Design Specifications for the Hanstholm WEPTOS Wave Energy Converter. *Energies* **2012**, *5*, 1001–1017.
23. Wan, L.; Gao, Z.; Moan, T. Experimental and numerical study of hydrodynamic responses of a combined wind and wave energy converter concept in survival modes. *Coast. Eng.* **2015**, *104*, 151–169.

24. Zurkinden, A.S.; Damkilde, L.; Gao, Z.; Moan, T. Structural modeling and analysis of a wave energy converter applying dynamical substructuring method. In Proceedings of the 32nd International Conference on Ocean, Offshore and Arctic Engineering (OMAE2013), Nantes, France, 9–14 June 2013.

25. Kenny, C.J.; Findlay, D.; Lazakis, I.; Shek, J.; Thies, P.R. Development of a Condition Monitoring System for an Articulated Wave Energy Converter. In Proceedings of the European Safety and Reliability Conference (ESREL 2016), Glasgow, Scotland, 25–29 September 2016.

26. Kenny, C.J.; Findlay, D.; Thies, P.R.; Shek, J.; Lazakis, I. Lessons Learned from 3 Years of Failure: Validating an FMEA with Historical Failure Data. In Proceedings of the 12th European Wave and Tidal Energy Conference (EWTEC2017), Cork, Ireland, 27 August–1 September 2017.

27. Okoro, U.; Kolios, A.; Cui, L. Multi-criteria risk assessment approach for components risk ranking—The case study of an offshore wave energy converter. *Int. J. Mar. Energy* **2017**, *17*, 21–39.

28. Coe, R.G.; Michelen, C.; Eckert-Gallup, A.; Sallaberry, C. Full long-term design response analysis of a wave energy converter. *Renew. Energy* **2018**, *116*, 356–366.

29. Muliawan, M.J.; Gao, Z.; Moan, T. Application of the Contour Line Method for Estimating Extreme Responses in the Mooring Lines of a Two-Body Floating Wave Energy Converter. *J. Offshore Mech. Arct. Eng.* **2013**, *135*, 031301.

30. Winterstein, S.R.; Ude, T.C.; Cornell, C.A.; Bjerager, P.; Haver, S. Environmental parameters for extreme response: Inverse FORM with omission factors. In Proceedings of the 6th International Conference on Structural Safety & Reliability (ICOSSAR), Innsbruck, Austria, 9–13 August 1993.

31. Haver, S.; Winterstein, S.R. Environmental contour lines: A method for estimating long term extremes by a short term analysis. *Trans. Soc. Nav. Archit. Mar. Eng.* **2009**, *116*, 116–127.

32. Yu, Y.H.; Van Rij, J.; Coe, R.; Lawson, M. Preliminary Wave Energy Converters Extreme Load Analysis. In Proceedings of the 35th International Conference on Ocean, Offshore and Arctic Engineering (OMAE2015), St. John's, NL, Canada, 19–24 June 2015.

33. Baarholm, G.S.; Haver, S.; Økland, O.D. Combining contours of significant wave height and peak period with platform response distributions for predicting design response. *Mar. Struct.* **2010**, *23*, 147–163.

34. Haver, S.; Sagli, G.; Gran, T. Long term response analysis of fixed and floating structures. In Proceedings of the 1998 International OTRC Symposium, Houston, TX, USA, 30 April–1 May 1998.

35. Rendon, E.A.; Manuel, L. Long-term loads for a monopile-supported offshore wind turbine. *Wind Energy* **2014**, *17*, 209–223.

36. Ren, N.; Gao, Z.; Moan, T.; Wan, L. Long-term performance estimation of the Spar-Torus-Combination (STC) system with different survival modes. *Ocean Eng.* **2015**, *108*, 716–728.

37. Michelen, C.; Coe, R. Comparison of Methods for Estimating Short-Term Extreme Response of Wave Energy Converters. In Proceedings of the OCEANS 2015—MTS/IEEE Washington, Washington, DC, USA, 19–22 October 2015.

38. Fonseca, N.; Pascoal, R.; Morais, T.; Dias, R. Design of a mooring system with synthetic ropes for the FLOW wave energy converter. In Proceedings of the 28th International Conference on Ocean, Offshore and Arctic Engineering (OMAE2009), Honolulu, HI, USA, 31 May–5 June 2009; pp. 1189–1198.

39. Agarwal, P. Structural Reliability of Offshore Wind Turbines. Ph.D. Thesis, The University of Texas at Austin, Austin, TX, USA, 2008.

40. Ruggiero, P.; Komar, P.D.; Allan, J.C. Increasing wave heights and extreme value projections: The wave climate of the U.S. Pacific Northwest. *Coast. Eng.* **2010**, *57*, 539–552.

41. Young, I.; Zieger, S.; Babanin, A.V. Global trends in wind speed and wave height. *Science* **2011**, *332*, 451–455.

42. Reeve, D.; Chen, Y.; Pan, S.; Magar, V.; Simmonds, D.; Zacharioudaki, A. An investigation of the impacts of climate change on wave energy generation: The Wave Hub, Cornwall, UK. *Renew. Energy* **2011**, *36*, 2404–2413.

43. Haver, S. On the joint distribution of heights and periods of sea waves. *Ocean Eng.* **1987**, *14*, 359–376.

44. Silva-González, F.; Heredia-Zavoni, E.; Montes-Iturrizaga, R. Development of environmental contours using Nataf distribution model. *Ocean Eng.* **2013**, *58*, 27–34.

45. Huseby, A.B.; Vanem, E.; Natvig, B. A new approach to environmental contours for ocean engineering applications based on direct Monte Carlo simulations. *Ocean Eng.* **2013**, *60*, 124–135.

46. Vanem, E.; Bitner-Gregersen, E.M. Alternative environmental contours for marine structural design—A comparison study. *J. Offshore Mech. Arct. Eng.* **2015**, *137*, 051601.

47. Eckert-Gallup, A.C.; Sallaberry, C.J.; Dallman, A.R.; Neary, V.S. Application of principal component analysis (PCA) and improved joint probability distributions to the inverse first-order reliability method (I-FORM) for predicting extreme sea states. *Ocean Eng.* **2016**, *112*, 307–319.

48. Coe, R.G.; Michelen, C.; Eckert-Gallup, A.C.; Yu, Y.H.; van Rij, J. WDRT: A toolbox for design-response analysis of wave energy converters. In Proceedings of the 4th Marine Energy Technology Symposium, Washington, DC, USA, 2016.

49. Ochi, M.K. *Ocean Waves: The Stochastic Approach*; Cambridge University Press: Cambridge, UK, 2005; Volume 6.

50. Faltinsen, O. *Sea Loads on Ships and Offshore Structures*; Cambridge University Press: Cambridge, UK, 1993; Volume 1.

51. Ochi, M.K. Wave statistics for the design of ships and ocean structures. *Trans. Soc. Nav. Archit. Mar. Eng.* **1978**, *86*, 47–76.

52. Tromans, P.S.; Anaturk, A.R.; Hagemeijer, P. A new model for the kinematics of large ocean waves-application as a design wave. In Proceedings of the 1st International Offshore and Polar Engineering Conference, Edinburgh, UK, 11–16 August 1991; pp. 64–71.

53. Orszaghova, J.; Rafiee, A.; Wolgamot, H.; Draper, S.; Taylor, P. Experimental Study of Extreme Responses of a Point Absorber Wave Energy Converter. In Proceedings of the 20th Australasian Fluid Mechanics Conference, Perth, Australia, 1–8 December 2016.

54. Weller, S.; Stallard, T.; Stansby, P. Experimental measurements of the complex motion of a suspended axisymmetric floating body in regular and near-focused waves. *Appl. Ocean Res.* **2013**, *39*, 137–145.

55. Rafiee, A.; Wolgamot, H.; Draper, S.; Orszaghova, J.; Fiévez, J.; Sawyer, T. Identifying the design wave group for the extreme response of a point absorber wave energy converter. In Proceedings of the Asian Wave and Tidal Energy Conference (AWTEC), Singapore, 24–28 October 2016.

56. Santo, H.; Taylor, P.; Moreno, E.C.; Stansby, P.; Taylor, R.E.; Sun, L.; Zang, J. Extreme motion and response statistics for survival of the three-float wave energy converter M4 in intermediate water depth. *J. Fluid Mech.* **2017**, *813*, 175–204.

57. Hann, M.; Greaves, D.; Raby, A.; Howey, B. Use of constrained focused waves to measure extreme loading of a taut moored floating wave energy converter. *Ocean Eng.* **2018**, *148*, 33–42.

58. Quon, E.; Platt, A.; Yu, Y.H.; Lawson, M. Application of Most Likely Extreme Response Method for Wave Energy Converters. In Proceedings of the 35th International Conference on Ocean, Offshore and Arctic Engineering (OMAE2016), Busan, Korea, 19–24 June 2016.

59. Taylor, P.H.; Jonathan, P.; Harland, L.A. Time domain simulation of jack-up dynamics with the extremes of a Gaussian process. *J. Vib. Acoust.* **1997**, *119*, 624–628.

60. Göteman, M.; Engström, J.; Eriksson, M.; Hann, M.; Ransley, E.; Greaves, D.; Leijon, M. Wave loads on a point-absorbing wave energy device in extreme waves. In Proceedings of the 25th International Ocean and Polar Engineering Conference, Kona, HI, USA, 21–26 June 2015.

61. Topper, M.B. *Guidance for Numerical Modelling in Wave and Tidal Energy*; SuperGen Marine; The University of Edinburgh: Edinburgh, Scotland, 2010.

62. Penalba, M.; Giorgi, G.; Ringwood, J.V. Mathematical modelling of wave energy converters: A review of nonlinear approaches. *Renew. Sustain. Energy Rev.* **2017**, *78*, 1188–1207.

63. Folley, M.; Babarit, A.; Child, B.; Forehand, D.; O'Boyle, L.; Silverthorne, K.; Spinneken, J.; Stratigaki, V.; Troch, P. A Review of Numerical Modelling of Wave Energy Converter Arrays. In Proceedings of the 31st International Conference on Ocean, Offshore and Arctic Engineering (OMAE2012), Rio de Janeiro, Brazil, 1–6 July 2012; pp. 535–545.

64. Li, Y.; Yu, Y.H. A synthesis of numerical methods for modeling wave energy converter-point absorbers. *Renew. Sustain. Energy Rev.* **2012**, *16*, 4352–4364.

65. Falnes, J. *Ocean Waves and Oscillating Systems*; Cambridge University Press: Cambridge, UK, 2002.

66. Newman, J.N. *Marine Hydrodynamics*; MIT Press: Cambridge, MA, USA, 1978.

67. WAMIT. *WAMIT User Manual*, 7th ed.; WAMIT: Chestnut Hill, MA, USA, 2012.

68. Babarit, A.; Delhommeau, G. Theoretical and numerical aspects of the open source BEM solver NEMOH. In Proceedings of the 11th European Wave and Tidal Energy Conference (EWTEC2015), Nantes, France, 6–11 September 2015.

69. Bacelli, G.; Coe, R.G.; Patterson, D.; Wilson, D. System Identification of a Heaving Point Absorber: Design of Experiment and Device Modeling. *Energies* **2017**, *10*, 472.

70. Kim, Y.; Kim, J.H.; Kim, Y. Time series prediction of nonlinear ship structural responses in irregular seaways using a third-order Volterra model. *J. Fluids Struct.* **2014**, *49*, 322–337.

71. Gkikas, G.; Athanassoulis, G. Development of a novel nonlinear system identification scheme for the pressure fluctuation inside an oscillating water column-wave energy converter Part I: Theoretical background and harmonic excitation case. *Ocean Eng.* **2014**, *80*, 84–99.

72. Cummins, W.E. *The Impulse Response Function and Ship Motions*; Technical Report DTNSDRC 1661; Department of the Navy, David Taylor Model Basin: Bethesda, MD, USA, 1962.

73. Lawson, M.; Yu, Y.H.; Nelessen, A.; Ruehl, K.; Michelen, C. Implementing Nonlinear Buoyancy and Excitation Forces in the WEC-Sim Wave Energy Converter Modeling Tool. In Proceedings of the 33rd International Conference on Ocean, Offshore and Arctic Engineering (OMAE2014), San Francisco, CA, USA, 8–13 June 2014.

74. Penalba, M.R.; Mérigaud, A.; Gilloteaux, J.C.; Ringwood, J. Nonlinear Froude-Krylov force modelling for two heaving wave energy point absorbers. In Proceedings of the 11th European Wave and Tidal Energy Conference (EWTEC2015), Nantes, France, 6–11 September 2015.

75. Merigaud, A.; Gilloteaux, J.C.; Ringwood, J. A nonlinear extension for linear boundary element methods in wave energy device modelling. In Proceedings of the International Conference on Offshore Mechanics and Arctic Engineering (OMAE2012), Rio de Janeiro, Brazil, 1–6 July 2012.

76. Gilloteaux, J.C.; Bacelli, G.; Ringwood, J. A non-linear potential model to predict large-amplitudes-motions: Application to a multi-body wave energy converter. In Proceedings of the 10th World Renewable Energy Congress (WREC), Glasgow, Scotland, 19–25 July 2008; Sayigh, A., Ed.; WREC: Brighton, UK, 2008; pp. 934–939.

77. Zurkinden, A.S.; Ferri, F.; Beatty, S.; Kofoed, J.P.; Kramer, M. Non-linear numerical modeling and experimental testing of a point absorber wave energy converter. *Ocean Eng.* **2014**, *78*, 11–21.

78. Sirnivas, S.; Yu, Y.H.; Hall, M.; Bosma, B. Coupled mooring analyses for the WEC-Sim wave energy converter design tool. In Proceedings of the 35th International Conference on Ocean, Offshore and Arctic Engineering (OMAE2016), Busan, Korea, 19–24 June 2016.

79. Giorgi, G.; Ringwood, J.V. Nonlinear Froude-Krylov and viscous drag representations for wave energy converters in the computation/fidelity continuum. *Ocean Eng.* **2017**, *141*, 164–175.

80. Bharath, A.; Nader, J.R.; Penesis, I.; Macfarlane, G. Nonlinear hydrodynamic effects on a generic spherical wave energy converter. *Renew. Energy* **2018**, *118*, 56–70.

81. Kring, D.C. Time Domain Ship Motions by a Three-Dimensional Rankine Panel Method. Ph.D. Thesis, Massachusetts Institute of Technology, Cambridge, MA, USA, 1994.

82. Shin, Y.; Belenky, V.; Lin, W.; Weems, K.; Engle, A. Nonlinear Time Domain Simulation Technology for Seakeeping and Wave-Load Analysis for Modern Ship Design. *Trans. Soc. Nav. Archit. Mar. Eng.* **2003**, *111*, 557–583.

83. Letournel, L.; Ferrant, P.; Babarit, A.; Ducrozet, G.; Harris, J.C.; Benoit, M.; Dombre, E. Comparison of fully nonlinear and weakly nonlinear potential flow solvers for the study of wave energy converters undergoing large amplitude motions. In Proceedings of the 33rd International Conference on Ocean, Offshore and Arctic Engineering (OMAE2014), San Francisco, CA, USA, 8–13 June 2014; p. 23912.

84. Greenhow, M.; Vinje, T.; Brevig, P.; Taylor, J. A theoretical and experimental study of the capsize of Salter's duck in extreme waves. *J. Fluid Mech.* **1982**, *118*, 221–239.

85. Guerber, E.; Benoit, M.; Grilli, S.T.; Buvat, C. A fully nonlinear implicit model for wave interactions with submerged structures in forced or free motion. *Eng. Anal. Bound. Elem.* **2012**, *36*, 1151–1163.

86. International Towing Tank Conference. *Practical Guidelines for Ship CFD Applications*; Technical Report 7.5-03-02-03; International Towing Tank Conference: Rio de Janeiro, Brazil, 2011.

87. Ransley, E.; Greaves, D.; Raby, A.; Simmonds, D.; Hann, M. Survivability of wave energy converters using CFD. *Renew. Energy* **2017**, *109*, 235–247.

88. Yu, Y.H.; Li, Y. Reynolds-Averaged Navier-Stokes simulation of the heave performance of a two-body floating-point absorber wave energy system. *Comput. Fluids* **2013**, *73*, 104–114.

89. Neary, V.S.; Lawson, M.; Previsic, M.; Copping, A.; Hallett, K.C.; Labonte, A.; Rieks, J.; Murray, D. *Methodology for Design and Economic Analysis of Marine Energy Conversion (MEC) Technologies*; Technical Report SAND2014-9040; Sandia National Laboratories: Albuquerque, NM, USA, 2014.

90. Sjökvist, L.; Göteman, M. Peak Forces on Wave Energy Linear Generators in Tsunami and Extreme Waves. *Energies* **2017**, *10*, 1323.
91. Westphalen, J.; Greaves, D.M.; Raby, A.; Hu, Z.Z.; Causon, D.M.; Mingham, C.G.; Omidvar, P.; Stansby, P.K.; Rogers, B.D. Investigation of Wave-Structure Interaction Using State of the Art CFD Techniques. *Open J. Fluid Dyn.* **2014**, *4*, 18–43.
92. Madhi, F.; Yeung, R.W. On survivability of asymmetric wave-energy converters in extreme waves. *Renew. Energy* **2017**, in press.
93. Holmes, B. *Tank Testing of Wave Energy Conversion Systems: Marine Renewable Energy Guides*; European Marine Energy Centre: Orkney, Scotland, 2009.
94. International Towing Tank Conference. *Wave Energy Converter Model Test Experiments*; 7.5-02-07-03.7; International Towing Tank Conference: Rio de Janeiro, Brazil, 2011.
95. International Towing Tank Conference. *Final Report and Recommendations to the 23rd ITTC—The Specialist Committee on Waves*; International Towing Tank Conference: Rio de Janeiro, Brazil, 2002.
96. Payne, G. *Guidance for the Experimental Tank Testing of Wave Energy Converters*; SuperGen Marine; University of Strathclyde: Glasgow, Scotland, 2008.
97. Henry, A.; Kimmoun, O.; Nicholson, J.; Dupont, G.; Wei, Y.; Dias, F. A two dimensional experimental investigation of slamming of an oscillating wave surge converter. In Proceedings of the Twenty-Fourth International Ocean and Polar Engineering Conference, Busan, Korea, 15–20 June 2014.
98. Blommaert, C.; Van Paepegem, W.; Dhondt, P.; De Backer, P.G.; Degrieck, J.; De Rouck, J.; Vantorre, M.; Van Slycken, J.; De Baere, I.; De Backer, H. Large scale slamming tests on composite buoys for wave energy applications. In Proceeding of the 17th International Conference on Composite Materials (ICCM), Edinburgh, UK, 27–31 July 2010.
99. Van Paepegem, W.; Blommaert, C.; De Baere, I.; Degrieck, J.; De Backer, G.; De Rouck, J.; Degroote, J.; Vierendeels, J.; Matthys, S.; Taerwe, L. Slamming wave impact of a composite buoy for wave energy applications: Design and large-scale testing. *Polym. Compos.* **2011**, *32*, 700–713.
100. Hughes, O.F.; Paik, J.K. *Ship Structural Analysis and Design*; Society of Naval Architects and Marine Engineers: Jersey City, NJ, USA, 2010.
101. Lewis, E. *Principles of Naval Architecture: Stability and Strength*; Principles of Naval Architecture, Society of Naval Architects and Marine Engineers: Jersey City, NJ, USA, 1989.
102. Veldkamp, D. Chances in Wind Energy: A Probabilistic Approach to Wind Turbine Fatigue Design. Ph.D. Thesis, Delft University of Technology (TU Delft), Delft, The Netherlands, 2006.
103. Downing, S.D.; Socie, D. Simple rainflow counting algorithms. *Int. J. Fatigue* **1982**, *4*, 31–40.
104. Ambühl, S.; Ferri, F.; Kofoed, J.P.; Sørensen, J.D. Fatigue reliability and calibration of fatigue design factors of wave energy converters. *Int. J. Mar. Energy* **2015**, *10*, 17–38.
105. Ambühl, S.; Kramer, M.; Sørensen, J.D. Reliability-based Calibration of Partial Safety Factors for Wave Energy Converters. In Proceedings of the International Conference on Applications of Statistics and Probability in Civil Engineering (ICASP12), Vancouver, BC, Canada, 12–15 July 2015.
106. Ambühl, S.; Kramer, M.; Sørensen, J.D. Reliability-Based Structural Optimization of Wave Energy Converters. *Energies* **2014**, *7*, 8178–8200.
107. Sørensen, J.D. Reliability and Maintenance for Offshore Wind Turbines and Wave Energy Devices. In *Renewable Energies Offshore*; CRC Press: Boca Raton, FL, USA, 2015; pp. 39–46.
108. Thies, P.R.; Johanning, L.; Harnois, V.; Smith, H.C.; Parish, D.N. Mooring line fatigue damage evaluation for floating marine energy converters: Field measurements and prediction. *Renew. Energy* **2014**, *63*, 133–144.
109. Ferri, F.; Ambühl, S.; Fischer, B.; Kofoed, J.P. Balancing power output and structural fatigue of wave energy converters by means of control strategies. *Energies* **2014**, *7*, 2246–2273.
110. Coles, S. *An Introduction to Statistical Modeling of Extreme Values*; Springer: London, UK, 2001.
111. Naess, A.; Moan, T. *Stochastic Dynamics of Marine Structures*; Cambridge University Press: Cambridge, UK, 2012.
112. Naess, A. Extreme value estimates based on the envelope concept. *Appl. Ocean Res.* **1982**, *4*, 181–187.
113. Naess, A.; Gaidai, O. Estimation of extreme values from sampled time series. *Struct. Saf.* **2009**, *31*, 325–334.

114. Vazquez-Hernandez, A.O.; Sagrilo, L.V.S.; Ellwanger, G.B. On the Extreme Analysis Applied to Moored Floating Platforms. In Proceedings of the 22nd International Conference on Ocean, Offshore and Arctic Engineering (OMAE2003), Cancun, Mexico, 8–13 June 2003; pp. 241–247.
115. Lian, G.; Haver, S.K. Estimating long term extreme slamming from breaking waves. In Proceedings of the 34th International Conference on Ocean, Offshore and Arctic Engineering (OMAE2015), St. Johns, NL, Canada, 31 May–5 June 2015.

Article

Assessment of the Potential of Energy Extracted from Waves and Wind to Supply Offshore Oil Platforms Operating in the Gulf of Mexico

Francisco Haces-Fernandez [1], Hua Li [2,*] and David Ramirez [1]

[1] Environmental Engineering Department, Texas A&M University-Kingsville, Kingsville, TX 78363, USA; fr.haces@gmail.com (F.H.-F.); david.ramirez@tamuk.edu (D.R.)

[2] Mechanical and Industrial Engineering Department, Texas A&M University-Kingsville, Kingsville, TX 78363, USA

* Correspondence: hua.li@tamuk.edu; Tel.: +1-361-593-4057

Received: 19 March 2018; Accepted: 24 April 2018; Published: 27 April 2018

Abstract: Offshore oil platforms operate with independent electrical systems using gas turbines to generate their own electricity. However, gas turbines operate very inefficiently under the variable offshore conditions, increasing fuel costs and air pollutant emissions. This paper focused on investigating the feasibility of implementing a hybrid electricity supply system for offshore oil platforms in the Gulf of Mexico, both for the United States and Mexico Exclusive Economic Zones. Geographic Information Systems methodologies were used to analyze the data from various sources. Three different scenarios were studied, including wind power only, wave power only, and wind and wave power combined. The results showed that all the offshore locations were within accepted feasible distance to the coast for connecting to the onshore grid. Most of the locations had acceptable power levels of either wind or wave energy while the combination of both resources can improve the overall energy harvesting efficiency and reduce the variability in a significant number of locations. The proposed methodology can be applied for specific locations with finer spatial and time resolution, which will allow stakeholders to improve the decision making process, generate important savings on the normal operation, reduce pollution, and potentially increase income by selling surplus energy from renewable sources.

Keywords: wind power; wave power; offshore oil platforms; Gulf of Mexico; Geographic Information Systems; WaveWatch III

---

1. Introduction

Production and ancillary activities on offshore oil platforms require electric power that ranges from 10 MW up to hundreds of MW, depending on the sizes of oil platforms [1,2]. Most of these platforms function with independent electrical systems, generating energy using gas turbines, which are expensive to operate [1]. The nature of the offshore production operations generates a variable system, with periods of low energy consumption followed by higher load requirements [1]. Gas turbine fuel efficiency is affected under variable operation conditions, considering that the energy consumption during idling conditions can be about 20% of what they would consume at full power [1,3]. Gas turbines in offshore oil platforms normally operate under 30% efficiency ranges, when the normal average efficiency should be about 55% considering a combined cycle gas power plant [3–8]. Furthermore, gas turbines increase emissions of NO_x, SO_2, Volatile Organic Compounds (VOC), Particulate Matter 10 micrometers or less (PM_{10}), Particulate Matter 2.5 micrometers or less ($PM_{2.5}$), CO, and CH_4 under these inefficient operation cycles [1,2,7,9,10]. Studies on offshore oil platforms in Texas and Norway indicated that gas turbines are the main contributors of several criteria air pollutants on the offshore areas [4,6,7,9,10].

Several approaches have been taken to ameliorate this problem [11,12]. Norway has required that new or major modifications on offshore oil and gas platforms must consider the use of onshore energy since 2007 [5,7,13–15]. In fact, several fields in that region use onshore power [5,7], such as Ormen Lange, Snøhvit, Gjøa, Valhall, and Goliat. Another alternative involves the use of offshore renewable energy, such as wind and wave energies [3,7,14,15]. Simulation and modeling studies have indicated that important savings and reduction on pollutant emissions can be obtained when developing isolated offshore systems with wind turbines and oil platforms. Simulation for the year 2009, performed by Korpås et al. [1] in the North Sea for the inclusion of wind energy on oil platform operation, resulted in 53,790 tons of CO_2 and 367 tons of NO_x reductions, with yearly savings of €5.73 Million assuming €64/MWh for wind energy generated [1].

Furthermore, it has been ascertained that the combination of the two alternatives, using offshore renewable energy and connecting to the onshore grid, could create important synergies for both oil and renewable energy industries [6,8,16–21]. Supplying offshore oil platforms with renewable energy reduces the load for the gas turbines, with the variability issue tackled by the use of coupled energy storage systems [3,17,22]. Including a connection to the onshore electric grid increases the reliability of the system by supplementing the electricity supply for longer than expected renewable energy interruptions without ramping up gas turbines, and allows for the excess energy generated from renewable sources to be sold to the grid [4,6,8]. Therefore, it is able to balance energy generation and consumption in a safe and efficient manner while generating important savings and synergies, potentially even generating profit for the facility [4,6,14,23].

Although offshore oil and ocean renewable energy industries have coexisted independently over the last two decades, it is becoming increasingly clear that they can benefit from each other in many ways [15]. Besides the previous stated yearly savings for the oil industry (€5.73 Million assuming €64/MWh for wind energy generated), there are many other short and long term advantages [1]. The reduction or elimination of heavy and bulky gas turbines will reduce the design and construction cost of offshore oil platforms, free spaces for other activities, and reduce the risk associated with gas turbine operations [24]. With continuous technological improvements, offshore wind energy has become a mature industry, which is competitive with traditional energy resources. Meanwhile, wave energy is considered as one of the biggest untapped renewable energy resource in the world with predicted potentials enough to provide ten percent of the world energy in the following decades [25–28]. It is desirable to introduce wave energy to the electrical systems of installations in oceanic environments, specially isolated systems such as islands, to reduce costs and pollution caused by fossil fuels [29–31]. Ocean renewable energy industry, especially wave energy, will also benefit from the oil industry and accelerate its development by applying all the knowhow from the offshore oil sector, accumulated through more than 50 years in the offshore environment [32]. In the long term, the possibility of cost sharing on offshore electrification will be beneficial for both industries. Electrification on the offshore environment is expensive and complex, and the proposed scheme will make the distribution of the installation and maintenance costs more efficient if the infrastructure can be shared by both offshore oil and renewable energy installations, providing each with important advantages [2,3,13,14,16,33,34]. The oil industry will not only be able to sell excess renewable energy to the grid, potentially generating extra income, but it can transform aging offshore infrastructures into wind installations. This could reduce decommissioning costs and allow the oil industry to continue investing on the renewable energy sector [35,36].

Motivated by these considerations, this paper focused on investigating the feasibility of using wind and wave energy to supply electricity to the U.S. offshore oil platforms and potential Mexico Exclusive Economic Zones (EEZ) in the Gulf of Mexico using Geographical Information Systems (GIS) methods. The locations of the areas for hydrocarbon exploration and extraction on Mexican EEZ were named as Comisión Nacional de Hidrocarburos (CNH) areas in this paper. Figure 1 shows the entire study area with small green dots representing offshore oil platforms and green rectangles representing CNH areas. In order to investigate and compare the wind and wave energy behaviors within the CNH areas, three major CNH regions (labeled as Regions I, II, and III in Figure 1) were grouped to conduct a detailed analysis.

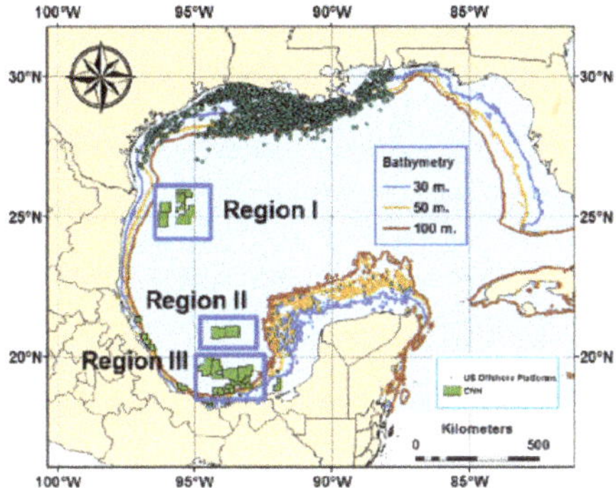

Figure 1. Study area with offshore platforms and Comisión Nacional de Hidrocarburos (CNH) areas identified.

Bathymetry data was obtained from [37], and included in Figure 1 to provide additional information for the decision making process on the selection and deployment of the energy harvesting equipment. Depending on the sea depth, three wind turbine foundations systems are available: monopole (up to 30 m), jacket-tripod (up to 50 m) and floating (deeper than 50 m) [38]. The average depth of existing offshore wind turbines is 22 m, and the largest depth for a jacket-tripod foundation is 44 m (EnBV Baltic 2 Wind Farm in Germany) [36,38]. The floating wind turbine concept was originally proposed in 1970, and Blue H technologies conducted a test on the Italian coast in 2008 followed by the Poseidon 37 project in 2009 [38]. Statoil also connected a floating wind turbine to the grid in 2009, while Repsol installed a floating 2 MW Vestas wind turbine on the Portuguese coast in 2011 [38]. Hywind offshore floating wind farm started operations in October 2017 in Scotland with five wind turbines [36]. The inclusion of bathymetry data in the GIS analysis will help selecting the most adequate foundation system for the wind turbine model to ensure the best fit for the meteorological and geographical conditions of each location. Both the jacked-tripod foundation [39] and the floating wind turbine system [26] might incorporate a hybrid wind-wave system that would bring a number of synergistic benefits to the project, including cost sharing to reduce operational and management expenditures [39,40].

2. Materials and Methods

This paper assessed the wind and wave energy potentials in the Gulf of Mexico for application in the oil and gas industry for both the United States and Mexico. GIS analysis was performed to ascertain the possibility of connecting offshore oil and gas facilities with the onshore electric grid, and to analyze the available wind and wave energy resources in each particular area considering historical data provided by the National Oceanic and Atmospheric Administration (NOAA) WaveWatch III system [41,42]. Big Data analysis was integrated into GIS to investigate the feasibility of supplying electricity to offshore oil facilities in the Gulf of Mexico with wave and wind energy. The locations of more than four thousand oil and gas platforms in the northern region of the Gulf of Mexico (U.S. oil platforms) were obtained from data published by the U.S. Geological Survey, Coastal and Marine Geology Program from information provided by the Minerals Management Service [43]. The locations of the areas for hydrocarbon exploration and extraction on Mexican EEZ were obtained from the National Commission for Hydrocarbons (CNH) of the Mexican Federal Government (CNH areas) [44]. The CNH areas are in the process of being assigned, through public bidding, for hydrocarbon exploration and extraction by Mexican and International companies, individually or in join projects with Petroleos Mexicanos (PEMEX), in accordance to legal changes to the Mexican Constitution of December 2013 [44].

This paper focused on developing a general methodology to evaluate the feasibility of using renewable energy to supply offshore oil platforms. The use of actual extracted electric power, rather that energy density values, was considered a more effective indicator to understand the energy capabilities of the region. Since two energy resources, wave and wind, were being simultaneously analyzed, the output data was used with the same measuring unit rather than wave density in kW/m and wind density in kW/m^2. Therefore, it was necessary to select equipment for the harvesting of wind and wave energy to develop and validate the methodology.

Wave energy was assumed to be extracted by one Pelamis P2 750 kW Wave Energy Converter (WEC). There is no WEC being commercially operated in large scale nowadays, and the developer of the Pelamis WEC went into administration with Wave Energy Scotland now owning their intellectual property and assets [45]. However, the Pelamis WEC was considered a good option because of several important reasons. First of all, it has been extensively used in previous researches [25,27,46–48], and its power curve was provided by previous research and the manufacturer (Figure 2) [49–52]. Secondly, the Pelamis P1 750 kW was the first WEC to operate commercially on a wave farm in Aguçadoura, Portugal that was connected to the grid [53]. The closing of this wave farm was mostly due to the financial collapse of the main shareholder of that project, Babcock & Brown infrastructure group from Australia [53]. At last, the second generation Pelamis P2 750 kW was successfully tested for 3 years in the Billia Croo wave test site [45,53].

Tp (s)	3	4	5	6	7	8	9	10	11	12	13	14	15	16	17	18	20
Hs (m)				Hs: wave height (meter) Tp: wave peak period (second)												
0.125	0	0	0	0	0	0	0	0	0	0	0	0	0	0	0	0	0
0.5	0	0	0	0	0	0	0	0	0	0	0	0	0	0	0	0	0
1	0	0	11	27	50	62	64	57	49	41	34	28	23	0	0	0	0
1.5	0	0	26	62	112	141	143	129	110	91	76	63	52	43	36	30	23
2	0	0	66	109	199	219	225	205	195	162	135	112	93	77	64	54	41
2.5	0	7	93	171	279	342	351	320	274	230	210	174	145	120	100	84	65
3	0	91	180	246	402	424	417	369	343	331	275	229	208	173	144	120	93
3.5	0	86	211	326	484	577	568	502	421	394	330	312	260	216	196	164	140
4	105	216	326	394	632	616	583	585	494	454	374	361	339	283	236	197	153
4.5	94	233	371	467	735	744	738	634	626	520	473	390	382	319	299	250	208
5	259	364	469	539	750	750	750	750	644	641	531	482	399	394	330	308	274
5.5	428	497	566	612	750	750	750	750	750	635	642	532	482	400	399	341	322
6	597	630	663	684	750	750	750	750	750	750	616	633	525	476	396	386	329
6.5	750	750	750	750	750	750	750	750	750	750	723	592	617	513	458	430	384
7	750	750	750	750	750	750	750	750	750	750	750	692	566	560	500	474	425
7.5	750	750	750	750	750	750	750	750	750	750	750	748	610	607	542	518	467
8	750	750	750	750	750	750	750	750	750	750	750	750	630	653	584	562	509
8.5	750	750	750	750	750	750	750	750	750	750	750	750	650	699	626	606	551
9	750	750	750	750	750	750	750	750	750	750	750	750	670	746	668	650	592
9.5	750	750	750	750	750	750	750	750	750	750	750	750	691	750	710	694	662
10	750	750	750	750	750	750	750	750	750	750	750	750	711	750	750	738	734

Figure 2. Electric power output (kW) table of Pelamis WEC [49,50].

Wind energy was assumed to be extracted by one Vestas V90 3 MW wind turbine. It is one of the most widely used offshore wind turbines in the world, and its power curve has been provided by the manufacturer (Figure 3) [33,34]. The Vestas V90 3 MW has a hub height of 80 m, diameter of 90 m, cut-in wind speed of 4 m/s, rated wind speed of 16 m/s, cut-out wind speed of 25 m/s, and restart (cut-back-in) wind speed of 20 m/s [54]. It is also possible to modify existing offshore wind turbines by changing their current foundation systems to floating system [55]. Manufacturers such as Vestas [56], Siemens [38] and General Electric [57] are installing its current offshore wind turbine models on floating foundations [38,58,59].

Wind Speed	Air Density [kg/m^3]											
[m/s]	0.97	1	1.03	1.06	1.09	1.12	1.15	1.18	1.21	1.225	1.24	1.27
0	0	0	0	0	0	0	0	0	0	0	0	0
4	53	56	59	61	64	67	70	72	75	77	78	81
5	142	148	153	159	165	170	176	181	187	190	193	198
6	271	281	290	300	310	319	329	339	348	353	358	368
7	451	466	482	497	512	528	543	558	574	581	589	604
8	691	714	737	760	783	806	829	852	875	886	898	921
9	995	1028	1061	1093	1126	1159	1191	1224	1257	1273	1289	1322
10	1341	1385	1428	1471	1515	1558	1602	1645	1688	1710	1732	1775
11	1686	1740	1794	1849	1903	1956	2010	2064	2118	2145	2172	2226
12	2010	2074	2137	2201	2265	2329	2392	2454	2514	2544	2573	2628
13	2310	2382	2455	2525	2593	2658	2717	2771	2817	2837	2856	2889
14	2588	2662	2730	2790	2841	2883	2915	2940	2958	2965	2971	2981
15	2815	2868	2909	2939	2960	2975	2984	2990	2994	2995	2996	2998
16	2943	2965	2979	2988	2993	2996	2998	2999	2999	3000	3000	3000
17	2988	2994	2997	2998	2999	3000	3000	3000	3000	3000	3000	3000
18	2998	2999	3000	3000	3000	3000	3000	3000	3000	3000	3000	3000
19	3000	3000	3000	3000	3000	3000	3000	3000	3000	3000	3000	3000
20	3000	3000	3000	3000	3000	3000	3000	3000	3000	3000	3000	3000
21	3000	3000	3000	3000	3000	3000	3000	3000	3000	3000	3000	3000
22	3000	3000	3000	3000	3000	3000	3000	3000	3000	3000	3000	3000
23	3000	3000	3000	3000	3000	3000	3000	3000	3000	3000	3000	3000
24	3000	3000	3000	3000	3000	3000	3000	3000	3000	3000	3000	3000
25	3000	3000	3000	3000	3000	3000	3000	3000	3000	3000	3000	3000

Figure 3. Electric Power output (kW) table of Vesta V90 3 MW [54].

Meteorological data over 36 years (1979–2015) in the Gulf of Mexico region generated by the NOAA's WaveWatch III system was used to first calculate wave and wind power output by considering one device in each geographical location. The resolution of NOAA data is one sixth longitude by one sixth latitude, which is also the dimension of each geographical location considered in this paper.

The electric power output generated on each location was calculated by applying meteorological data with the power curves (Figures 2 and 3). The significant wave height-*Hs* (in meters) and the dominant wave period-*Tp* (in seconds) were applied to the Pelamis electric power curve (Figure 2) to estimate its power output. Dominant wave period (*Tp*) was calculated from the energy wave period provided by NOAA's WaveWatch III by multiplying factor α. The value of α approaches to one as the spectral width decreases, and it has been considered as 0.86 for a fully developed ocean [60,61]. In this paper, 0.9 was selected as α value as indicated by previous research [60–64], which is also the equivalent of assuming a standard JONSWAP (Joint North Sea Wave Observation Project) spectrum (Deutsches Hydrographisches Institut—Hamburg—1973 Hasselmann et al.) [35,37,61,65–68].

Electric power generated by the Vestas V90 3MW was calculated based on wind speeds provided by the NOAA from the WaveWatch III system at the same locations and time periods as for wave. The provided wind speed vectors were at a height of 10 m above sea level, which was converted to wind speeds at wind turbine hub height (80 m for Vestas V90 3MW) applying the wind profile power law formula Equation (1), considering near-neutral stability conditions at locations in the Gulf of Mexico [69,70]:

$$\frac{U_2}{U_1} = \left(\frac{Z_2}{Z_1}\right)^P \tag{1}$$

where, U_2 is the wind speed at height Z_2, U_1 is the wind speed at height Z_1, and $P = 0.10$ for offshore wind turbine [70].

The maps created in this paper to analyze wave power and wind power generated by the Pelamis 750 kW and the Vestas V90 3 MW are represented by a color scale applying Jenks Natural Breaks classification method. This method classifies values by minimizing each category average deviation from the category mean, and simultaneously maximizes each category deviation from the means of the other categories in the same array aiming to minimize variance on each category while maximizing variance between categories. It is a very good fit to evaluate geospatial data that has high variation on temporal and spatial criteria, such as wave energy and wind energy. It is a good analysis tool for arrays with relatively big differences on the data values [71–74]. The color bars presented on each of the maps in this paper ensured a good contrast between the diverse electric power extracted to perform geospatial analysis and gain better understanding on their temporal and spatial behavior and variability. These different classification classes allow finding the best fit for every particular location and time period. Maps in this paper were designed to provide additional information on the behavior of electric power from wind and wave in addition to the results presented on the corresponding tables and graphics. However, since the main purpose of the maps is to show contrast of the renewable energy behaviors (in light of its high geo temporal variability), comparison between maps should always consider that the color bar in each map may be in different scale since the Jenks Natural Breaks classification method was used.

Figure 4 represents the average wave power generated by one Pelamis 750 kW WEC over the 36 year period on each location in the Gulf of Mexico. It shows a high wave power concentration on the western central location and the Yucatan Strait. It can be observed that the CNH region I in the north of the Gulf of Mexico is one of highest wave power regions, and a significant number of U.S. oil platforms and CNH regions II and III in the Gulf of Campeche are located in the yellow ring surrounding the wave power map, which indicates commercially acceptable power levels while not being subjected to the harsher marine conditions caused by more energetic waves.

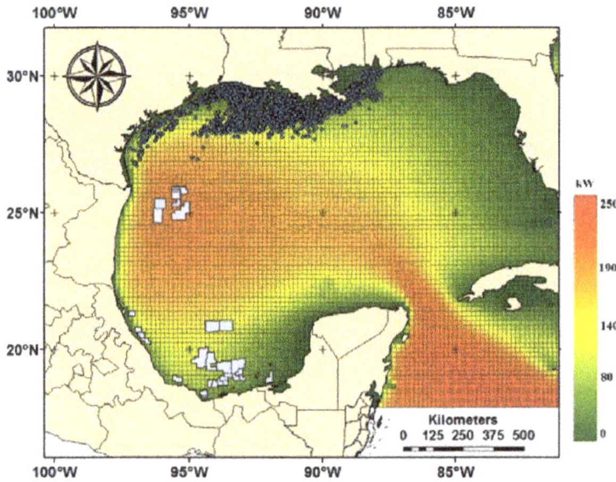

Figure 4. Yearly average wave power generated by one Pelamis 750 kW WEC over 36 years.

The average wind power generated by one Vestas V90 3 MW over the 36 year period in the Gulf of Mexico is presented in Figure 5, showing a high wind power concentration on the northwestern Gulf of Mexico region, the western Yucatan Peninsula coast, and the Florida Strait. It can be observed that the CNH region I in the northern region of the Gulf of Mexico is one of the highest wind power regions similarly with wave power. In addition, some U.S. oil platforms are located on high wind power locations along the Texas coast, and a significant number of U.S. oil platforms and CNH regions II and III in the Gulf of Campeche are located in the yellow band surrounding the wind power map, which indicates acceptable power levels.

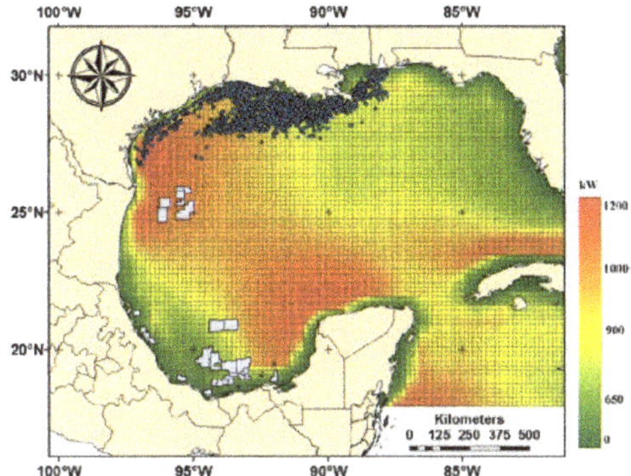

Figure 5. Yearly average wind power generation by one Vestas V90 3 MW over 36 years.

3. Results and Discussion

The distance to the coast is an important factor when considering the feasibility of combining offshore renewable energy with oil platforms and the onshore grid. Longer distances increase the capital and maintenance costs and energy losses due to transmission. Three hundred (300) kilometers has been considered an acceptable distance from the coast to oil rigs when connecting to the onshore grid [7,13]. The Troll A offshore oil platform, located to the west of Bergen, Norway, has been successfully connected to the onshore grid over a distance of 65 km [7]. In this paper, two distinctive analyses related to distance to the coast were performed to calculate the distance from each installation to its closest coastal location, considering the U.S. oil platforms and the CNH areas separately. For the CNH areas, the distance was calculated from each of the WaveWatch III data locations that are relevant to a particular CNH area.

The distance analysis results are presented in Figure 6, which shows the cumulative percentage of U.S. oil platforms and CNH areas according to different ranges of distance to the coast, indicating an acceptable distance range (less than 300 km) to the coast on both cases. Almost 80% of the U.S. oil platforms and CNH areas are located at a distance less than 80 km and 90 km from the coast, respectively. The maximum distance is 230 km for U.S. oil platforms and 240 km for the CNH areas, which is less than 300 km as the acceptable distance.

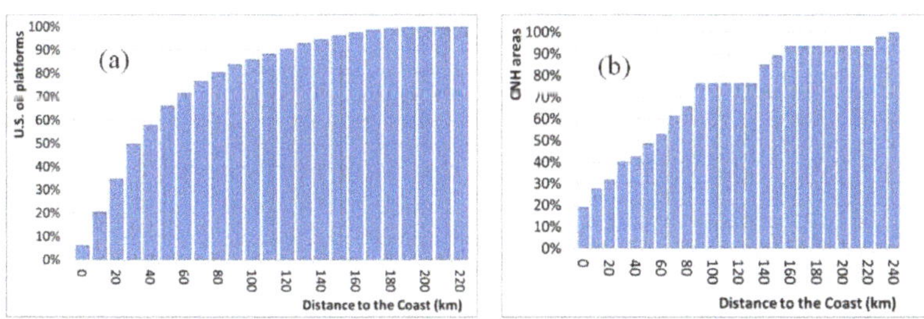

Figure 6. Cumulative histograms of distance to the coast: (**a**) U.S. oil platforms, and (**b**) CNH areas.

Three different scenarios were analyzed for in this paper, including electric power from: (1) wind only, (2) wave only, and (3) wind and wave combined. The results and discussion of each scenario are presented below.

Previous research [66–68,75] has indicated that both wind and wave energy in this area has a distinctive seasonal and monthly behavior. Since wave energy is particularly dependent on local weather patterns in the Gulf of Mexico, both wind and wave energy share the seasonal and monthly variability behavior. In the Gulf of Mexico, summer has been ascertained as the lowest wind and wave energy periods, with July and August having the lowest power outputs in the entire year. On the other hand, both resources have higher power output during fall and winter seasons with January and December providing the highest power output in the entire year. Spring season behaves as a transition period to the lower summer months [66–68]. Inter-year variations are also important, and should be considered when performing particular analysis for an installation or cluster of installations [67].

3.1. Wind Power Only

The wind energy assessments for the U.S. oil platforms and the three CNH areas were done monthly, considering that the Gulf of Mexico presents important temporal variability on both wind and wave energy resources throughout the year [66–68]. In addition, the assessments were done separately for both regions due to several reasons. First, the U.S. oil platforms were considered as fixed point installations while the CNH areas were considered as polygons, with larger area including more WaveWatch III data locations. Second, the map shown in Figure 5 indicates that the wind energy characteristics of both locations are different, and it was considered important to differentiate the wind energy behavior in both locations. Last, the assessment objectives were different, considering that the U.S. oil platforms are already in operation while the CNH areas are mostly geared for planning and future installations.

Table 1 shows the percentages of total U.S. oil platforms and CNH areas corresponding to different wind power output levels. Wind energy in the U.S. oil platforms areas had a seasonal behavior with each month of the year showing a distinctive pattern. In the U.S. oil platforms areas, the wind power was high from November to April. It can also be observed that most U.S. oil platform locations were in higher wind power levels between December and March (curves skewed to the higher levels), where the curves of November and April tended to be similar as normal distribution curve. On the other hand, most U.S. oil platforms were in lower wind power areas from June to August (curves skewed to the lower levels) with low overall power levels, which match with the expected summer low offshore renewable energy levels [66–68,75]. The months of May and September showed moderate power levels with curves tending to be normally distributed as well.

Table 1. Percentage of U.S. oil platforms and CNH areas generating the indicated electric power from one Vestas V90 3 MW.

Power (kW)	Wind Power U.S. Oil Platforms (%)												Wind Power CNH Total Areas (%)											
	Jan	Feb	Mar	Apr	May	Jun	Jul	Aug	Sep	Oct	Nov	Dec	Jan	Feb	Mar	Apr	May	Jun	Jul	Aug	Sep	Oct	Nov	Dec
150	–	–	–	–	–	–	–	–	–	–	–	–	–	–	–	–	–	2	3	5	2	1	–	–
200	–	–	–	–	–	–	–	1	–	–	–	–	–	–	–	–	–	3	3	4	3	–	–	–
250	–	–	–	–	–	–	4	4	–	–	–	–	1	–	–	2	3	2	4	3	2	2	2	2
300	–	–	–	–	–	1	15	27	–	–	–	–	–	1	2	2	–	3	2	5	3	2	–	–
350	–	–	–	–	–	3	50	44	–	–	–	–	2	2	2	–	2	2	3	7	5	2	2	2
400	–	–	–	–	1	3	15	14	1	–	–	–	2	1	1	2	2	4	4	5	3	2	2	2
450	–	–	–	–	–	24	6	7	1	–	–	–	1	–	1	2	4	4	3	12	5	2	2	2
500	–	–	–	–	–	29	6	3	3	–	–	–	2	2	–	2	3	2	4	19	4	2	1	2
550	–	–	–	–	3	16	2	–	10	–	–	–	1	1	2	3	3	5	4	11	18	4	2	–
600	–	–	–	1	2	7	2	–	5	1	–	–	2	4	3	3	5	4	8	6	18	3	3	3
650	–	–	–	–	14	7	–	–	14	2	–	–	4	2	2	2	5	4	13	5	9	5	2	3
700	–	–	–	–	25	7	–	–	31	5	–	–	2	2	2	5	3	2	11	9	23	4	3	2
750	–	–	–	–	18	2	–	–	33	6	–	–	2	4	3	3	3	4	7	4	5	17	4	3
800	–	–	1	2	10	1	–	–	2	5	1	–	3	2	2	4	4	27	6	3	–	13	4	4
850	–	2	2	3	7	–	–	–	–	12	2	–	5	7	5	4	6	15	5	–	–	12	15	11
900	1	2	4	4	6	–	–	–	–	16	2	2	10	9	6	5	8	10	4	–	–	3	14	13
950	–	5	7	13	9	–	–	–	–	18	8	1	10	10	5	8	9	5	6	2	–	2	11	8
1000	3	5	11	16	4	–	–	–	–	30	5	3	8	7	8	5	8	–	5	–	–	21	6	8
1050	6	10	17	19	1	–	–	–	–	5	9	6	10	9	7	5	7	2	3	–	–	3	2	9
1100	5	10	13	19	–	–	–	–	–	–	10	5	9	8	6	7	18	–	–	–	–	–	–	–
1150	10	13	28	16	–	–	–	–	–	–	14	12	1	2	8	5	7	–	2	–	–	–	–	–
1200	13	16	14	6	–	–	–	–	–	–	19	13	–	–	6	3	–	–	–	–	–	–	–	–
1250	17	20	3	1	–	–	–	–	–	–	17	14	–	–	7	19	–	–	–	–	–	–	9	–
1300	18	12	–	–	–	–	–	–	–	–	11	21	–	6	22	9	–	–	–	–	–	–	11	–
1350	18	5	–	–	–	–	–	–	–	–	2	16	5	20	–	–	–	–	–	–	–	–	5	3
1400	7	–	–	–	–	–	–	–	–	–	–	6	16	1	–	–	–	–	–	–	–	–	–	14
1450	2	–	–	–	–	–	–	–	–	–	–	1	4	–	–	–	–	–	–	–	–	–	–	9

The capacity factor of wind turbine (actual power output divided by nameplate capacity) is normally 30%, but it can vary between 20% and 30% due to time-varying influences, such as wind resource inter-year variations [69,76]. Vestas V90 3MW installed in the Barrow Offshore Wind Farm (UK) reported a 24.1% capacity factor while a capacity factor of 27.7% was reported for those installed in Kentish Flats Offshore Wind Farm (Thames River Estuary in the Kent coast, UK) [77]. Different offshore wind farms in other European locations reported different capacity factors, depending on the installed wind turbine model and the local meteorological conditions [77]. Considering the nameplate capacity of one Vestas V90 is 3 MW, it will reach 20% or higher capacity factor if its output power is 600 kW or higher, while it will reach 30% or higher capacity factor if its output power is 900 kW or higher. For most of the U.S. oil platforms areas, the capacity factors of seven months fall above the 30% range and nine months above the 20% range with only three months (June to August) having less than 20% capacity factor. For most of the CNH areas capacity factor are above 20–30% for all months of the year, except August. However, CNH areas do not have distinctive pattern of the wind energy behavior within the same month. Instead, different patterns in the CNH areas overlapped in the same month.

To complement and further explain the results listed in Table 1, maps of January and August were created as shown in Figure 7 applying Jenks natural breaks classification method for the color bars. January (Figure 7a) showed several distinctive patterns overlapped or combined in which higher wind power was concentrated in the northern section of the Gulf of Mexico, explaining why the U.S. oil platform locations performed better than CNH in January. In addition, Figure 7a also directly shows that 25% of CNH areas fell in the high power generation range (1350–1450 kW) in January in Table 1, which is disjointed from the main body of data and performed better than the rest of the CNH areas. This can be explained by CNH Region I being engulfed by the higher northern wind pattern. Meanwhile, Figure 7b allows understanding that some sections of the CNH areas performed better than the US oil platforms during August. Sections of CNH Regions II and III are part of the southern Gulf of Mexico higher wind pattern during August. Furthermore, the maps indicate that one possible reason for CNH areas lacking of uniform behavior is that CNH areas spread over larger geographic areas, which experienced different wind energy patterns.

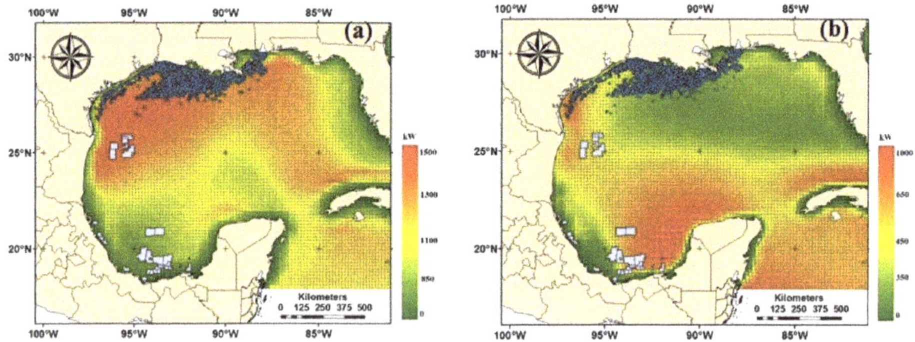

Figure 7. Monthly average wind power generated by one Vestas V90 3 MW over 36 years applying Jenks natural breaks classification method for color bar: (**a**) January, and (**b**) August.

The maximum wind power level in January was 1500 kW, while the maximum level in August was 1000 kW. In January, CNH region I was in the highest wind power level, while wind power in CNH regions II and III was in lower level. In August, it shows an almost opposite scenario. CNH regions II and III produced higher wind power than CNH region I did.

Individual histogram data of different portions of the CNH areas was then considered to better understand the wind energy behavior in these areas. Figure 8 shows that different percentages of CNH region I produced different monthly wind power levels. The ranges of monthly wind power level of CNH region I were within 200 kW, with small variation within the same month. Summer months had lower wind power levels. As for capacity factor, only August showed a poor performance below 20% (600 kW). June, July and September had capacity factors ranging between 20% (600 kW) and 30% (900 kW), while the other months performed above 30%, with best power levels from November to April.

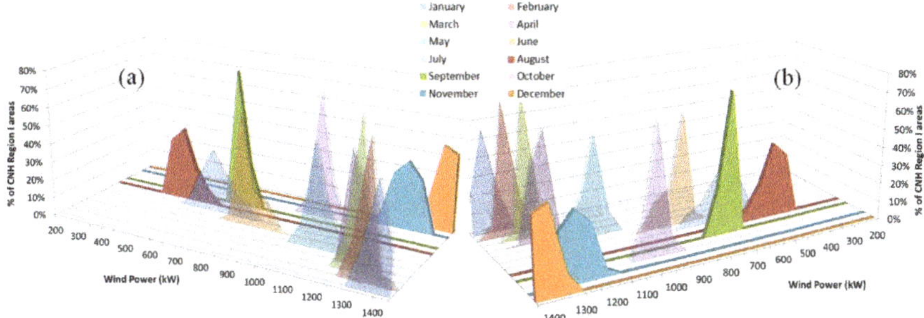

Figure 8. Percentage of the area of CNH region I generating the indicated wind power from one Vestas V90 3 MW: (**a**) front view, and (**b**) back view.

The CNH region II (Figure 9) had lower overall wind power levels than the CNH region I, with a high seasonal behavior. Similarly, with the exception of August, the other months had wind power capacity factor higher than 20%. There were seven months, from November to May, having power performance higher than 30% of nameplate capacity. Furthermore, Table 2 shows that the CHN region III, located in the south of the Gulf of Campeche, had overall low wind power levels and less compact distribution, indicating that wind power in CNH region III had very high variation within a given month. Similarly, most locations in CNH region III have 20% or above capacity factor over the entire year except August and September, and almost all the locations had 30% or above capacity factor from November to May.

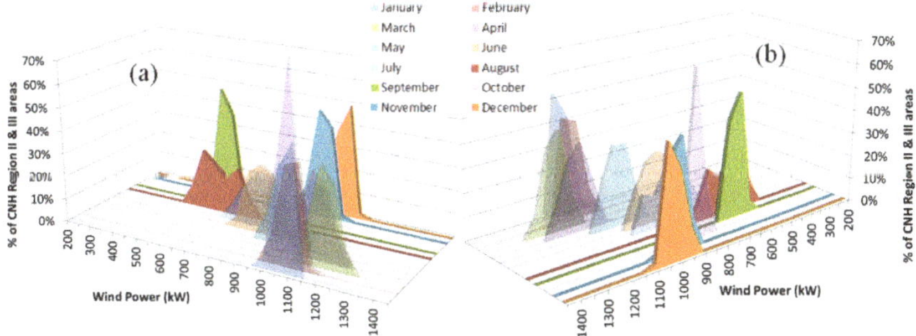

Figure 9. Percentage of the area of CNH region II generating the indicated wind power from one Vestas V90 3 MW: (**a**) front view, and (**b**) back view.

Table 2. Percentage of the CNH region III area that would generate the indicated electric power from one Vestas V90 3 MW.

Power (kW)	Wind Power in CNH Region III (% of Areas)											
	Jan	Feb	Mar	Apr	May	Jun	Jul	Aug	Sep	Oct	Nov	Dec
500	–	–	–	–	–	–	–	8	2	–	–	–
550	–	–	–	–	–	–	2	11	4	–	–	–
600	–	–	–	–	–	3	5	12	6	2	–	–
650	–	–	–	–	3	6	6	8	6	2	–	–
700	–	–	–	2	5	5	7	12	2	2	2	–
750	–	–	–	3	3	9	10	8	40	3	2	2
800	2	3	3	5	8	9	6	6	32	6	2	3
850	3	3	4	3	13	11	5	5	8	6	3	2
900	5	5	3	6	8	6	10	17	–	3	5	6
950	2	6	4	8	8	10	6	7	–	44	4	–
1000	5	2	6	10	9	12	6	6	–	26	3	5
1050	5	18	10	10	12	13	5	–	–	6	38	22
1100	26	19	11	13	14	10	5	–	–	–	29	33
1150	19	23	7	18	8	6	7	–	–	–	9	15
1200	17	9	18	9	6	–	15	–	–	–	3	9
1250	10	6	16	6	3	–	5	–	–	–	–	3
1300	6	6	9	5	–	–	–	–	–	–	–	–
1350	–	–	6	2	–	–	–	–	–	–	–	–
1400	–	–	3	–	–	–	–	–	–	–	–	–

3.2. Wave Power Only

A similar histogram analysis related to the wave energy extracted by one Pelamis 750 kW was conducted to investigate the feasibility of using wave energy to supply electricity to the U.S. oil platforms and potential CNH areas (Table 3). By comparing to wind energy results above, it is possible to assess which renewable energy source has better potential for each location. Previous research has estimated that the capacity factor of the Pelamis 750 kW in several locations in Canada would fluctuate from the lowest value of 14.3% in the Tofino Ucluelet location to the highest estimation of 26.2% at the Hibernia Oil Platform [78]. Different research indicated that the Pelamis 750 kW would operate at 20% capacity factor in San Francisco, California [49,77]. In addition, the performances of two other WECs (AquaBuOY and WaveDragon) were evaluated alongside the Pelamis 750 kW in the same Canadian locations. Results indicated that the lowest annual performance among the WECs was the AquaBuOY in Tofino Ucluelet with 9.8% capacity factor, while the highest WEC performance was the WaveDragon with 32.1% at Hibernia Oil Platform [79]. Considering 20% (150 kW) as acceptable capacity factor for Pelamis WEC [69,79], it can be observed from Tables 4 and 5 that most U.S. oil platform areas appears to be underperforming with capacity factor less than 20%. The proximity of the U.S. oil platforms to the northern coast of the Gulf of Mexico and the dependence of wave energy on northerly weather patterns can explain the low wave energy levels. However, a small number of these oil platforms perform above 20% capacity factor during eight months over the years. It is important to further segment the U.S. oil platform areas or perform individual analysis for each platform to assess the feasibility of supplying electricity with wave energy to a particular platform.

Table 3. Wave power generated by one Pelamis 750 kW WEC presented by the percentage of study areas under different power levels.

Power (kW)	Wave Power U.S. Oil Platforms (%)												Wave Power CNH Total Areas (%)											
	Jan	Feb	Mar	Apr	May	Jun	Jul	Aug	Sep	Oct	Nov	Dec	Jan	Feb	Mar	Apr	May	Jun	Jul	Aug	Sep	Oct	Nov	Dec
10	16	16	14	14	15	23	30	26	13	16	15	15	–	–	–	–	3	13	17	17	1	–	–	–
20	6	6	4	5	14	20	29	31	11	8	10	6	–	–	–	1	7	10	12	21	3	–	–	–
30	9	8	4	5	12	12	14	23	12	11	8	9	–	–	–	–	7	11	14	22	6	–	–	–
40	8	6	6	8	10	12	9	18	12	4	7	5	–	–	–	3	9	11	13	39	8	–	–	–
50	11	12	8	6	9	11	12	2	11	13	11	13	–	–	–	4	10	9	10	1	11	–	–	–
60	6	4	7	9	6	8	6	–	7	6	5	4	–	–	2	4	9	8	34	–	12	2	–	–
70	3	5	7	3	8	7	–	–	9	4	4	3	–	–	2	5	7	8	–	–	18	2	–	–
80	4	4	4	4	5	5	–	–	10	3	3	4	–	–	4	8	5	4	–	–	12	2	–	–
90	2	4	4	6	5	2	–	–	8	4	3	3	–	–	2	8	6	12	–	–	2	3	–	3
100	4	4	5	4	3	–	–	–	6	8	3	4	–	2	3	9	7	14	–	–	–	2	2	3
110	3	3	4	5	3	–	–	–	1	4	4	4	1	3	3	8	4	–	–	–	11	2	2	–
120	4	5	3	6	3	–	–	–	–	5	6	4	3	2	5	5	–	–	–	–	16	5	2	4
130	5	5	4	5	2	–	–	–	–	4	4	4	2	2	8	5	–	–	–	–	–	7	1	3
140	3	2	6	4	–	–	–	–	–	5	3	3	1	2	8	6	–	–	–	–	–	11	2	3
150	2	4	5	5	1	–	–	–	–	5	3	3	2	3	9	5	21	–	–	–	–	10	2	3
160	3	3	2	3	–	–	–	–	–	–	2	3	1	4	7	2	5	–	–	–	–	15	4	3
170	2	2	4	3	–	–	–	–	–	–	3	3	2	6	5	–	–	–	–	–	–	10	5	6
180	2	2	4	3	–	–	–	–	–	–	2	2	4	8	7	–	–	–	–	–	–	2	7	5
190	2	3	3	2	–	–	–	–	–	–	2	2	4	8	5	–	–	–	–	–	–	19	6	8
200	3	2	2	–	–	–	–	–	–	–	2	4	7	10	12	–	–	–	–	–	–	8	8	7
210	1	–	–	–	–	–	–	–	–	–	–	2	6	12	–	8	–	–	–	–	–	–	9	9
220	1	–	–	–	–	–	–	–	–	–	–	–	9	7	–	19	–	–	–	–	–	–	14	13
230	–	–	–	–	–	–	–	–	–	–	–	–	14	2	11	–	–	–	–	–	–	–	9	12
240	–	–	–	–	–	–	–	–	–	–	–	–	12	–	17	–	–	–	–	–	–	–	–	6
250	–	–	–	–	–	–	–	–	–	–	–	–	5	–	–	–	–	–	–	–	–	–	–	–
260	–	–	–	–	–	–	–	–	–	–	–	–	–	6	–	–	–	–	–	–	–	–	27	–
270	–	–	–	–	–	–	–	–	–	–	–	–	–	23	–	–	–	–	–	–	–	–	–	–
280	–	–	–	–	–	–	–	–	–	–	–	–	27	–	–	–	–	–	–	–	–	–	–	12

Table 4. Percentage of the area of CNH region III generating the indicated electric power from one Pelamis 750 kW.

Power (kW)	Wave Power in CNH Region III (% of Areas)											
	Jan	Feb	Mar	Apr	May	Jun	Jul	Aug	Sep	Oct	Nov	Dec
10	–	–	–	–	–	4	6	6	–	–	–	–
20	–	–	–	–	–	16	23	39	–	–	–	–
30	–	–	–	–	8	27	30	50	–	–	–	–
40	–	–	–	–	19	27	29	5	12	–	–	–
50	–	–	–	–	25	18	12	–	25	–	–	–
60	–	–	–	2	22	8	–	–	27	–	–	–
70	–	–	–	6	17	–	–	–	28	–	–	–
80	–	–	–	17	9	–	–	–	8	–	–	–
90	–	–	–	21	–	–	–	–	–	–	–	–
100	–	–	2	20	–	–	–	–	–	2	–	–
110	–	–	6	19	–	–	–	–	–	4	–	–
120	–	–	10	11	–	–	–	–	–	10	–	–
130	–	–	18	4	–	–	–	–	–	15	–	–
140	–	3	20	–	–	–	–	–	–	23	–	–
150	–	5	20	–	–	–	–	–	–	18	3	–
160	–	9	17	–	–	–	–	–	–	28	7	4
170	4	14	7	–	–	–	–	–	–	–	11	9
180	9	17	–	–	–	–	–	–	–	–	15	10
190	10	18	–	–	–	–	–	–	–	–	17	15
200	14	17	–	–	–	–	–	–	–	–	14	16
210	17	17	–	–	–	–	–	–	–	–	16	16
220	16	–	–	–	–	–	–	–	–	–	17	19
230	21	–	–	–	–	–	–	–	–	–	–	11

Table 5. Percentages of areas on U.S. oil platforms and CNH areas under different power levels generated by four Pelamis 750 kW WEC and one Vestas V90 3 MW.

Power (kW)	Wave Wind Combined Power U.S. Oil Platforms (%)												Wave Wind Power CNH Total Areas (%)											
	Jan	Feb	Mar	Apr	May	Jun	Jul	Aug	Sep	Oct	Nov	Dec	Jan	Feb	Mar	Apr	May	Jun	Jul	Aug	Sep	Oct	Nov	Dec
100	–	–	–	–	–	–	–	–	–	–	–	–	–	–	–	–	–	–	–	1	–	–	–	–
200	–	–	–	–	–	–	–	–	–	–	–	–	–	–	–	–	–	3	5	6	1	–	–	–
300	–	–	–	–	–	–	10	13	–	–	–	–	–	–	–	1	1	3	3	6	4	–	–	–
400	–	–	–	–	–	4	39	38	–	–	–	–	–	–	1	1	3	4	5	10	3	1	–	–
500	–	–	–	–	1	20	28	42	4	–	–	–	–	–	–	2	2	7	5	8	5	1	–	–
600	–	–	–	–	4	29	17	4	10	2	–	–	–	–	1	2	5	4	8	14	8	2	1	1
700	–	–	–	–	12	17	3	2	9	4	–	–	1	2	2	2	4	7	4	33	9	1	–	–
800	–	–	2	2	23	17	2	1	20	9	2	–	–	2	2	3	6	7	8	15	22	3	2	1
900	2	2	3	5	17	6	–	–	20	7	2	2	1	1	2	4	6	10	25	6	17	4	–	1
1000	3	8	10	10	12	4	1	–	16	11	9	4	1	2	2	3	12	12	19	1	4	5	3	2
1100	8	8	8	11	13	2	–	–	15	18	9	9	2	2	5	5	13	10	15	–	5	5	3	3
1200	9	9	8	13	8	–	–	–	6	11	5	8	2	3	4	9	8	30	3	–	22	7	3	2
1300	8	14	14	9	4	1	–	–	–	7	19	11	3	4	6	13	5	3	–	–	–	13	3	4
1400	16	14	11	13	3	–	–	–	–	13	11	14	4	6	4	10	7	–	–	–	–	19	5	4
1500	11	7	7	8	2	–	–	–	–	9	8	9	5	4	13	6	–	–	–	–	–	10	6	6
1600	7	7	7	10	1	–	–	–	–	7	6	8	3	9	10	2	–	–	–	–	–	2	9	6
1700	6	8	10	8	–	–	–	–	–	2	9	6	7	18	9	6	21	–	–	–	–	–	21	18
1800	8	8	8	5	–	–	–	–	–	–	7	9	18	6	1	3	7	–	–	–	–	27	6	11
1900	7	6	7	4	–	–	–	–	–	–	5	6	11	8	7	–	–	–	–	–	–	–	10	6
2000	5	4	5	2	–	–	–	–	–	–	4	5	8	5	3	–	–	–	–	–	–	–	–	7
2100	5	3	–	–	–	–	–	–	–	–	3	4	6	–	–	20	–	–	–	–	–	–	–	–
2200	3	2	–	–	–	–	–	–	–	–	1	3	–	–	27	8	–	–	–	–	–	–	–	–
2300	2	–	–	–	–	–	–	–	–	–	–	2	–	–	1	–	–	–	–	–	–	–	19	–
2400	–	–	–	–	–	–	–	–	–	–	–	–	–	28	–	–	–	–	–	–	–	–	9	–
2500	–	–	–	–	–	–	–	–	–	–	–	–	20	–	–	–	–	–	–	–	–	–	–	10
2600	–	–	–	–	–	–	–	–	–	–	–	–	8	–	–	–	–	–	–	–	–	–	–	18

The wave energy behavior in the CNH areas did not show distinctive pattern within the same month. Some locations in the CNH areas had 20% or lower capacity factor, and a considerable number of locations had high wave power levels above 20% capacity factor. From October to May, a large number of locations had capacity factor higher than 20%, and some of them produced 200 kW during five months of a year, which is encouraging. However, it would be necessary to further segment CNH areas to determine the best locations with wave energy potential.

Figure 10 indicates the average wave power in January and August applying the Jenks natural breaks classification method to provide further understanding of the results listed in Table 3. This provides a good contrast for analysis this high variability (geo temporal) renewable energy resource. Figure 10) indicates that the maximum power level was 350 kW in January concentrated mostly in Strait of Yucatan and the northwest coastal section of the Gulf of Mexico, encompassing the CNH areas closer to the U.S.–Mexico EEZ border. The southern section showed low wave power levels, affecting the CNH regions II and III. It directly indicates that CNH areas performed better than U.S. oil platforms during January, and also explains that more than 25% of the CNH areas performed above the 285 kW level as shown in Table 3. According to Figure 10, it can be ascertained that the better performing area mainly belongs to CNH region I and probably some sections of region II. Figure 10b indicates that the highest wave power levels in August occurred in Strait of Yucatan, the central Gulf of Mexico and the southern Texas Coast. It allows discovering of CNH region I as the main CNH area under the influence of the highest wave energy patterns at the northern Gulf of Mexico. The information provided by these maps indicates that it would be beneficial to further segment the CNH areas to better understand wave energy harvesting feasibility.

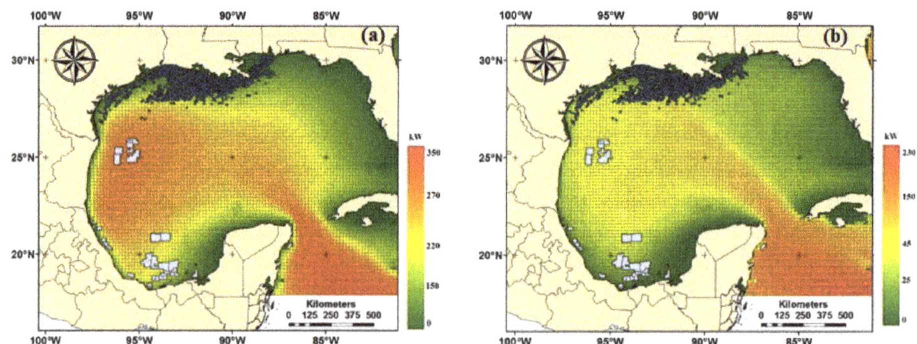

Figure 10. Monthly average wave power generated by one Pelamis WEC over 36 years applying Jenks natural breaks classification method for color bar: (**a**) January, and (**b**) August.

Figure 11 shows the monthly average wave power in CNH region I and II indicating a strong seasonal behavior in region I. The capacity factor was above 20% (150 kW) from October to May, while it shows much more energetic wave behavior from November to April. However, the wave power levels were lower from June to September, due to the seasonal meteorological conditions in the area, leading to the possibility of adjusting the Pelamis WEC to perform better under these different meteorological conditions.

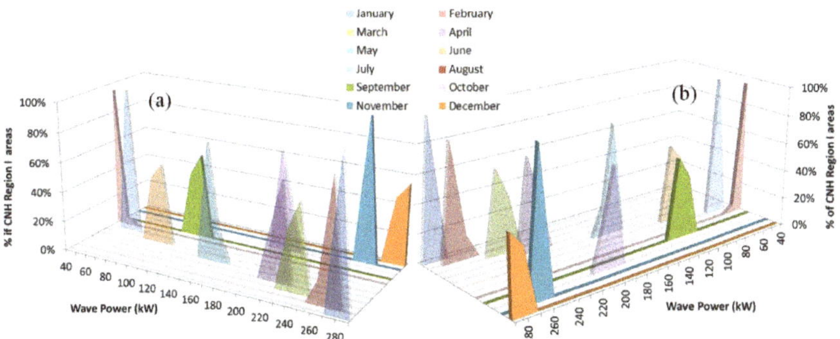

Figure 11. Percentage of the area of CNH region I generating the indicated wave power from one Pelamis 750 kW: (**a**) front view, and (**b**) back view.

In CNH region II, as shown in Figure 12, the overall wave power was lower than CNH region I, but it was still above the 20% capacity factor from October to March and for almost 50% of the locations in April. The rest of the calendar year showed lower wave power levels.

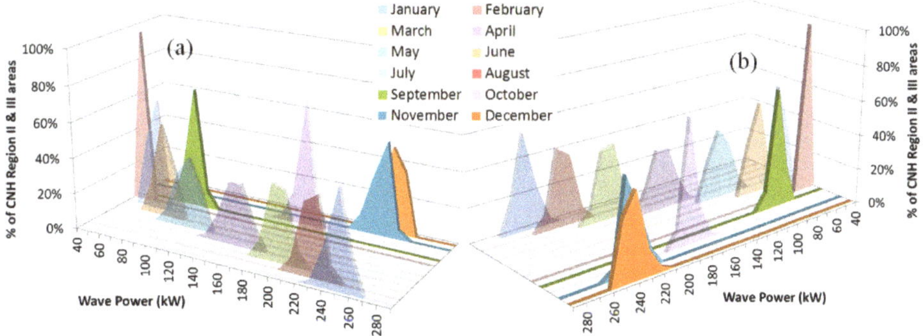

Figure 12. Percentage of the area of CNH region II generating the indicated wave power from one Pelamis 750 kW: (**a**) front view, and (**b**) back view.

On the other hand, results of the CNH region III, as shown in Table 4, indicates acceptable performance at most locations from November to February and almost 50% of locations from October to April. The wider histogram distribution indicates less similar behavior in this region in regards to wave power. It is possible to analyze particular areas leading to finding promising locations for longer periods of time with acceptable performance.

3.3. Wind and Wave Power Combined

After independently assessing the wave and wind power potential, it was important to combine them, considering simultaneously extracting both renewable energy resources. As previously discussed the Gulf of Mexico is not uniform in its spatial and time distribution of wind and wave energy, with a large number of locations having high variability and diverse behavior on the wind and wave power potentials. Therefore, the combination of both resources could aid in reducing variability and enhance energy production at those locations [26,48,69,80,81].

Since the equipment considered for this research have different nameplate capacities, a combination of one Vesta V90 3 MW wind turbine with four Pelamis 750 kW WEC was applied, creating an installation with total nameplate capacity of 6 MW, equally divided between both resources. Therefore, an array of four Pelamis 750 kW and one Vestas V90 3 MW was considered for each location of US oil platforms and for the locations of the CNH regions. Figure 13 shows yearly average wind and wave power combinations with the proposed installation, indicating a pattern of high energy in the U.S.–Mexico border coast extending to the central Gulf of Mexico to the western section of the Yucatan Peninsula. When compared with the individual patterns of wind and wave power, this map indicates that the variability of average renewable energy was reduced and the areas above the 20% capacity factor threshold (1.2 MW) was extended to larger sections of the Gulf of Mexico, creating more adequate potential areas.

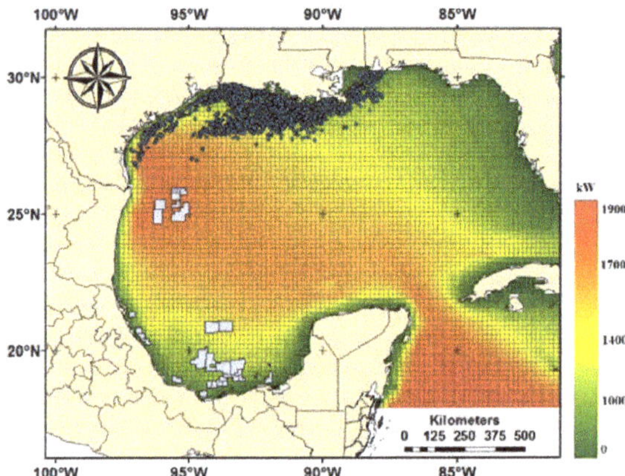

Figure 13. Yearly average of combined power generated by four Pelamis 750 kW WEC and one Vestas V90 3 MW with Jenks natural breaks classification method for color bar.

Table 5 indicates that a large number of locations in U.S. oil platforms areas between October and April have 20% or higher capacity factor, with power reduction occurring during the summer months. It will be important to perform specific analysis for desired locations to ascertain if the use of one of the available resources or its combination is the optimal alternative, depending on the obtained variability reduction and the behavior of the resources during the year. On the other hand, Table 5 also shows that at least two distinctive behaviors were presented in the CNH areas when combining wind and wave power. One of the potential behaviors is having high combined power output with potentially only underperforming on the months of June to August. Overall, a large number of locations in the CNH areas performed over the threshold limit from October to April.

Maps for the months of January and August were created to better understand the diverse seasonal and geographical performance of the combined power output (Figure 14). The Jenks natural breaks classification method was used to provide further analysis tools on results previously presented. Figure 14a shows that the highest power output in January could reach up to 2600 kW concentrated on the northwest region of the Gulf of Mexico, benefiting the CNH region I and a number of U.S. oil platforms. It explains that the CNH areas had better performance than the U.S. oil platforms during January. CNH region I and part of region II were under moderate power levels, shown as the yellow ring surrounding the high red power areas. The power range bin 2500–2600 kW in Table 5 contains more than 25% of the CNH areas, and is disconnected from the general performance data on the CNH table. It indicates that different sections of the CNH areas are under diverse wind and wave energy influences, which is validated by Figure 14a. On the other hand, the highest power level in August, the lowest performing month of the year, was 1500 kW, and it was concentrated on the coastal region on the U.S.–Mexico border and on the south of the Gulf of Mexico (Figure 14b), benefiting some of the U.S. oil platforms and the CNH region III.

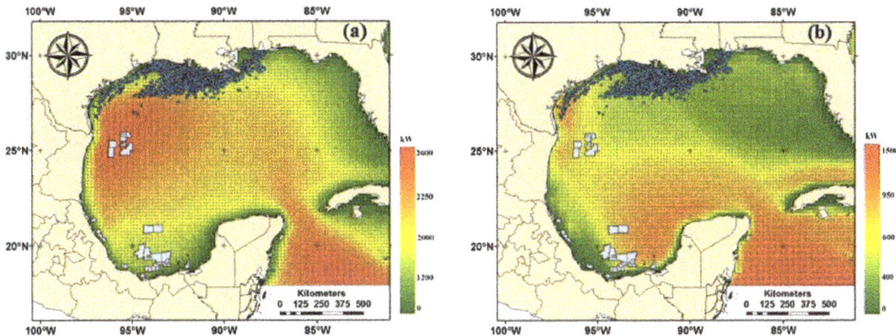

Figure 14. Monthly average of combined wave and wind power generated by four Pelamis and one Vestas with Jenks natural breaks classification method for color bar: (**a**) January, and (**b**) August.

A better understanding of the wind and wave energy combined behavior was obtained by creating histograms segmenting the CNH areas in three major regions. The CNH region I as shown in Figure 15 had a very clear seasonal behavior with small variation in each month, where the capacity factor was over 20% from October to May. The CNH region II as shown in Figure 16 had higher overall power levels than the CNH region I, and it had only three months, from July to September, underperforming below the 20% threshold with six months above 2000 kW.

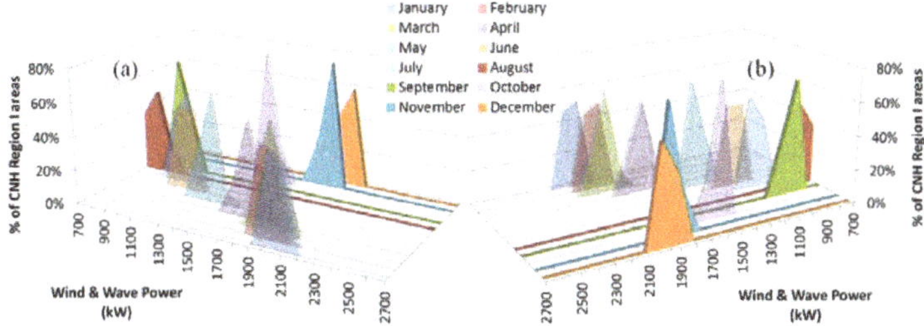

Figure 15. Percentage of the area of CNH region I generating the indicated wave and wind power by four Pelamis 750 kW and one Vestas V90 3 MW: (**a**) front view, and (**b**) back view.

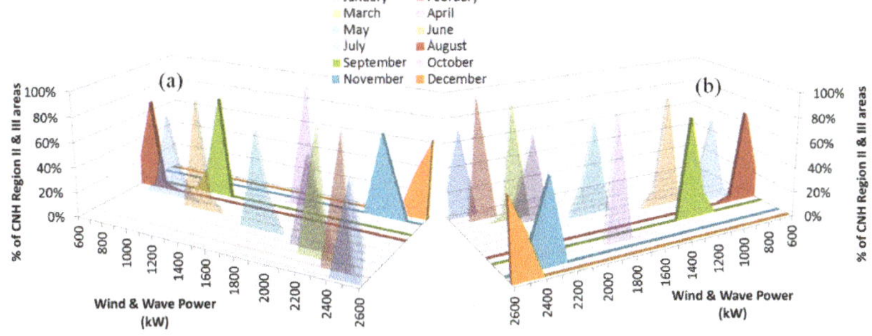

Figure 16. Percentage of the area of CNH region II generating the indicated wave and wind power by four Pelamis 750 kW and one Vestas V90 3 MW: (**a**) front view, and (**b**) back view.

Table 6 shows that the results of the CNH region III for each month spread over a larger range of power levels, indicating that the behavior of the different areas was not consistent. It shows a right skewedness curve or trend for most months, except August. The power output in most locations remained above the 20% threshold in seven months, which indicates a significant number of locations in the CNH region III having good power levels.

Table 6. Percentage of the CNH region III area generating the indicated electric power by four Pelamis 750 kW and one Vestas V90 3 MW.

Power (kW)	Wind and Wave Power Combined in CNH Region III (% of Areas)											
	Jan	Feb	Mar	Apr	May	Jun	Jul	Aug	Sep	Oct	Nov	Dec
400	–	–	–	–	–	–	–	13	–	–	–	–
500	–	–	–	–	–	7	5	18	5	–	–	–
600	–	–	–	–	3	5	13	21	7	–	–	–
700	–	–	–	–	8	15	12	19	11	–	–	–
800	–	–	–	3	9	10	15	20	52	3	–	–
900	–	–	–	6	14	24	12	9	25	2	–	–
1000	–	–	3	6	21	29	17	–	–	3	3	–
1100	–	2	3	4	28	10	21	–	–	5	–	2
1200	2	3	6	18	17	–	5	–	–	12	3	2
1300	2	3	6	31	–	–	–	–	–	26	2	2
1400	3	6	3	22	–	–	–	–	–	49	5	3
1500	3	7	30	10	–	–	–	–	–	–	11	9
1600	7	17	24	–	–	–	–	–	–	–	16	10
1700	10	49	25	–	–	–	–	–	–	–	55	42
1800	46	13	–	–	–	–	–	–	–	–	5	30
1900	27	–	–	–	–	–	–	–	–	–	–	–

4. Conclusions

The assessment of power extracted from wave and wind in the Gulf of Mexico for its application in offshore oil industry showed promising results. The concept of connecting offshore oil installations to wave and wind harvesters and simultaneously connecting them to the onshore grid is feasible for both U.S. and Mexico EEZ in the Gulf of Mexico. Research results indicated that the distance from the coast to current and planned offshore facilities is mostly on the shortest ranges of the feasibility threshold, which is encouraging.

Analysis performed for the assessment wind and wave power output in the Gulf of Mexico showed a lack of spatial and temporal uniformity, with high geo temporal variability on both wind and wave resources. Results provided by the maps and statistical tools indicated that most of the U.S. oil platforms and CNH areas have very good potential for the extraction of either wind or wave energy. Furthermore, there are a significant number of locations in the Gulf of Mexico where renewable energy extraction for the combined two sources is feasible, generating significant economic and environmental advantages.

The maps generated by the GIS and statistical tools allows for better understanding of the statistical results generated by the methodology. In addition, these maps show that the combination of wind and wave energy promoted the advantages in many locations, increasing the energy extraction levels and reducing its variability. Synergies generated by the proposed system, considering each resource individually or combined, could be maximized in an important number of locations on the Gulf of Mexico.

Considering that some of the locations evaluated are better suited for the extraction of just one renewable energy resource or for the combination of both resources, it would be important to perform individual analysis for particular areas (regional analysis) applying the proposed methodology for each location with better spatial resolution if possible. This will help the decision making process of the design of the best system in each particular oil installation location.

Regional analysis will be of special importance for both future projects and existing installations, which economic and environmental characteristics would be enhanced by including renewable energy to its overall operation, allowing for savings on electricity consumption, potential extra income from sales of energy to the onshore grid and reduction on the emission of pollutants to the atmosphere.

It should be noted that the equipment selected in this paper is incidental to the methodology. The main requirement is to have power curve from the selected equipment that allows calculating its power output while operating under different geographic and temporal meteorological conditions. Since power curves of different equipment can be incorporated to the methodology on a plug-in basis to calculate the required power output, the possibility of choosing different equipment with the same methodology is one of the strengths, allowing researchers and developers to perform sensitivity analysis on the selection of the best technology for each particular location. In addition, the methodology could be expanded to other geographical regions with the inclusion of the corresponding historical meteorological data.

Future research will evaluate the correlation between wind and wave in the Gulf of Mexico, particularly considering the dependence of wave to localized weather patterns and monthly as well as seasonal variations of both resources. As previously indicated, both wind and wave resources are higher during fall and winter seasons with a transitional period during spring and low energy during summer [66–68]. Previous research has also indicated that the combination of both resources reduces the variability of the wave resource and increases overall power output [66]. Future research will evaluate the complementarity of both resources and the possible synergies related to variability reduction achieved by its combination.

Future research will also include comparative analysis between different equipment to harvest wind and wave energy, evaluating the most adequate technology for each particular location. The plug in capability for diverse power curves offered by the methodology will allow for the development of these comparative analyses, applying the same underlying meteorological and geographical data. New insights on the development of equipment could be derived from these comparative analyses. Evaluating several designs it could be possible to gain better understanding on the characteristics that perform better on each location and to find feasible combinations.

Author Contributions: Francisco Haces-Fernandez developed the research method and completed the research activities under the supervision of Hua Li and David Ramirez. The initial paper was written by Francisco Haces-Fernandez, while Hua Li made major revision on the draft paper and David Ramirez also revised the draft paper. Both Hua Li and David Ramirez approved the final version to be published.

Acknowledgments: The authors are thankful to the support from CONACYT (Consejo Nacional de Ciencia y Tecnología) and COTACYT (Consejo Tamaulipeco de Ciencia y Tecnologia) through the scholarship number 218681, Texas A&M University-Kingsville, Eagle Ford Center for Research and Outreach, the Center for Research Excellence in Science and Technology Research on Environmental Sustainability in Semi-Arid Coastal Areas (CREST-RESSACA) and National Science Foundation (award # EEC 1359414).

Abbreviations

CNH	Comisión Nacional de Hidrocarburos
EEZ	Exclusive Economic Zones
EMEC	European Marine Energy Centre
GIS	Geographical Information Systems
NOAA	National Oceanic and Atmospheric Administration
PEMEX	Petroleos Mexicanos
PM10	Particulate Matter 10 micrometers or less
PM2.5	Particulate Matter 2.5 micrometers or less
VOC	Volatile Organic Compounds
WEC	Wave Energy Converter

References

1. Korpås, M.; Warland, L.; He, W.; Tande, J.O.G. A case-study on offshore wind power supply to oil and gas rigs. *Energy Procedia* **2012**, *24*, 18–26. [CrossRef]

2. He, W.; Uhlen, K.; Hadiya, M.; Chen, Z.; Shi, G.; del Rio, E. Case study of integrating an offshore wind farm with offshore oil and gas platforms and with an onshore electrical grid. *J. Renew. Energy* **2013**, *2013*, 607165. [CrossRef]

3. Årdal, A.R.; Undeland, T.; Sharifabadi, K. Voltage and frequency control in offshore wind turbines connected to isolated oil platform power systems. *Energy Procedia* **2012**, *24*, 229–236. [CrossRef]

4. Årdal, A.R. Feasibility Studies on Integrating Offshore Wind Power with Oil Platforms. Master's Thesis, Institutt for Elkraftteknikk, Trondheim, Norway, 2011.

5. Kolstad, M.L.; Sharifabadi, K.; Årdal, A.R.; Undeland, T.M. Grid integration of offshore wind power and multiple oil and gas platforms. In Proceedings of the 2013 MTS/IEEE OCEANS-Bergen, Bergen, Norway, 10–13 June 2013; pp. 1–7.

6. Kolstad, M.L. Integrating Offshore Wind Power and Multiple Oil and Gas Platforms to the Onshore Power Grid Using VSC-HVDC Technology. *Mar. Technol. Soc. J.* **2014**, *48*, 31–44. [CrossRef]

7. Marvik, J.I.; Øyslebø, E.V.; Korpås, M. Electrification of offshore petroleum installations with offshore wind integration. *Renew. Energy* **2013**, *50*, 558–564. [CrossRef]

8. Aardal, A.R.; Marvik, J.I.; Svendsen, H.; Tande, J.O.G. Study of Offshore Wind as Power Supply to Oil and Gas Platforms. In Proceedings of the Offshore Technology Conference, Houston, TX, USA, 30 April–3 May October 2012.

9. TCEQ—Eastern Research Group, Inc. Emissions from Oil and Gas Production Facilities. Texas Commission on Environmental Quality. TCEQ. Final Report. TCEQ Contract 582-7-84003 August 31, 2007. ERG No. 0227.00.001.001. Available online: https://www.tceq.texas.gov/assets/public/implementation/air/am/contracts/reports/ei/5820784003FY0701-20090831-ergi-ei_from_old_gas_facilities.pdf (accessed on 4 July 2017).

10. Villasenor, R.; Magdaleno, M.; Quintanar, A.; Gallardo, J.C.; López, M.T.; Jurado, R.; Vallejo, C.J. An air quality emission inventory of offshore operations for the exploration and production of petroleum by the Mexican oil industry. *Atmos. Environ.* **2003**, *37*, 3713–3729. [CrossRef]

11. O'Connor, P.A.; Cleveland, C.J. US Energy Transitions 1780–2010. *Energies* **2014**, *7*, 7955–7993. [CrossRef]

12. Han, S.; Zhang, B.; Sun, X.; Han, S.; Höök, M. China's Energy Transition in the Power and Transport Sectors from a Substitution Perspective. *Energies* **2017**, *10*, 600. [CrossRef]

13. Pierobon, L. *Novel Design Methods and Control Strategies for Oil and Gas Offshore Power Systems*; DTU Mechanical Engineering: Lyngby, Denmark, 2014.

14. Svendsen, H.G.; Hadiya, M.; Øyslebø, E.V.; Uhlen, K. Integration of offshore wind farm with multiple oil and gas platforms. In Proceedings of the 2011 IEEE PES Trondheim PowerTech, Trondheim, Norway, 19–23 June 2011; pp. 1–3.

15. Boxem, T.; Koornneef, J.; van Dijk, R.; Halter, M.; EBN, E.N.; Breunesse, E.; Prins, J. *System Integration Offshore Energy: Innovation Project North Sea Energy*; The Netherlands Organization for Applied Scientific Research (TNO): Hague, The Netherlands, 2016.

16. Jepma, C.J. On the Economics of Offshore Energy Conversion: Smart Combinations. Converting Offshore Wind Energy into Green Hydrogen on Existing Oil and Gas Platforms in the North Sea. Energy Delta Institute (EDI). 3 February 2017. Available online: https://www.gasmeetswind.eu/wp-content/uploads/2017/05/EDI-North-Sea-smart-combinations-final-report-2017.pdf (accessed on 3 July 2017).

17. Mundheim, K. Monotower production platform swith renewable energy. In Proceedings of the Offshore Mediterranean Conference and Exhibition, Ravenna, Italy, 28–30 March 2007.

18. Nielsen, J.S.; Sørensen, J.D. Methods for risk-based planning of O&M of wind turbines. *Energies* **2014**, *7*, 6645–6664.

19. Halabi, M.A.; Al-Qattan, A.; Al-Otaibi, A. Application of solar energy in the oil industry—Current status and future prospects. *Renew. Sustain. Energy Rev.* **2015**, *43*, 296–314. [CrossRef]

20. Yang, M.; Khan, F.I.; Sadiq, R.; Amyotte, P. A rough set-based quality function deployment (QFD) approach for environmental performance evaluation: A case of offshore oil and gas operations. *J. Clean. Prod.* **2011**, *19*, 1513–1526. [CrossRef]

21. Wilson, J.C.; Elliott, M.; Cutts, N.D.; Mander, L.; Mendão, V.; Perez-Dominguez, R.; Phelps, A. Coastal and Offshore Wind Energy Generation: Is It Environmentally Benign? *Energies* **2010**, *3*, 1383–1422. [CrossRef]

22. Lu, S.Y.; Jason, C.S.; Wesnigk, J.; Delory, E.; Quevedo, E.; Hernández, J.; Anastasiadis, P. Environmental aspects of designing multi-purpose offshore platforms in the scope of the FP7 TROPOS Project. In Proceedings of the OCEANS 2014, Taipei, Taiwan, 7–10 April 2014.

23. Smailes, M.; Ng, C.; Mckeever, P.; Shek, J.; Theotokatos, G.; Abusara, M. Hybrid, Multi-Megawatt HVDC Transformer Topology Comparison for Future Offshore Wind Farms. *Energies* **2017**, *10*, 851. [CrossRef]

24. Hyttinen, M.; Lamell, J.O.; Nestli, T.F. New application of voltage source converter (VSC) HVDC to be installed on the gas platform Troll A. In Proceedings of the Cigre Session, New Delhi, India, 17–18 November 2004; p. B4-210.

25. Antonio, F.d.O. Wave energy utilization: A review of the technologies. *Renew. Sustain. Energy Rev.* **2010**, *14*, 899–918.

26. Perez-Collazo, C.; Greaves, D.; Iglesias, G. A review of combined wave and offshore wind energy. *Renew. Sustain. Energy Rev.* **2015**, *42*, 141–153. [CrossRef]

27. Silva, D.; Rusu, E.; Soares, C.G. Evaluation of various technologies for wave energy conversion in the Portuguese nearshore. *Energies* **2013**, *6*, 1344–1364. [CrossRef]

28. Silva, D.; Rusu, E.; Guedes Soares, C. High-Resolution Wave Energy Assessment in Shallow Water Accounting for Tides. *Energies* **2016**, *9*, 761. [CrossRef]

29. Franzitta, V.; Curto, D. Sustainability of the renewable energy extraction close to the Mediterranean Islands. *Energies* **2017**, *10*, 283. [CrossRef]

30. Franzitta, V.; Catrini, P.; Curto, D. Wave energy assessment along Sicilian coastline, based on deim point absorber. *Energies* **2017**, *10*, 376. [CrossRef]

31. Franzitta, V.; Curto, D.; Milone, D.; Rao, D. Assessment of renewable sources for the energy consumption in Malta in the Mediterranean Sea. *Energies* **2016**, *9*, 1034. [CrossRef]

32. New Plans to Increase Clean Power Are Ambitious and Expensive Offshore Wind Power. Oil Rigs to Whirligigs. Available online: https://www.economist.com/node/15287732 (accessed on 14 January 2010).

33. Montilla-D Jesus, M.E.; Arnaltes, S.; Castronuovo, E.D.; Santos-Martin, D. Optimal Power Transmission of Offshore Wind Power Using a VSC-HVdc Interconnection. *Energies* **2017**, *10*, 1046. [CrossRef]

34. Poulsen, T.; Hasager, C.B. How Expensive Is Expensive Enough? Opportunities for Cost Reductions in Offshore Wind Energy Logistics. *Energies* **2016**, *9*, 437. [CrossRef]

35. Shankleman, J. Big Oil Replaces Rigs with Wind Turbines. Bloomberg Markets Online. Available online: https://www.bloomberg.com/news/articles/2017-03-23/oil-majors-take-a-plunge-in-industry-that-may-hurt-fossil-fuel (accessed on 30 November 2017).

36. 4C Submarine Cable Consultancy. Offshore Wind Turbine Database. Available online: http://www.4coffshore.com/windfarms/enbw-baltic-2-germany-de52.html (accessed on 30 March 2018).

37. Texas A&M University. The Gulf of Mexico Coastal Ocean Observing System. Bathymetry Related Data in Shapefile Format. Available online: http://gcoos.tamu.edu/products/topography/Shapefiles.html (accessed on 13 December 2016).

38. European Wind Energy Association. Deep Water: The Next Step for Offshore Wind Energy. Available online: http://www.ewea.Org (accessed on 18 March 2018).

39. Perez-Collazo, C.; Greaves, D.; Iglesias, G. A novel hybrid wind-wave energy converter for jacket-frame substructures. *Energies* **2018**, *11*, 637. [CrossRef]

40. Antonio, F.d.O. First-generation wave power plants: Current status and R&D requirements. In Proceedings of the ASME 2003 22nd International Conference on Offshore Mechanics and Arctic Engineering, Cancun, Mexico, 8–13 June 2003; American Society of Mechanical Engineers: New York, NY, USA, 2003; pp. 723–731.

41. Shih, H.J.; Chang, C.H.; Chen, W.B.; Lin, L.Y. Identifying the Optimal Offshore Areas for Wave Energy Converter Deployments in Taiwanese Waters Based on 12-Year Model Hindcasts. *Energies* **2018**, *11*, 499. [CrossRef]

42. Wu, W.; Yang, Z.; Wang, T. Wave Resource Characterization Using an Unstructured Grid Modeling Approach. *Energies* **2018**, *11*, 605. [CrossRef]

43. U.S. Geological Survey, Coastal and Marine Geology Program. *Locations of Oil and Gas Platforms in the Gulf of Mexico*; Woods Hole Science Center: Woods Hole, MA, USA, 2017.

44. National Commission for Hydrocarbons (Comisión Nacional de Hidrocarburos—CNH). Mexican Federal Government. Available online: http://rondasmexico.gob.mx/ (accessed on 3 July 2017).

45. EMEC. Pelamis Wave Power. EMEC. The European Marine Energy Center LTD: Orkney, Scotland. Available online: http://www.emec.org.uk/about-us/wave-clients/pelamis-wave-power/ (accessed on 30 March 2018).

46. Rusu, E. Evaluation of the wave energy conversion efficiency in various coastal environments. *Energies* **2014**, *7*, 4002–4018. [CrossRef]

47. Wan, Y.; Fan, C.; Zhang, J.; Meng, J.; Dai, Y.; Li, L.; Zhang, X. Wave Energy Resource Assessment off the Coast of China around the Zhoushan Islands. *Energies* **2017**, *10*, 1320. [CrossRef]

48. Wan, Y.; Fan, C.; Dai, Y.; Li, L.; Sun, W.; Zhou, P.; Qu, X. Assessment of the Joint Development Potential of Wave and Wind Energy in the South China Sea. *Energies* **2018**, *11*, 398. [CrossRef]

49. Bedard, R.; Hagerman, G.; Siddiqui, O. *System Level Design, Performance and Costs for San Francisco California Pelamis Offshore Wave Power Plant*; Report WP-006-SFa; EPRI: San Francisco, CA, USA, 2004.

50. Reikard, G.; Robertson, B.; Buckham, B.; Bidlot, J.R.; Hiles, C. Simulating and forecasting ocean wave energy in western Canada. *Ocean Eng.* **2015**, *103*, 223–236. [CrossRef]

51. Previsic, M. *Deployment Effects of Marine Renewable Energy Technologies: Wave Energy Scenarios (No. DOE/GO18175)*; RE Vision Consulting, LLC: Sacramento, CA, USA, 2010.

52. Pelamis Wave Power. Pelamis Wave Power. Brochure. Technical Details. Patents: US6476511, AU754950, ZA20012008, EP1115976B. Available online: http://ctp.lns.mit.edu/energy/files/pelamisbrochure.pdf (accessed on 30 March 2018).

53. Bahaj, A.S. Generating electricity from the oceans. *Renew. Sustain. Energy Rev.* **2011**, *15*, 3399–3416. [CrossRef]

54. Vestas Wind Systems A/S. General Specification for V90—3.0 MW.V. (2006, October). Available online: http://report.nat.gov.tw/ReportFront/report_download.Jspx?sysId=C09503816&fileNo=002 (accessed on 4 October 2016).

55. Sclavounos, P.D.; Lee, S.; DiPietro, J.; Potenza, G.; Caramuscio, P.; De Michele, G. Floating offshore wind turbines: Tension leg platform and taught leg buoy concepts supporting 3–5 MW wind turbines. In Proceedings of the European Wind Energy Conference EWEC, Warsaw, Poland, 20–23 April 2010; pp. 20–23.

56. MHI Vestas Offshore Wind A/S. Vestas Wind Float (PT). Denmark. Available online: http://www.mhivestasoffshore.com/windfloat/ (accessed on 3 April 2018).

57. Mark, E. Ship Shape: This Floating Offshore Wind Farm Could Be the Future of Renewable Energy. Available online: https://www.ge.com/reports/ship-shape-this-floating-offshore-wind-farm-could-be-the-future-of-renewable-energy/ (accessed on 30 August 2016).

58. IRENA. Floating Foundations: A Game Changer for Offshore Wind Power, International. Renewable Energy Agency: Abu Dhabi, 2016. Available online: http://www.irena.org/-/media/Files/IRENA/Agency/Publication/2016/IRENA_Offshore_Wind_Floating_Foundations_2016.pdf (accessed on 3 April 2018).

59. U.S. Department of Energy. *Wind on the Waves: Floating Wind Power Is Becoming a Reality*; Office of Energy Efficiency and Renewable Energy U.S. Department of Energy: Washington, DC, USA, 2017.

60. Hagerman, G. Southern New England wave energy resource potential. In *Proceedings of the Building Energy*; Tufts University: Boston, MA, USA, 2001.

61. Contestabile, P.; Ferrante, V.; Vicinanza, D. Wave energy resource along the coast of Santa Catarina (Brazil). *Energies* **2015**, *8*, 14219–14243. [CrossRef]

62. Musial, W.D. *Status of Wave and Tidal Power Technologies for the United States*; National Renewable Energy Laboratory: Golden, CO, USA, 2008.

63. Kamranzad, B.; Etemad-Shahidi, A.; Chegini, V. Sustainability of wave energy resources in southern Caspian Sea. *Energy* **2016**, *97*, 549–559. [CrossRef]

64. Mirzaei, A.; Tangang, F.; Juneng, L. Wave energy potential along the east coast of Peninsular Malaysia. *Energy* **2014**, *68*, 722–734. [CrossRef]

65. Ramírez, G.; de Dios, J.; Mejía Fuentes, O.I.; Menjívar Pino, P.J. Evaluacion del Potential Energetico del Oleaje en las Costas de El Salvador. Ph.D. Thesis, Universidad de El Salvador, San Salvador, El Salvador, 2007.

66. Haces-Fernandez, F. Investigation on the Possibility of Extracting Wave Energy from the Texas Coast. Master's Thesis, Texas A&M University-Kingsville, Kingsville, TX, USA, December 2014.

67. Haces-Fernandez, F.; Li, H.; Ramirez, D. Wave Energy Characterization and Assessment in the US Gulf of Mexico, East and West Coasts with Energy Event Concept. *Renew. Energy* **2018**, *123*, 312–322. [CrossRef]

68. Haces-Fernandez, F.; Martinez, A.; Ramirez, D.; Li, H. Characterization of wave energy patterns in Gulf of Mexico. In Proceedings of the IIE Annual Conference on Institute of Industrial and Systems Engineers, Pittsburgh, PA, USA, 20–23 May 2017; pp. 1532–1537.

69. Stoutenburg, E.D.; Jenkins, N.; Jacobson, M.Z. Power output variations of co-located offshore wind turbines and wave energy converters in California. *Renew. Energy* **2010**, *35*, 2781–2791. [CrossRef]

70. Hsu, S.A.; Meindl, E.A.; Gilhousen, D.B. Determining the power-law wind-profile exponent under near-neutral stability conditions at sea. *J. Appl. Meteorol.* **1994**, *33*, 757–765. [CrossRef]

71. Chen, J.; Yang, S.; Li, H.; Zhang, B.; Lv, J. Research on geographical environment unit division based on the method of natural breaks (Jenks). In Proceedings of the International Archives of the Photogrammetry, Remote Sensing and Spatial Information Sciences XL-4 (W3), Beijing, China, 5–6 December 2013; pp. 47–50.
72. Wong, M.S.; Zhu, R.; Liu, Z.; Lu, L.; Peng, J.; Tang, Z.; Ho Lo, C.; Chan, W.K. Estimation of Hong Kong's solar energy potential using GIS and remote sensing technologies. *Renew. Energy* **2016**, *99*, 325–335. [CrossRef]
73. Mellino, S.; Ulgiati, S. Mapping the evolution of impervious surfaces to investigate landscape metabolism: An Emergy–GIS monitoring application. *Ecol. Inform.* **2015**, *26*, 50–59. [CrossRef]
74. Miller, A.; Li, R. A geospatial approach for prioritizing wind farm development in Northeast Nebraska, USA. *ISPRS Int. J. Geo-Inf.* **2014**, *3*, 968–979. [CrossRef]
75. Rusu, E.; Onea, F. Joint Evaluation of the Wave and Offshore Wind Energy Resources in the Developing Countries. *Energies* **2017**, *10*, 1866. [CrossRef]
76. Wiser, R.; Bolinger, M.; Barbose, G.; Darghouth, N.; Hoen, B.; Mills, A.; Widiss, R. *2015 Wind Technologies Market Report*; U.S. Department of Energy (DOE): Washington, DC, USA, 2015.
77. Feng, Y.; Tavner, P.J.; Long, H. Early experiences with UK Round 1 offshore wind farms. *Proc. Inst. Civ. Eng. Energy* **2010**, *163*, 167–181. [CrossRef]
78. Dunnett, D.; Wallace, J.S. Electricity generation from wave power in Canada. *Renew. Energy* **2009**, *34*, 179–195. [CrossRef]
79. Dalton, G.J.; Alcorn, R.; Lewis, T. Case study feasibility analysis of the Pelamis wave energy convertor in Ireland, Portugal and North America. *Renew. Energy* **2010**, *35*, 443–455. [CrossRef]
80. Astariz, S.; Iglesias, G. Enhancing wave energy competitiveness through co-located wind and wave energy farms. A review on the shadow effect. *Energies* **2015**, *8*, 7344–7366. [CrossRef]
81. Lee, H.; Poguluri, S.K.; Bae, Y.H. Performance Analysis of Multiple Wave Energy Converters Placed on a Floating Platform in the Frequency Domain. *Energies* **2018**, *11*, 406. [CrossRef]

Article

Cost-Based Design and Selection of Point Absorber Devices for the Mediterranean Sea

Vincenzo Piscopo [1],*, Guido Benassai [2], Renata Della Morte [2] and Antonio Scamardella [1]

[1] Department of Science and Technology, Centro Direzionale Isola C4, The University of Naples "Parthenope", 80143 Naples, Italy; antonio.scamardella@uniparthenope.it

[2] Department of Engineering, Centro Direzionale Isola C4, The University of Naples "Parthenope", 80143 Naples, Italy; guido.benassai@uniparthenope.it (G.B.); renata.dellamorte@uniparthenope.it (R.D.M.)

* Correspondence: vincenzo.piscopo@uniparthenope.it; Tel.: +39-081-5476946

Received: 26 March 2018; Accepted: 13 April 2018; Published: 16 April 2018

Abstract: Sea wave energy is one of the most promising renewable sources, even if relevant technology is not mature enough for the global energy market and is not yet competitive if compared with solar, wind and tidal current devices. Particularly, among the variety of wave energy converters developed in the last decade, heaving point absorbers represent one of the most feasible and studied technologies, as shown by the small-scale testing and full-scale prototypes, deployed in the last years throughout the world. Nevertheless, the need for further reduction of the energy production costs requires a specialized design of wave energy converters, accounting for the restraints provided by the power take-off unit and the device operational profile. Hence, actual analysis focuses on a new cost-based design procedure for heaving point absorbers. The device is equipped with a floating buoy with an optional fully submerged mass connected, by means of a tensioned line, to the power take-off unit. It consists of a permanent magnet linear generator, lying on the seabed and equipped with a gravity-based foundation. The proposed procedure is applied to several candidate deployment sites located in the Mediterranean Sea; the incidence of the power take-off restraint and the converter operational profile is fully investigated and some recommendations for preliminary design of wave energy converter devices are provided. Current results show that there is wide scope to make the wave energy sector more competitive on the international market, by properly selecting the main design parameters of point absorbers, on the basis of met-ocean conditions at the deployment site.

Keywords: point absorber; power take-off; hydrodynamic optimization; levelised cost of energy; Mediterranean Sea

---

1. Introduction

In the last decade the interest in harnessing wave power has grown fast, thanks to several research activities funded by national administrations and private companies, with the main aim of facilitating and coordinating ocean energy research, development and demonstration, through international cooperation and information exchanges [1]. Starting from the first pioneering work by Salter [2], different wave energy conversion technologies have been developed and investigated [3–6], moving the research activities from small-scale testing to full-scale prototypes. Nevertheless, the wave energy sector actually experiences a moderate slow-down in the rate of progress, mainly due to the high investment needs, combined with challenging environmental conditions and technical risks [7]. In this respect, it is quite challenging to achieve the 7/8 Technology Readiness Level [8], with reference to system prototype demonstration in the operational environment, and move the sector towards competitive manufacturing systems, at least in the case of key enabling technologies.

The issue of moving the wave energy sector towards full effective systems, ready for the international market, is mainly due to technical and financial difficulties in increasing the rated power of wave energy converter (WEC) devices [7] and consistently slow-down the levelised cost of energy (LCoE). Hence, all these factors make the wave energy sector not yet competitive enough on the international market, if compared with other renewable sources, such as hydroelectric, solar, wind and tidal stream technologies. Nevertheless, it is predictable that in the future the wave energy sector will reach competitive LCoE levels, with the increase of the total installed capacity that will help to reduce uncertainties and costs, as well as improve system reliability and availability [7,9]. These targets comply with the EU Strategic Energy Technology Plan, according to which the LCoE of wave energy devices is expected to decrease up to 0.15 €/kWh by 2030 and 0.10 €/kWh by 2035.

One of the most promising pathway to reduce the LCoE consists of optimizing the hydrodynamic performances of WEC devices [3,4,9–13] by a proper design of the energy converter and the power take-off (PTO) unit. Furthermore, different motion control techniques, such as latching, unclutching or phase control methods [14], can be also applied in order to increase the power production. In this respect, the control strategies for optimum design of sea structures, widely applied in the last decade in the offshore wind energy sector among others [15–17], play a fundamental role also in the wave energy field, to reduce capital, operating and decommissioning costs. Hence, following the variety of attempts, devoted to making the wave energy sector competitive on the international market, a new cost-based design procedure of WEC devices, consisting of heaving point absorbers with an optional fully submerged mass and connected to a permanent magnet linear generator lying on the seabed, is outlined and investigated, focusing on the following main subjects:

(i) Two point absorber configurations, mainly consisting of a floating buoy plus an optional fully submerged mass, required to properly tune the device heave natural frequency with dominant sea states at the candidate deployment site, are investigated by a purposely developed programme in Matlab [18]. Particularly, the hydrodynamic model of the floating buoy is developed accounting for the WEC operational profile and the PTO mechanical restraint, in terms of free stroke length, provided by the permanent magnet linear generator.

(ii) A new procedure to select the optimum WEC design, that allows minimizing the LCoE, is outlined and applied to several candidate deployment sites, located in the Mediterranean Sea, to investigate the incidence of available energy and metocean conditions on the attained power production and cost.

(iii) The incidence of PTO free stroke length and WEC operational profile on power production and LCoE is analysed, to investigate the impact on annual power production and costs.

(iv) Some guidelines to select, in the preliminary design phase, the optimum point absorber configuration that allows minimizing the LCoE, are provided.

Actual efforts are undertaken to furnish a possible pathway to speed up the mass production of WEC devices and push the energy sector towards the commercial phase. In this respect, several cost reduction opportunities in the wave energy sector can arise from: (i) increasing scale/volume productions; (ii) improving experience in manufacturing and operational phases and (iii) providing innovative design strategies, resulting in increased device ratings and reduced capital costs.

2. Review of Point Absorber Technology

It is well-known that wave energy technologies vary widely in both mechanical and electrical conversion modes, mainly depending on relevant working principle [1]. In current analysis, the point absorber is assumed as a reference WEC device, provided that it represents one of the most promising technologies, as proved by the variety of concept designs and full scale prototypes, developed in the last decade by several research administrations and private companies. Anyway, it must be pointed out that even if the working principle of point absorbers is always the same and mainly consists of harnessing the heave motion of a floating buoy in a seaway, the overall layout may substantially differ. In fact, it mainly depends on the applied motion transmission and power conversion technologies, as proved by the PB3 PowerBuoy and the Uppsala wave energy converters, among others. The first device, with a nominal power equal to 3 kW, was developed by the Company "Ocean Power Technologies" and recently deployed off the New Jersey coastline, under the US Navy "Littoral Expeditionary Autonomous PowerBuoy (LEAP) Program". It consists of a floating buoy with spar-type configuration and a damping heave plate on the bottom, connected to the seabed by means of mooring lines [19]. The plate maintains the spar in a stationary position, while the relative motion of the floating buoy drives a mechanical system that converts the linear motion into a rotary one, moving an electrical generator and producing electricity. The second device, with a nominal power of 10 kW, was developed by the Uppsala University and widely tested in open sea conditions at the Lysekil research site on the Swedish west coast. It mainly consists of a floating buoy connected, by means of a tensioned line, to a gravity-based foundation, lying on the seabed and equipped with a permanent magnet linear generator, converting the heave motion into electricity [20].

The conversion of wave energy into electrical power can be performed by means of commonly applied technologies, such as linear or rotary electrical generators, the latter in conjunction with oil-hydraulic transformer units [21]. In this respect, in the last decade several research activities have been also performed to optimize the PTO layout and maximize the power production, by means of active control techniques, such as latching or unclutching [14]. They are mainly based on putting the device velocity in phase with the incoming wave one by: (i) holding the floater at its extreme excursion and releasing it before the incoming wave occurs or (ii) switching on/off the PTO system [22]. Furthermore, different speed control techniques have been also investigated for oil-hydraulic PTO units, to minimize spring, oscillatory and transient effects due to valve operations, which could lead to possible damages of pipelines, pumps and accessories [23]. Finally, permanent magnet linear generators have been widely studied at Uppsala University [24], that deployed more than 10 WEC devices at the Lysekil site since 2006, to optimize the whole system and decrease the unit production costs.

Finally, a new point absorber type, with a fully submerged mass, has been recently proposed and investigated [3,12,25,26], to increase the power production by phase control. The method mainly consists of shifting the device heave natural frequency towards the incoming wave one, so that the WEC operates at near-resonance conditions. In this respect, several geometries have been also investigated [9], to increase the floating buoy hydrodynamic efficiency and maximize the power production. Hence, starting from actual state of art, in current analysis a new cost-based design and selection procedure of point absorber devices is outlined and applied based on the following main design data:

(i)   The WEC device consists of a floating buoy, with a bullet-type configuration, plus an optional fully submerged mass, connected by means of a tensioned line to a PTO unit, lying on the seabed and equipped with a gravity-based foundation;

(ii) The PTO consists of a direct driven permanent magnet linear generator, modelled in the frequency domain by an additional damping and a spring stiffness;

(iii) Phase control is applied, to properly tune the device heave natural frequency, maximize the power production and minimize the LCoE, with reference to dominant sea states at candidate deployment sites.

Finally, a preliminary cost analysis is also performed before carrying out the optimization procedure, to evaluate the incidence of buoy geometry and WEC size on capital, maintenance and decommissioning costs.

3. Method

3.1. Hydrodynamic Model

The point absorber, consisting of a floating buoy and an additional fully submerged mass as depicted in Figure 1, is connected by means of a tensioned line to a three-phase permanent magnet linear generator. It lies on the seabed and converts the relative motion between the stator and translator into electricity. The floating buoy is characterized by a bullet-type configuration, consisting of a submerged hemisphere with diameter D, and a cylindrical body above the water line with height equal to 20% of relevant diameter. The fully submerged body, having a sphere-shaped configuration with diameter D_{ext}, allows properly tuning the device heave natural frequency, so that wave power extraction is maximum. In this respect, it must be pointed out that it allows increasing the heave mass, without varying the radiation damping and the heave exciting force of the floating buoy, if the draught of the fully submerged mass is sufficiently high, i.e., at least five times its diameter [26].

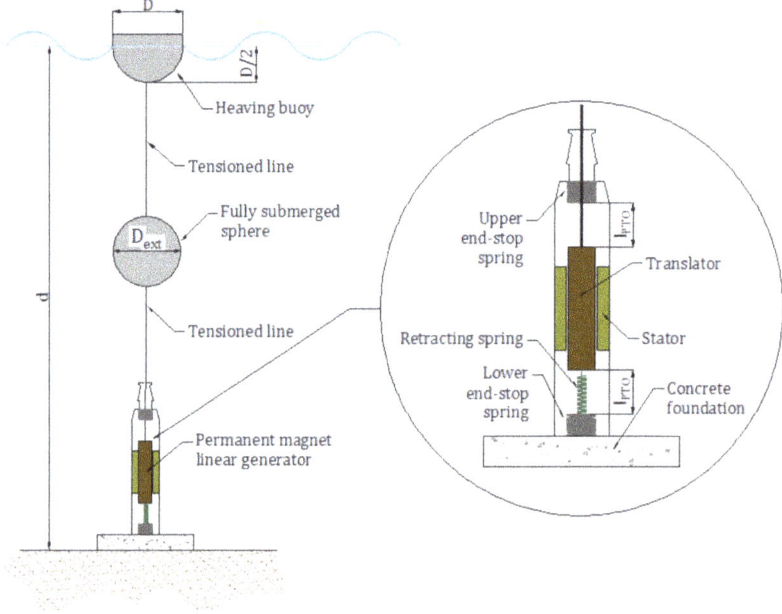

Figure 1. Heaving point absorber with permanent magnet linear generator.

Besides, the permanent magnet linear generator provides additional damping b_{PTO} and restoring stiffness k_{PTO}, as further discussed in Section 3.2. The former mainly depends on the magnetic flux generated in the coils on the oscillating mass, the latter is required to keep tensioned the connection rope with the heaving buoy when the actuator moves downwards. Hence, the point absorber is modelled as a linear mass-spring-damper system in the frequency domain [3]:

$$[m + m_a(\omega) + m_{ext}]\ddot{z} + \left[b_{hyd}(\omega) + b_{PTO}\right]\dot{z} + [k + k_{PTO}]z = \varsigma_a Re\left\{f_{ext}(\omega)e^{-i\omega t}\right\} \qquad (1)$$

having denoted by: (i) m and $m_a(\omega)$ the floating buoy mass and heave added mass respectively; (ii) m_{ext} the additional mass provided by the fully submerged sphere; (iii) $b_{hyd}(\omega)$ the floating buoy radiation damping; (iv) k the restoring stiffness and (v) $f_{ext}(\omega)$ the heave exciting force per unit wave amplitude ς_a. Based on Equation (1), non-linear heave drag forces, exerted by the floating buoy and the fully submerged mass, are not considered, provided that the WEC device is modelled as a linear mass-spring-damper system in the frequency domain [3,9,12,13]. If time-domain analysis is performed, non-linear drag forces can be simply added in the hydrodynamic model, as carried out by Lok et al. [27], among others. Nevertheless, the incidence of drag forces on power production is expected to be moderate and less than 5%, as gathered by a preliminary comparative analysis between frequency and time-domain codes. Based on current results, the frequency domain model reveals to be suitable at least in the preliminary design phase, when several alternative configurations need to be investigated and compared, in order to detect the optimum one for the candidate deployment site. Besides, the heave added mass and radiation damping are determined according to the theoretical solution provided by Hulme [28] for a floating hemisphere, while the heave exciting force per unit wave amplitude is assessed in compliance with the numerical solution provided by Chen et al. [29]. Hence, the heave motion z of the floating buoy is determined as follows:

$$z(\omega, t) = \text{Re}\left\{ H(\omega) e^{-i\omega t} \right\} \tag{2}$$

having denoted by $H(\omega)$ the heave motion transfer function:

$$H(\omega) = \frac{f_{ext}}{-\omega^2 [m + m_a(\omega) + m_{ext}] + i\omega \left[b_{hyd}(\omega) + b_{PTO} \right] + k + k_{PTO}} \tag{3}$$

whose modulus is the response amplitude operator (RAO) of the WEC device. Finally, it is noticed that the additional mass provided by the fully submerged sphere m_{ext} is the sum of the displaced water and the heave added mass [30].

3.2. Permanent Magnet Linear Generator Model

As previously said, the three-phase permanent magnet linear generator is modelled in the frequency domain by a constant damping b_{PTO} and a spring stiffness k_{PTO}. The former is related to the electromechanical force exerted by the PTO, the latter is required to keep tensioned the connection line with the floating buoy during the downward motion of the translator mass. Hence, assuming that the inductive impedance of the generator equivalent circuit is negligible, as regards the resistive one in the range of wave frequencies [31], the damping b_{PTO}^i exerted by any of the three coils can be easily determined. In fact, it mainly depends on: (i) the voltage $e(t) = -d\phi/dt$, generated by Faraday's law and opposite to the magnetic flux first time derivative; (ii) the circuit external resistance R_l and (iii) the translator velocity $\dot{z}$, according to the following equation [32]:

$$b_{PTO}^i = \frac{1}{R_l} \left[\frac{e(t)}{\dot{z}} \right]^2 \tag{4}$$

If full overlap between translator and stator areas exists, the magnetic flux generated by each coil connected in parallel is a sinusoidal function of the translator vertical position z [33], so that the total damping coefficient b_{PTO} is the sum of three terms, each one with the same sinusoidal pattern and a phase leg equal to $w_p/6$:

$$b_{PTO} = \sum_{i=1}^{3} b_{PTO}^i = \frac{3}{2} \left(\frac{2\pi}{w_p} \right)^2 \frac{\phi_t^2}{R_l} \tag{5}$$

In Equation (5) w_p is the generator polar pair width, namely the distance between one north pole to the next on the translator mass, and $\phi_t = B_t w_t dpqc$ is the magnetic flux amplitude [33]. It depends on: (i) the magnetic field in a tooth B_t; (ii) the tooth width w_t; (iii) the width of stator side d; (iv) the total number of poles p; (v) the winding ratio q and (vi) the number of cables in a slot c. Hence, denoted by R_g the linear generator reactive resistance, the electrical efficiency η_{ele} is determined as follows in order to account for copper/eddy current and hysteresis power losses [20,34]:

$$\eta_{ele} = \frac{R_l}{R_l + R_g} \tag{6}$$

Finally, if partial overlapping between translator and stator areas occurs, a drop-off of the PTO damping, is caused [35]. Nevertheless, in current analysis the PTO damping dependence on the translator mass vertical position is not considered, provided that the Equation (1) would become non-linear and the frequency domain analysis would be no more applicable. Anyway, the negligence of this dependence does not significantly affect the analysis as, when partial overlapping occurs, the translator mass hits the upper/lower end-stop springs, with a drop-off of the instantaneous vertical velocity and, consequently, of the vertical force $b_{PTO}\dot{z}$ exerted by the PTO on the wave energy converter.

3.3. Power Production Assessment

In order to properly assess the WEC power production, relevant operational profile needs to be preliminarily assessed. In fact, beyond a certain sea state the device shifts to a survival mode, to avoid possible damages to the PTO system and excessive slamming loads, due to harsh weather conditions [36,37]. Based on previous studies, devoted to the hydrodynamic optimization of point absorbers with hydraulic PTOs [12], the maximum significant wave height is set equal to 3 m. This value allows achieving a good availability of the WEC device at the candidate deployment site, as further discussed in Section 6.2. Hence, the yearly available energy E_{ava} in kWh is determined on the basis of wave statistics at the deployment site, according to the following equation [9,37–39]:

$$E_{ava} = 8760 \sum_{i=1}^{N} \sum_{j=1}^{M} P_{ij} \int_{0}^{\infty} c_g(\omega) S_\varsigma(\omega, H_{s,i}, T_{p,j}) d\omega \tag{7}$$

having denoted by: (i) p_{ij} the probability of occurrence of the sea state obtained by combining the significant wave height $H_{s,i}$ with the peak period $T_{p,j}$; (ii) $c_g(\omega)$ the wave group celerity and (iii) S_ς the wave spectrum. As concerns the point absorber power production, it is mainly limited by the linear generator free stroke length. In fact, when the translator mass hits the upper or lower end-stop springs, a sudden drop-off of the instantaneous velocity occurs that affects the power conversion. Hence, the choice of the most suitable stroke length is quite challenging and would be performed on the basis of a cost/benefit analysis [40], as further discussed in Section 6.1. In fact, a longer stroke requires a taller WEC, with increased economical and mechanical issues, while a shorter stroke may limit too much the power production. Hence, in current analysis the Annualised Energy Production (AEP) in kWh, with reference to a total amount of 8760 h in one year, is assessed by the following equation, that accounts for the heave motion restraint provided by the PTO unit, as further discussed in the Appendix A:

$$AEP = 8760 \eta_{ele} b_{PTO} \sum_{i=1}^{N} \sum_{j=1}^{M} P_{ij} \int_{0}^{\infty} |H(\omega)|^2 S_\varsigma(\omega, H_{s,i}, T_{p,j}) \varphi(\omega, l_{PTO}) d\omega; \; H_{s,i} \leq 3m \tag{8}$$

In Equation (8) l_{PTO} is the upper/lower stroke length, namely the free spacing between the translator mass and the upper/lower end-stop springs, while the corrective function $\varphi(\omega, l_{PTO})$ is discussed in the Appendix. Finally, the point absorber capacity factor CF $= AEP/8760R$, namely the ratio of effective AEP to theoretical energy at rated power R during one year, is determined. In this respect, it must be pointed out that electrical production is not significantly affected by the PTO rated power, as permanent magnet linear generators are capable of enduring long and short term overloads up to 5 and 10 times the rated power respectively. Particularly, long term overloads are mainly due to the most extreme wave conditions at the deployment site, while the short term ones are due to the sudden peaks of the translator's instantaneous velocity, resulting from steep wave fronts [20].

Finally, it must be pointed out that Equation (8) furnishes the AEP of the WEC device with no control strategy of the heave motion, provided that current research focuses on the selection of optimum buoy dimensions that allow to minimize the LCoE, depending on met-ocean conditions at the deployment site. Nevertheless, it is predictable that the expected power production can be further increased if the proper motion control strategy is applied, as investigated by Belmont [41] and Ringwood et al. [42], among others. In this respect, the average power output of the WEC device can be increased by simple control strategies, by determining the optimum velocity profile of the WEC device as a frequency-dependent function of the heave excitation force, or by more refined predictive control methods, such as latching or declutching among others.

3.4. Levelised Cost of Energy Assessment

The LCoE represents the most important factor to efficiently compare different energy sources and evaluate the economic feasibility of new investments in the renewable energy sector. It accounts for capital, operating and decommissioning costs, experienced from the preliminary design up to the end of the power production plant lifetime. It is mainly based on the "present value" approach that relates the risk perceived by the investors to the rate of return of the investment. Hence, projects perceived by the investors as having a high overspend risk, or a lower than predicted revenue, require a high rate of return to become attractive on the international market. Assuming that the yearly maintenance costs are constant during the plant lifetime, the LCoE is determined according to the following formula [9]:

$$\text{LCoE} = \frac{\text{SCI}(1 + r)^n + \text{SDC}}{8760\text{CF}} \cdot \frac{r}{(1 + r)^n - 1} + \frac{\text{OM}}{8760\text{CF}} \tag{9}$$

having denoted by: (i) SCI, SDC and OM the ratios of capital, decommissioning and annual operating costs to the device rated power R in €/kW; (ii) r the discount rate; (iii) n and CF the device expected lifetime and capacity factor, respectively. As concerns the capital costs, they are mainly related to material and manufacturing expenses required for assembling the linear generator and the floating buoy with the fully submerged mass. As concerns the linear generator, the translator is provided with Neodymium–Iron–Boron magnets, mounted on a steel plate between aluminium spacers, while the stator is made of non-oriented laminated electrical steel, with thin insulating coating to reduce eddy current losses [20].

In this respect, it must be pointed out that the unit cost of rare earth materials continues to show stability since the peaks of August 2011, caused by the high mismatch between supply and demand, coming from the high and clean technology sector. Besides, the re-design of many products, to reduce or eliminate the usage of rare earth materials, as well as the increased recycling efforts and the openings of new rare earth ore mines around the world, contributed to put downward the pressure on pricings in the last few years.

As concerns the stator winding, it is made of standard cables with circular cross-section, that can be purchased at relative low prices on the international market. Besides, the concrete foundation is realized by sea-water resistant materials, whose unit price on the international market is about twice higher than the one of ordinary materials used in the civil building sector. Furthermore, manufacturing costs are assessed on the basis of the average hourly total pay for a specialized assembler, inclusive of social and insurance fees [43]. Finally, the floating buoy and the fully submerged mass are assumed to be realised in normal strength steel, with a unit cost equal to 4 €/kg [42]. Hence, based on a comprehensive analysis of current prices on the international market, Table 1 reports the unit prices of the most significant item costs for the analysed WEC device, together with the expected range. In this respect, it must be pointed out that the unit cost of fabricated steel is assumed equal to 4 €/kg [44], even if a reduction up to 3 €/kg is certainly achievable, as suggested by Stansby et al. [45] on the basis of the 2016 UK construction rates for fabricated steel. Nevertheless, in current analysis the upper bound value of the unit cost range for fabricated steel is assumed on safe side, provided that it includes all manufacturing expenses that are characterized by a great variability on the international market.

As concerns the decommissioning expenses, it is assumed that both the linear electric generator and the floating buoy are not reusable, but rather recycled and sold for scrap. Hence, in current analysis decommissioning costs are taken equal to 20% of the capital ones, on safe side, provided that they have not been experienced for wave energy plants and a correct estimate is still challenging. Similarly, annual operating expenses are assumed equal to 5% of capital costs, which is a reasonable value for wave energy converters [9].

Table 1. Material and manufacturing costs.

Item	Material	Current Unit Cost	Unit Cost Range
Permanent magnets	Neodymium-Iron-Boron	80 €/kg	60–100 €/kg
Stator	Electrical steel	2 €/kg	1.5–2.5 €/kg
Translator	Electrical steel	2 €/kg	1.5–2.5 €/kg
Rim	Aluminium alloy	5 €/kg	4–6 €/kg
Cables (16 mm^2)	Electrical copper	5 €/m	4–6 €/m
Foundation	Marine concrete	300 €/m^3	200–400 €/m^3
Manufacturing [1]	–	25 €/h	—
Floating buoy [2]	Normal Strength Steel	4 €/kg	3–4 €/kg

[1] It refers to the only assembly costs of the linear generator; [2] It includes manufacturing costs.

4. Main Data for Cost-Based Design

4.1. Selection of Candidate Deployment Sites

During the last years, several studies were performed to assess the wave energy potential in the Mediterranean Sea, along the Italian coastlines. Really, its wave climate is quite mild, if compared to the Atlantic Ocean, off the Spanish coastlines, and the North Sea. Nevertheless, it still represents an attractive choice for possible deployment of WEC devices. In fact, it is predictable that all technical issues, mainly related to the WEC survival in extreme sea state conditions, can be more easily solved, so that the power production is expected to be competitive, in terms of LCoE, on the international market [46]. Particularly, the most energetic sites are located in the Tyrrhenian Sea at Alghero, off the western coast of Sardinia, while the mean wave energy decreases in the Ionian and Adriatic Seas. Nevertheless, as highlighted by Liberti et al. [46], the mean wave power alone is not sufficient to identify the most promising areas for possible deployment of wave arrays. In fact, it arises from the contribution of individual sea states, distributed over a wide range of wave heights, periods and directions, so that contributions due to most energetic sea states cannot be efficiently harnessed, since relevant exploitation requires an over-sized WEC device.

Based on previous remarks, in current analysis five candidate locations are selected in the Mediterranean Sea along the Italian coastlines, as depicted in Figure 2. Besides, Table 2 reports: (i) the coordinates of candidate deployment sites; (ii) the JONSWAP spectrum peak enhancement factor γ; (iii) the mean J and available J_{ava} wave power per unit length of wave front. Particularly, the mean wave power accounts for all sea states at the deployment site, each one with a certain probability of occurrence. The available wave power per unit length, instead, accounts for all environmental conditions up to the cut-out sea state, corresponding in current analysis to a significant wave height equal to 3 m, as discussed in Section 3.3. Really, this parameter reveals to be very effective, as only part of the available wave power at the deployment site can be effectively harnessed, depending on the WEC operational profile. In fact, beyond the cut-out sea state the WEC device switches to the survival mode and the power production is stopped. Finally, the JONSWAP spectrum peak enhancement factor is site-dependent, as it ranges from 1.84 up to 2.20 with reference to the five candidate sites. In all cases, wave data were taken from the website of the Italian governmental institution ISPRA, namely "Istituto Superiore per la Protezione e la Ricerca Ambientale".

Based on the data reported in Table 2, Alghero is the most promising site, with a mean wave power per unit length of wave front equal to 12.53 kW/m. On the contrary, the remaining candidate sites, which are characterized by a lower wave energy potential ranging from 2.58 up to 5.25 kW/m, seem to represent a less attractive choice. Nevertheless, if the comparative analysis is performed on the basis of the available power up to 3 m significant wave height, the previous gap consistently reduces in favour of less energetic sites. Hence, the main outcomes of Liberti et al. [46] are confirmed, provided that the only mean wave power per unit length of wave front is not sufficient to compare different deployment sites, if the WEC operational profile is not taken into due consideration. Finally, wave histograms are reported in Figure 3, based on 0.5 s and 0.5 m mean wave period T_1 and significant wave height H_s classes, respectively. It is noticed that T_1 is the ratio of 0-order to 1-order sea state spectral moments [47].

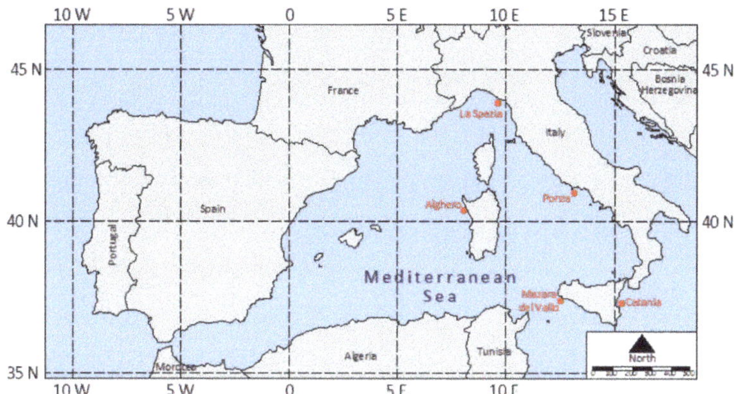

Figure 2. Location of candidate deployment sites in the central Mediterranean Sea.

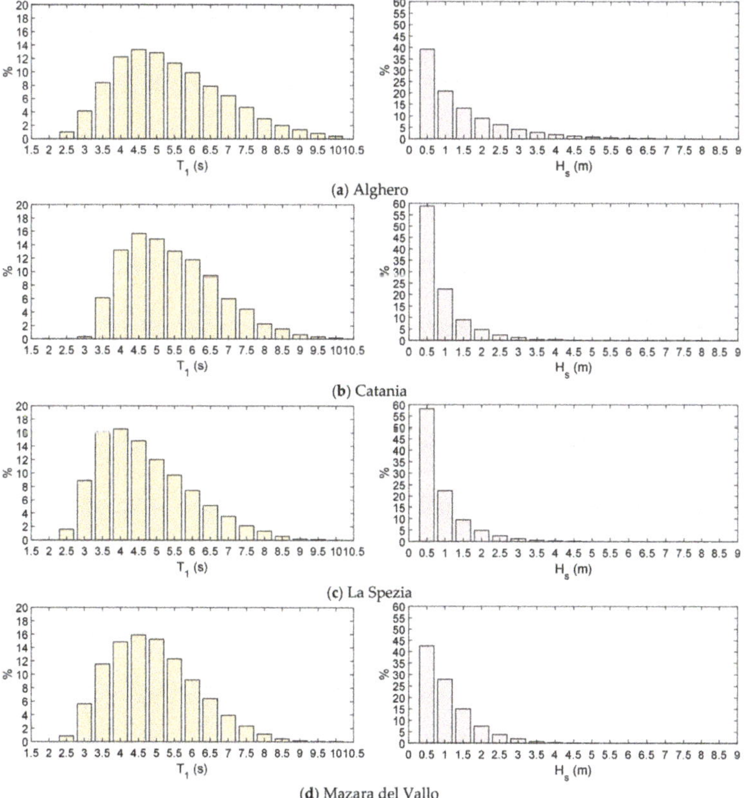

(**a**) Alghero

(**b**) Catania

(**c**) La Spezia

(**d**) Mazara del Vallo

Figure 3. *Cont.*

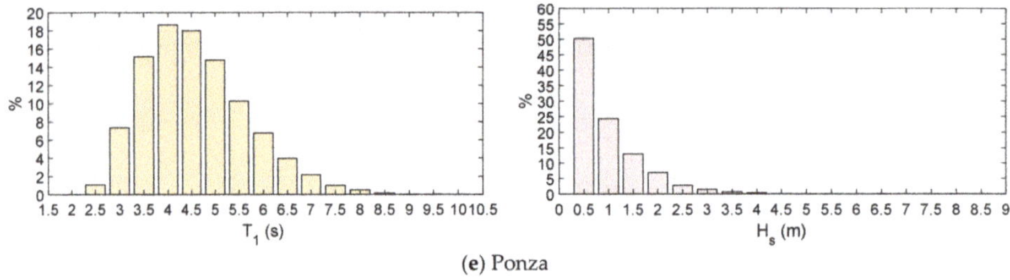

(**e**) Ponza

Figure 3. Frequency histograms of wave heights and periods at candidate deployment sites.

Table 2. Main data of candidate sites along the Italian coastlines.

Deployment Site	Latitude	Longitude	γ (—)	J (kW/m)	J_{ava} (kW/m)
Alghero	40°32′54″ N	08°06′24″ N	1.86	12.53	5.79
Catania	37°26′18″ N	15°08′48″ N	2.20	2.58	2.16
La Spezia	43°55′12″ N	09°49′06″ N	2.02	3.75	2.98
Mazara del Vallo	37°31′00″ N	12°32′00″ N	1.84	5.25	4.52
Ponza	40°52′00″ N	12°57′00″ N	2.15	4.46	3.73

4.2. Capital, Maintenance and Decommisioning Costs

As discussed in Section 3.4, capital costs are mainly due to the linear generator and the floating buoy. In current analysis a 10 kW linear generator, with a mechanical damping b_{PTO} and a spring stiffness k_{PTO} equal to 23.75 kNs/m and 6.20 kN/m respectively [48], is assumed as a reference PTO unit. Relevant data are listed in Table 3, while Table 4 reports the list of material and manufacturing costs. Particularly, Table 3 reports all quantities required to assess the mechanical damping and the electrical efficiency of the permanent magnet linear generator based on Equations (5) and (6). In this respect, it provides: the magnetic field in a tooth B_t, the width of stator side d, the total number of poles p, the winding ratio q, the number of cables in a slot c, the resistive load R_l and the generator polar pair width w_p. As concerns the manufacturing costs reported in Table 4, Danielson et al. [48] provide all material quantities required for the 10 kW generator, while the weight of the concrete foundation is assumed equal to 35 ton, corresponding to 15 m^3 [24]. Finally, the total manufacturing effort, required for assembling the permanent magnet linear generator, is assumed to be about 100 h, even if some cost reductions are achievable in view of a PTO mass-production. According to the data reported in Table 4, the base-cost of the 10 kW linear generator is equal to about 25 k€, even if other expenditure items, mainly related to the profit of the manufacturing plant and additional expenses for minor mechanical parts, are likely to occur. Nevertheless, as point absorbers are still in a pre-commercial phase, it is quite challenging to carry out a reliable estimate of these additional cost items. Hence they are roughly assumed equal to 20% of the base-cost that, in turn, increases up to 35 k€. As concerns the cost of the floating buoy and the fully submerged mass, it is determined on the basis of the following assumptions: (i) all structures are made of 10 mm normal strength steel; (ii) the weight of the internal reinforcements is equal to 30% of the external platings' one, which is a typical value for structures with spar-type configuration.

Finally, the deployment costs are equal to 5 k€, which is a reasonable value for chartering of small tug boats and barges, as there is no need for the employment of more expensive units, such as anchor handling vessels. Finally, decommissioning and annual operating costs are determined as a fraction of capital expenditures, in conjunction with a 7% discount rate and a 25-year lifetime of the WEC array [49], as listed in Table 5, where main data for the assessment of LCoE are reported.

Table 3. Linear generator main data.

Rated power	R	10	kW
Design speed	v	0.70	m/s
Efficiency	η_{ele}	0.86	—
Upper/lower stroke length	l_{PTO}	0.8	m
Magnetic field in a tooth	B_t	1.76	T
Tooth width	w_t	8	mm
Width of stator side	d	400	mm
Total number of poles	p	100	—
Winding ratio	q	1.2	—
Number of cables in a slot	c	6	—
Resistive load	R_l	4.1	Ω
Generator polar pair width	w_p	100	mm

Table 4. Linear generator preliminary cost analysis.

Item	Material	Quantity	Unit Cost	Cost (€)
Permanent magnets	Neodymium-Iron-Boron	115 kg	80 €/kg	9200
Stator	Electrical steel	766 kg	2 €/kg	1532
Translator	Electrical steel	432 kg	2 €/kg	864
Rim	Aluminium alloy	13 kg	15 €/kg	195
Cables (16 mm^2)	Electrical copper	1096 m	5 €/m	5480
Foundation	Marine concrete	15 m^3	300 €/m^3	4500
Manufacturing	—	100 h	25 €/h	2500

Table 5. Main data for LCoE assessment.

Item	Symbol	Equation/Value	Unit
Capital cost to power ratio	SCI	$3500 + 500 + 40\pi\left(D^2 + D_{ext}^2\right)$	€/kW
Decommissioning cost to power ratio	SDC	0.20SCI	€/kW
Annual operating cost to power ratio	OM	0.05SCI	€/kW
Discount rate	r	0.07	—
Device expected lifetime	n	25	years

5. Results

5.1. WEC Sizing without the Fully Submerged Mass

Optimum dimensions of the floating buoy at the deployment sites listed in Table 2 are preliminarily assessed without the fully submerged mass. Figure 4 reports the LCoE distribution versus the buoy diameter D at the candidate deployment sites, varying the buoy diameter in the range from 2 up to 7 m. In all cases, the curves are characterized by a left descending and a right ascending branch. Main data of optimum configurations are further detailed in Table 6. Based on current results, it is gathered that: (i) maximum power production occurs at Alghero and Mazara del Vallo, where the LCoE is equal to about 0.60 €/kWh; (ii) the energy production cost increases by about 10% and 25% at Ponza and La Spezia respectively; (iii) the LCoE almost doubles at Catania, where a consistent decrease of power production occurs. As concerns the floating buoy dimensions, the diameter corresponding to the minimum LCoE is site-dependent and ranges from 3.7 m at Alghero up to 4.2 m at La Spezia.

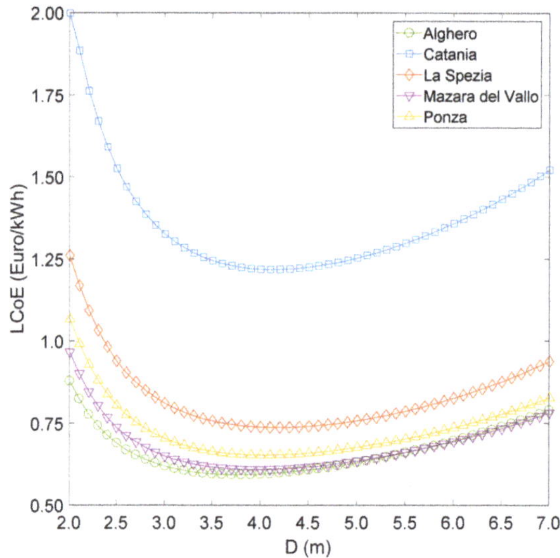

Figure 4. LCoE distribution versus the diameter of the buoy without the fully submerged mass.

Table 6. Main data of WEC devices without the fully submerged mass.

Location	D (m)	AEP (kWh)	CF (—)	SCI (€/kWh)	SDC (€/kWh)	OM (€/kWh)	LCoE (€/kWh)
Alghero	3.7	13,289	0.152	5668	1134	283	0.593
Catania	4.1	6900	0.079	6048	1210	302	1.218
La Spezia	4.2	11,587	0.132	6149	1230	307	0.738
Mazara del Vallo	4.0	13,587	0.155	5949	1190	297	0.609
Ponza	4.1	12,871	0.147	6048	1210	302	0.653

5.2. WEC Sizing with the Fully Submerged Mass

Sizing of WEC devices is performed by adding the fully submerged mass, to properly tune the device heave natural frequency, on the basis of dominant sea states at the deployment site. The LCoE distribution versus the buoy diameter D is plotted in Figure 5, while main data of optimum WECs are listed in Table 7. Hence, for each candidate configuration, the diameter D_{ext} of the fully submerged sphere is iteratively varied, with 0.1 m step, to detect for any dimension of the floating buoy, the optimum configuration, characterized by the minimum LCoE. Based on current results, the fully submerged sphere yields to a consistent increase of the AEP and the WEC capacity factor. Besides, the optimum buoy diameter is equal to 3.8 m, with the only exception of Catania, where it is equal to 4.0 m. Concerning the fully submerged mass diameter, it lies in the range 3.7–4.2 m, depending on the dominant wave periods at the deployment site. Finally, from Table 7 it is also gathered that: (i) power production is comparable at Alghero, Mazara del Vallo and Ponza, where the LCoE is equal to about 0.40 €/kWh; (ii) the energy production cost increases by about 25% and 80% at La Spezia and Catania, respectively.

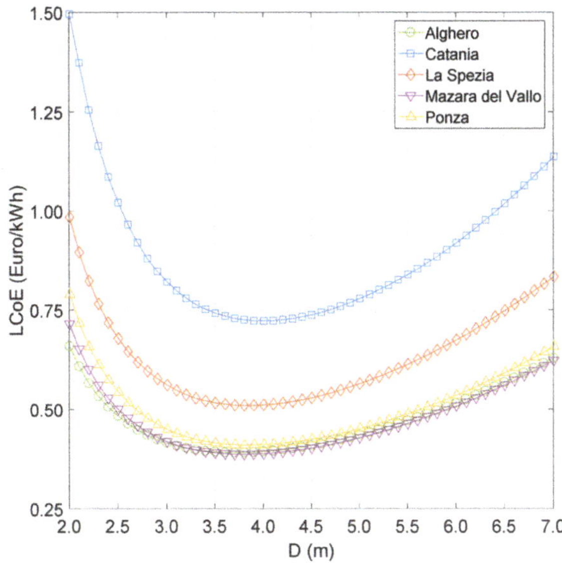

Figure 5. LCoE distribution versus the diameter of the buoy with the fully submerged mass.

Table 7. Main data of WEC devices with the fully submerged mass.

Location	D (m)	AEP (kWh)	CF (—)	SCI (€/kWh)	SDC (€/kWh)	OM (€/kWh)	LCoE (€/kWh)	D_{ext} (m)
Alghero	3.8	28,501	0.325	8021	1604	401	0.387	4.2
Catania	4.0	15,592	0.178	8105	1621	405	0.722	4.1
La Spezia	3.8	20,468	0.234	7515	1503	376	0.510	3.7
Mazara del Vallo	3.8	28,437	0.325	7915	1583	396	0.387	4.1
Ponza	3.8	26,480	0.302	7811	1562	391	0.410	4.0

6. Discussion

6.1. Incidence of Fully Submerged Mass on Cost-Based Design

The incidence of the fully-submerged mass on the AEP and LCoE is preliminarily investigated. Figure 6a,b report the comparative analysis between the two WEC configurations, on the basis of the data listed in Tables 6 and 7. Based on current results, the fully submerged mass allows to increase the AEP and decrease the LCoE. In fact, the AEP almost doubles, while the LCoE decreases by about 30–40%, depending on the deployment site. This outcome is mainly due to the presence of the fully submerged mass, that allows properly tuning the device heave natural frequency, based on wave periods of dominant sea states at the deployment site. In this respect, the WEC device with the fully submerged mass operates closer to the resonance condition with the incoming waves, as regards the first configuration, so as a consistent increase of the AEP is gained. Nevertheless, the AEP increase is not counterbalanced by an equivalent decrease of the LCoE. This outcome is mainly due to the increase of capital costs, due to the fully submerged mass, as well as of operating and decommissioning expenses that, in turn, are expected to slightly increase as gathered from relevant values reported in Tables 6 and 7.

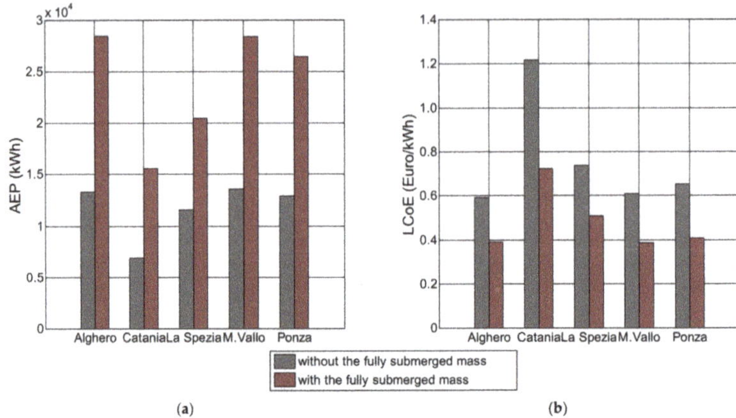

Figure 6. Comparative analysis among optimum WEC configurations. (**a**) AEP at candidate deployment sites; (**b**) LCoE at candidate deployment sites.

6.2. PTO Free Stroke Length

Previous results show that power production can be consistently increased by adding a fully submerged mass to the floating buoy, that allows properly tuning the device heave natural frequency, with reference to dominant sea states at the candidate deployment site. Nevertheless, as discussed in Section 3.3, power production is mainly affected by the PTO free stroke length. In fact, in the most energetic sea states only part of the wave cycle is effective, provided that a sudden drop off of the instantaneous power production occurs, when the translator mass hits the PTO upper and lower end-stop springs. Hence, it is conceivable that if the PTO free stroke length increases, the AEP grows up, so yielding to a further decrease of the LCoE.

Assuming that the capital cost of the permanent magnet linear generator remains almost unchanged for small variations of the free stroke length, Figure 7 reports a comparative analysis between the LCoE of WECs without and with the fully submerged mass at Alghero. Particularly, the free stroke length l_{PTO} ranges from 0.5 up to 2.0 m, with 0.5 m step. Based on current results, reported in Tables 8 and 9, it is gathered that if l_{PTO} increases from 0.8 m up to 2.0 m, the LCoE decreases by 8.6% and 40.7% for WECs without and with the fully submerged mass, respectively. This outcome is mainly due to the different operational profile of WEC devices. In fact it is overdamped without the additional sphere, while it operates at near resonance conditions if the fully submerged mass is provided. Hence, it is gathered that no variations of the PTO free stroke length are required for WECs without the fully submerged mass, as the expected decrease of LCoE is moderate. On the contrary, a new PTO design, with an increased free stroke length, can yield to a consistent decrease of LCoE for point absorbers equipped with the fully submerged mass.

Table 8. Incidence of PTO free stroke length on main data of WEC devices without the fully submerged mass at Alghero.

l_{PTO} (m)	D (m)	AEP (kWh)	CF (—)	SCI (€/kWh)	SDC (€/kWh)	OM (€/kWh)	LCoE (€/kWh)
0.5	3.8	10,652	0.122	5579	1116	279	0.728
0.8	4.0	13,289	0.152	5668	1134	283	0.593
1.0	3.8	14,521	0.166	5853	1171	293	0.560
1.5	3.8	14,997	0.171	5853	1171	293	0.542
2.0	3.8	14,997	0.171	5853	1171	293	0.542

Table 9. Incidence of PTO free stroke length on main data of WEC devices with the fully submerged mass at Alghero.

l_{PTO} (m)	D (m)	AEP (kWh)	CF (—)	SCI (€/kWh)	SDC (€/kWh)	OM (€/kWh)	LCoE (€/kWh)	D_{ext} (m)
0.5	3.6	18,997	0.217	7529	1506	376	0.551	3.9
0.8	3.8	28,501	0.325	8021	1604	401	0.387	4.2
1.0	3.8	34,002	0.388	8242	1648	412	0.337	4.4
1.5	4.0	46,479	0.531	8904	1781	445	0.266	4.8
2.0	4.1	55,356	0.632	9254	1851	463	0.232	5.0

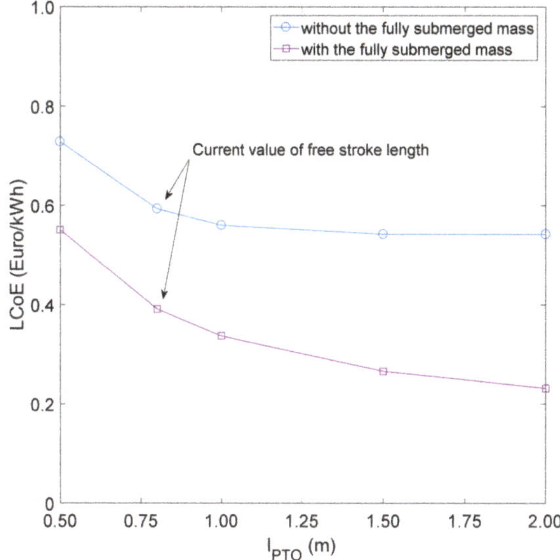

Figure 7. LCoE distribution versus the PTO free stroke length at Alghero.

6.3. WEC Operational Profile

It is well-known that survivability in extreme weather conditions represents a challenging issue for WECs [1,50] and that the selection of the most suitable survival mode mainly depends on the layout of the device and the PTO equipment [35]. In this respect, two different survival mode strategies can be applied. The former mainly consists of ballasting and submerging the floating buoy, so that no significant wave interactions occur; the latter is based on short-circuiting the permanent magnet linear generator [35], to increase its damping and reduce the heave motion amplitude of the WEC device. Nevertheless, independently of the selected survival mode, the choice of the "cut-out" sea state represents a basic issue for the reliable design of WEC devices, as both AEP and LCoE are affected by relevant operational profile.

In the previous analysis, it was assumed that the point absorber switches to the survival mode when the significant wave height exceeds 3 m. Hence, Figure 8 reports a comparative analysis, in terms of LCoE, for WEC devices deployed at Alghero, where H_s is varied in the range from 2.0 up to 4.0 m, with 0.5 m step. Current results, listed in Tables 10 and 11, show that if H_s increases from the design value of 3.0 m up to 4.0 m, the LCoE decreases by 14.5% and 11.5% for WEC devices without and with the fully submerged mass, respectively. On the contrary, the opposite holds true if H_s decreases up to 2.0 m, so yielding to a corresponding increase of LCoE equal to 44.7% and 33.2% for the two WEC configurations. Current results confirm that the proper selection of the "cut-out" sea state represents a basic issue for the power production assessment. Really, it has to be selected in view of a proper balancing between the device availability and the occurrence of undesirable phenomena, such as slamming events, in harsh weather conditions. Hence, the 3 m "cut-out" sea state seems to be a reasonable choice at least for WEC devices deployed in the Mediterranean Sea.

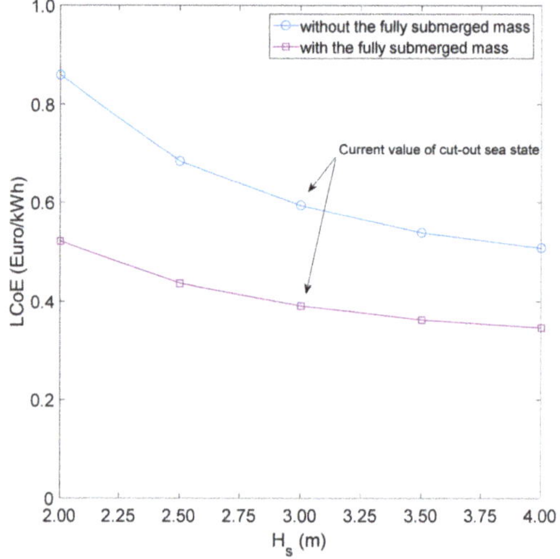

Figure 8. LCoE distribution versus maximum significant wave height at Alghero.

Table 10. Incidence of maximum significant wave height on main data of WEC devices without the fully submerged mass at Alghero.

H_S (m)	D (m)	AEP (kWh)	CF (—)	SCI (€/kWh)	SDC (€/kWh)	OM (€/kWh)	LCoE (€/kWh)
2.0	4.0	9636	0.110	5949	1190	297	0.858
2.5	3.9	11,910	0.136	5853	1171	293	0.683
3.0	3.7	13,289	0.152	5668	1134	283	0.593
3.5	3.7	14,640	0.167	5668	1134	283	0.538
4.0	3.7	15,527	0.177	5668	1134	283	0.507

Table 11. Incidence of maximum significant wave height on main data of WEC devices with the fully submerged mass at Alghero.

H_S (m)	D (m)	AEP (kWh)	CF (—)	SCI (€/kWh)	SDC (€/kWh)	OM (€/kWh)	LCoE (€/kWh)	D_{ext} (m)
2.0	3.8	21,401	0.244	8021	1604	401	0.521	4.2
2.5	3.8	25,545	0.292	8021	1604	401	0.436	4.2
3.0	3.8	28,501	0.325	8021	1604	401	0.387	4.2
3.5	3.7	30,407	0.347	7930	1586	396	0.362	4.2
4.0	3.7	31,821	0.363	7930	1586	396	0.346	4.2

6.4. Some Recommendations for WEC Design

Based on previous results, some recommendations for the preliminary design and optimization of the WEC device can be furnished:

(i)   Power production mainly depends on metocean conditions at the deployment site. Nevertheless, actual results show that the decrease rate of LCoE diminishes with the available wave power per unit length of wave front, as reported in Figure 9, with reference to WEC devices without and with the fully submerged mass. In fact, the LCoE is almost constant from 3.5 up to about 6.0 kW/m^2, which implies that the expected cost of energy is almost comparable at Ponza, Mazara del Vallo and Alghero, even if J_{ava} ranges from 3.73 up to 5.79 kW/m^2. This outcome is mainly due to the heave motion restraint provided by the PTO unit, so that the incoming wave energy cannot be completely harnessed in the most energetic sea states.

(ii)   Power production is mainly affected by the free stroke length of the permanent magnet linear generator, as in the most energetic sea states only part of the wave cycle is effective. In this respect, this vertical motion restraint has to be properly accounted in the most energetic sea states, as detailed in the Appendix, to correctly estimate the AEP and the LCoE.

(iii)   Floating buoy dimensions have to be properly selected by minimizing the LCoE, on the basis of wave statistics at the candidate deployment site. Particularly, the power production depends on the "cut-out" sea state that has to be selected by properly balancing the WEC operational profile and all technical issues related to the extreme loads in harsh weather conditions on the floating buoy, the mooring connection line and the PTO unit. Actual results show that a 3 m "cut-out" significant wave height is a reasonable design value for point absorbers in the Mediterranean Sea, as a further increase of the maximum significant wave height yields to a moderate decrease of the LCoE.

(iv)   There is a strong relation between the PTO free stroke length and the fully submerged mass. In fact, a low free stroke length of about 0.8 m is suitable for WEC devices without the fully submerged mass. On the contrary, if the floating buoy is equipped with a fully submerged mass, in order to properly tune the device heave natural frequency depending on dominant sea states at the deployment site, consistent reductions of LCoE can be achieved by increasing the PTO free stroke length. Current results show that for point absorbers equipped with the fully submerged mass and deployed in the Mediterranean Sea the free stroke length should be increased up to at least 1.5 m, i.e., one half the "cut-out" significant wave height.

(v)   Further reductions of LCoE can be achieved by increasing the PTO rated power from the current value of 10 kW up to 20/30 kW, which is the rated power of the last generators designed and tested at the Uppsala University [24,43].

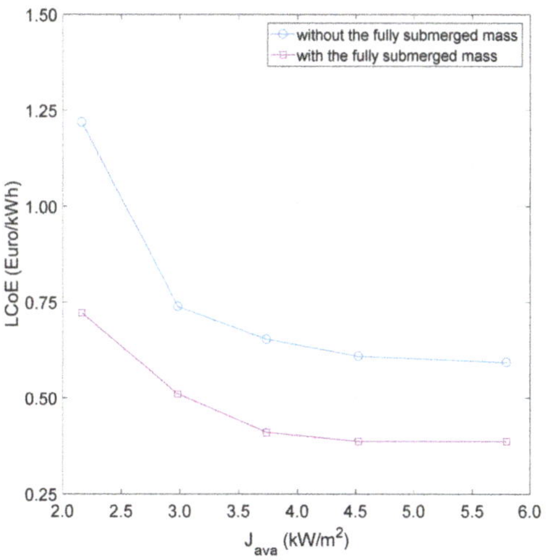

Figure 9. LCoE distribution versus J_{ava}.

Current recommendations refer to WEC devices deployed in the Mediterranean Sea. Nevertheless, it is predictable that almost similar outcomes can be stressed for other deployment sites, with reference to the incidence of the PTO free stroke length and the WEC operational profile on the attained LCoE.

7. Conclusions

A new-cost based design procedure for point absorbers, equipped with an optional fully submerged mass and connected to a permanent magnet linear generator lying on the seabed, was developed to minimize the LCoE. Based on current results, the following main outcomes were achieved:

(i) A new hydrodynamic model for point absorbers with a fully submerged mass was developed. It accounts for the WEC operational profile, in terms of "cut-out" sea state, and the heave motion restraint provided by the permanent magnet linear generator when the translator mass hits the upper/lower end-stop springs, as detailed in the Appendix A.

(ii) The new procedure, devoted to select the most suitable WEC device, characterized by the minimum LCoE, was applied to several candidate deployment sites, verifying that the only wave power per unit length of wave front is not sufficient to correctly characterize the site, in terms of power production and costs.

(iii) The incidence on the AEP and the LCoE of the PTO free stroke length and the WEC operational profile was investigated. It was found that current PTOs, with a free stroke length equal to 0.8 m, are suitable for point absorbers without the fully submerged mass. On the contrary, it is suggested to increase the free stroke length up to 1.5 m for WEC devices equipped with the fully submerged mass, due to the increased heave motion amplitude in harsh weather conditions.

(iv) Some guidelines to reduce the LCoE, to properly tune the PTO mechanical properties and select, at least in a preliminary project phase, the most suitable point absorber configuration are provided.

Current outcomes, stressed for the Mediterranean Sea, need to be further investigated focusing on other deployment sites, in order to reduce the LCoE of WEC devices and make the wave energy sector more competitive on the international market, as regards other renewable sources.

Acknowledgments: The authors wish to thank the anonymous reviewers for their useful comments and insights. Besides, the research activities were partly financed by the Individual Research Programme—Year 2017—funded by the University of Naples "Parthenope".

Author Contributions: All authors have contributed to the research. Vincenzo Piscopo developed, applied and discussed the mathematical model, while Guido Benassai, Renata Della Morte and Antonio Scamardella conceived the theoretical framework of the research activity. All authors took part in the writing of the paper.

Conflicts of Interest: The authors declare no conflict of interest.

Nomenclature

List of Abbreviations

AEP	Annualized Energy Production
LCoE	Levelised Cost of Energy
PTO	Power Take-Off
RAO	Response Amplitude Operator
WEC	Wave Energy Converter

List of Symbols

b_{hyd}	Frequency-dependent hydrodynamic damping coefficient
b_{PTO}	Power Take-Off damping coefficient
d	Width of stator side
B_t	Magnetic field in a tooth
CF	Capacity factor
c_g	Wave group celerity
D	Floating buoy diameter
D_{ext}	Diameter of fully submerged mass

E_{ava}	Yearly available energy
e	Voltage
f_{ext}	Heave exciting force
H	Heave motion transfer function
$H_{s,i}$	Significant wave height of the i-th sea state
J	Wave power per unit length of wave front
J_{ava}	Available wave power per unit length of wave front
k	Restoring stiffness
k_{PTO}	Power Take-Off restoring stiffness
l_{PTO}	Power Take-Off free stroke length
m	Mass of floating buoy
m_a	Frequency-dependent added mass of floating buoy
m_{ext}	Mass plus added mass of the fully submerged body
n	Device expected lifetime
OM	Ratio of annual operating costs to device rated power
P_{abs}	Absorbed power
$P_{abs,full}$	Absorbed power if the entire wave cycle is effective
$P_{abs,part}(\omega)$	Absorbed power if only part of the wave cycle is effective
p	Total number of poles
P_{ij}	Probability of occurrence of the sea state
q	Winding ratio
R	Device rated power
Re	Real part of a complex number
R_g	Generator reactive resistance
R_l	Circuit external resistance
r	Discount rate
SCI	Ratio of capital costs to device rated power
SDC	Ratio of decommissioning costs to device rated power
S_ς	Wave spectrum
$T_{p,j}$	Peak period of the j-th sea state
T_1	Mean wave period
t	Time
w_p	Generator polar pair width
w_t	Tooth width
z	Heave motion amplitude
$\dot{z}$	Heave motion velocity
$\ddot{z}$	Heave motion acceleration

List of Greek Symbols

γ	Peak enhancement factor of wave spectrum
ς	Sea surface elevation
ς_a	Wave amplitude
η_{ele}	Electrical efficiency
ϕ	Magnetic flux
ϕ_t	Magnetic flux amplitude
φ	Corrective factor of the absorber power
ω	Wave circular frequency

Appendix A. Power Production Assessment

In the following a new procedure is outlined to evaluate the power production of a point absorber, when the heave motion of translator mass is restrained by the PTO free stroke length and only part of the wave cycle is effective. Let us consider a regular sea state with: (i) wave amplitude ς_a; (ii) wave period T and (iii) circular frequency ω. The sea surface elevation $\varsigma(\omega, t)$ is expressed as follows:

$$\varsigma(\omega,t) = \varsigma_a \sin(\omega t) \tag{A1}$$

Denoted by $H(\omega)$ the WEC heave transfer function, the vertical motion $z(\omega,t)$ and the squared velocity $\dot{z}^2(\omega,t)$ of the linear generator translator mass can be determined:

$$z(\omega,t) = |H(\omega)|\varsigma_a \sin(\omega t) \tag{A2}$$

$$\dot{z}^2(\omega,t) = |H(\omega)|^2\omega^2\varsigma_a^2 \cos^2(\omega t) \tag{A3}$$

so that the instantaneous absorbed power is derived, as a function of PTO damping:

$$P_{abs}(\omega,t) = b_{PTO}\dot{z}^2(\omega,t) \tag{A4}$$

If the vertical motion amplitude of the translator mass is lower than the PTO free stroke length, as depicted in Figure A1, the wave cycle is fully effective and the mean absorber power is determined as follows:

$$P_{abs,full}(\omega) = \frac{1}{T}\int_0^T P_{abs}(\omega,t)dt = \frac{1}{2}b_{PTO}|H(\omega)|^2\omega^2\varsigma_a^2 \tag{A5}$$

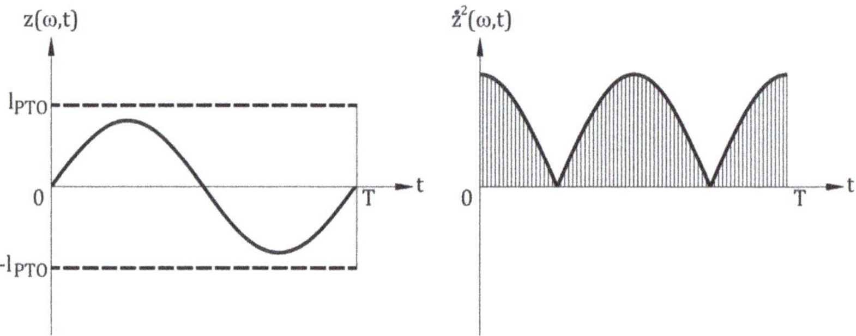

Figure A1. Power absorption without exceedance of PTO upper/lower stroke length (graphs are purely qualitative).

On the contrary, if the motion amplitude of translator mass is higher than the PTO free stroke length, as depicted in Figure A2, only part of the wave cycle is effective, as when the translator mass hits the upper or lower end-stop springs a sudden drop-off of the heave velocity occurs. Hence, if the end-stop spring stiffness is high enough, the heave velocity drop-off can be assumed almost instantaneous, so that no power production occurs in the time intervals $[T_1, T_2]$ and $[T_3, T_4]$, as depicted in Figure A2. Hence, the absorbed power in is determined as follows:

$$P_{abs,part}(\omega) = \frac{1}{T}\left[\int_0^{T_1} P_{abs}(\omega,t)dt + \int_{T_2}^{T_3} P_{abs}(\omega,t)dt + \int_{T_4}^T P_{abs}(\omega,t)dt\right] \tag{A6}$$

and can be further simplified:

$$P_{abs,part}(\omega) = \frac{2}{T}\left[\int_0^{T_1} P_{abs}(\omega,t)dt + \int_{T_2}^{T/2} P_{abs}(\omega,t)dt\right] \tag{A7}$$

Based on previous results, the following corrective function $\varphi(\omega)$ can be introduced. It is unitary, if the heave motion amplitude of translator mass is lower than the l_{PTO}, while it is equal to the ratio of partial to fully effective absorber power, if the translator mass heave motion amplitude exceeds the free stroke length:

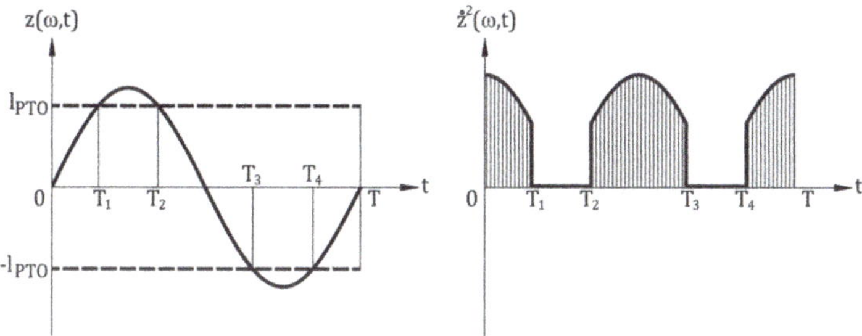

Figure A2. Power absorption with exceedance of PTO upper/lower stroke length (both graphs are purely qualitative).

$$\varphi(\omega) = \begin{cases} 1 & \text{if } |H(\omega)|\varsigma_a \leq l_{PTO} \\ \dfrac{P_{abs,part}(\omega)}{P_{abs,full}(\omega)} & \text{if } |H(\omega)|\varsigma_a >_{PTO} \end{cases} \tag{A8}$$

After some manipulations, Equation (A8) is resembled by the following one:

$$\varphi(\omega) = \begin{cases} 1 & \text{if} |H(\omega)|\varsigma_a \leq l_{PTO} \\ \dfrac{4}{T}\left[\displaystyle\int_0^{T_1} \cos^2(\omega t)dt + \int_{T_2}^{T/2} \cos^2(\omega t)dt\right] & \text{if} |H(\omega)|\varsigma_a > l_{PTO} \end{cases} \tag{A9}$$

so that the absorber power can be determined as follows, with reference to both full and partly effective wave cycles:

$$P_{abs}(\omega) = \frac{1}{2}b_{PTO}|H(\omega)|^2\omega^2\varsigma_a^2\varphi(\omega) \tag{A10}$$

Similar results can be extended to a seaway, by means of the wave spectrum $S_\varsigma(\omega)$, so that the following equality holds:

$$P_{abs}(\omega) = b_{PTO}\int_0^\infty |H(\omega)|^2\omega^2\varphi(\omega)S_\varsigma(\omega)d\omega \tag{A11}$$

Finally, the electrical power production $P_{ele}(\omega) = \eta_{ele}P_{abs}(\omega)$ depends on the efficiency of the permanent magnet linear generator η_{ele} and the absorbed power $P_{abs}(\omega)$, based on Equation (A11).

References

1. Falcão, A.F.O. Wave energy utilization: A review of the technologies. *Renew. Sustain. Energy Rev.* **2010**, *14*, 899–918. [CrossRef]
2. Salter, S.H. Wave power. *Nature* **1974**, *249*, 720–724. [CrossRef]
3. Vantorre, M.; Banasiak, R.; Verhoeven, R. Modelling of hydraulic performance and wave energy extraction by a point absorber in heave. *Appl. Ocean Res.* **2004**, *26*, 61–72. [CrossRef]
4. Falnes, J. A review of wave-energy extraction. *Mar. Struct.* **2007**, *20*, 185–201. [CrossRef]
5. Vicente, P.C.; António, F.D.; Gato, L.M.C.; Justino, P.A.P. Dynamics of arrays of floating point-absorber wave energy converters with inter-body and bottom slack-mooring connections. *Appl. Ocean Res.* **2009**, *31*, 267–281. [CrossRef]

6.   Nunes, G.; Valério, D.; Beirão, P.; Sá da Costa, J. Modelling and control of a wave energy converter. *Renew. Energy* **2011**, *36*, 1913–1921. [CrossRef]

7.   Ocean Energy Systems. *International Levelised Cost of Energy for Ocean Energy Technologies*; Report; International Energy Agency: Paris, France, 2015.

8.   European Commission. *European Commission Decision C (2014) 4995*; European Commission: Brussels, Belgium, 2014.

9.   De Backer, G. Hydrodynamic Design Optimization of Wave Energy Converters Consisting of Heaving Point Absorbers. Ph.D. Thesis, Ghent University, Ghent, Belgium, 2009.

10.   Bachynski, E.E.; Young, Y.L.; Yeung, R.W. Analysis and optimization of a tethered wave energy converter in irregular waves. *Renew. Energy* **2012**, *48*, 133–145. [CrossRef]

11.   Yu, Y.-H.; Li, Y.; Hallet, K.; Hotimsky, C. *Design and Analysis for a Floating Oscillating Surge Wave Energy Converter*; National Renewable Energy Laboratory Report NREL/CP-5000-61283; National Renewable Energy Laboratory: Golden, CO, USA, 2014; pp. 1–11.

12.   Piscopo, V.; Benassai, G.; Cozzolino, L.; Della Morte, R.; Scamardella, A. A new optimization procedure of heaving point absorber hydrodynamic performances. *Ocean Eng.* **2016**, *116*, 242–259. [CrossRef]

13.   Piscopo, V.; Benassai, G.; Della Morte, R.; Scamardella, A. Towards a cost-based design of heaving point absorbers. *Int. J. Mar. Energy* **2017**, *18*, 15–29. [CrossRef]

14.   Salter, S.H.; Taylor, J.R.M.; Caldwell, N.J. Power conversion mechanisms for wave energy. *Proc. Inst. Mech. Eng. Part M J. Eng. Marit. Environ.* **2002**, *216*, 1–27. [CrossRef]

15.   Benassai, G.; Campanile, A.; Piscopo, V.; Scamardella, A. Mooring control of semi-submersible structures for wind turbines. *Procedia Eng.* **2014**, *70*, 132–141. [CrossRef]

16.   Benassai, G.; Campanile, A.; Piscopo, V.; Scamardella, A. Ultimate and accidental limit state design for mooring systems of floating wind turbines. *Ocean. Eng.* **2014**, *92*, 64–74. [CrossRef]

17.   Benassai, G.; Campanile, A.; Piscopo, V.; Scamardella, A. Optimization of Mooring Systems for Floating Offshore Wind Turbines. *Coast. Eng. J.* **2015**, *57*, 1550021-1–1550021-19. [CrossRef]

18.   MathWorks. Matlab User Guide R2017b. Available online: www.mathworks.com (accessed on 4 December 2017).

19.   Mekhiche, M.; Edwards, K.A. Ocean power technologies PowerBuoy®: System-level design, development and validation methodology. In Proceedings of the 2nd Marine Energy Technology Symposium, Seattle, WA, USA, 15–18 April 2014; pp. 1–9.

20.   Danielson, O. Wave Energy Conversion—Linear Synchronous Permanent Magnet Generator. Ph.D. Thesis, Acta Universitatis Upsaliensis, Uppsala, Sweden, 2006.

21.   Drew, B.; Plummer, A.R.; Sahinkaya, M.N. A review of wave energy converter technology. *Proc. Inst. Mech. Eng. A J. Power Energy* **2009**, *223*, 887–902. [CrossRef]

22.   Falnes, J.; Budal, K. Wave-power conversion by point absorbers. *Norw. Mar. Res.* **1978**, *6*, 2–11.

23.   Gaspar, J.F.; Calvário, M.; Kamarlouei, M.; Guedes Soares, C. Power take-off concept for wave energy converters based on oil-hydraulic transformer units. *Renew. Energy* **2016**, *86*, 1232–1246. [CrossRef]

24.   Lejerskog, E.; Boström, C.; Hai, L.; Waters, R.; Leijon, M. Experimental results on power absorption from a wave energy converter at Lysekil wave energy research site. *Renew. Energy* **2015**, *77*, 9–14. [CrossRef]

25.   Bozzi, S.; Moreno Miquel, A.; Antonini, A.; Passoni, G.; Archetti, R. Modelling of a Point Absorber for Energy Conversion in Italian Seas. *Energies* **2013**, *6*, 3033–3051. [CrossRef]

26.   Archetti, R.; Moreno Miquel, A.; Antonini, A.; Passoni, G.; Bozzi, S.; Gruosso, G.; Scarpa, F.; Bizzozero, F.; Giassi, M. Designing a point-absorber wave energy converter for the Mediterranean Sea. In *Energia, Ambiente e Innovazione—Speciale Ocean Energy: Ongoing Research in Italy*; ENEA Agenzia Nazionale per le Nuove Tecnologie, L'energia e lo Sviluppo Economico Sostenibile: Rome, Italy, 2015; pp. 76–85.

27.   Lok, K.; Stallard, T.; Stansby, P.; Jenkins, N. Optimisation of a Clutch-Rectified Power Take Off System for a Heaving Wave Energy Device in Irregular Waves with Experimental Comparison. *Int. J. Mar. Energy* **2014**, *8*, 1–16. [CrossRef]

28.   Hulme, A. The wave forces acting on a floating hemisphere undergoing forced periodic oscillations. *J. Fluid. Mech.* **1982**, *121*, 443–463. [CrossRef]

29.   Chen, X.B.; Diebold, L.; Doutreleau, Y. New Green-Function Method to Predict Wave-Induced Ship Motions and Loads. In Proceedings of the Twenty-Third Symposium on Naval Hydrodynamics, Val de Reuil, France, 17–22 September 2000; pp. 66–81.

30.   Newman, J.N. *Marine Hydrodynamics*; MIT Press: Cambridge, MA, USA, 1977.

31. El-hami, M.; Glynne-Jones, P.; White, N.M.; Hill, M.; Beeby, S.; James, E.; Brwon, A.D.; Ross, J.N. Design and fabrication of a new vibration-based electromechanical power generator. *Sensor. Actuator A Phys.* **2001**, *92*, 335–342. [CrossRef]

32. Guizzi, G.L.; Manno, M.; Manzi, G.; Salvatori, M.; Serpella, D. Preliminary study on a kinetic energy recovery system for sailing yachts. *Renew. Energy* **2014**, *62*, 216–225. [CrossRef]

33. Thorburn, K.; Leijon, M. Farm size comparison with analytical model of linear generator wave energy converters. *Ocean. Eng.* **2007**, *34*, 908–916. [CrossRef]

34. Ekström, R.; Ekergård, B.; Leijon, M. Electrical damping of linear generators for wave energy converters—A review. *Renew. Sustain. Energy Rev.* **2015**, *42*, 116–128. [CrossRef]

35. Ulvgård, L.; Sjökvist, L.; Göteman, M.; Leijon, M. Line Force and Damping at Full and Partial Stator Overlap in a Linear Generator for Wave Power. *J. Mar. Sci. Eng.* **2016**, *4*, 81. [CrossRef]

36. Bachynski, E.E.; Moan, T. Point absorber design for a combined wind and wave energy converter on a tension-leg support structure. In Proceedings of the 32nd ASME International Conference on Ocean, Offshore and Arctic Engineering, Nantes, France, 9–14 June 2013; pp. 1–10.

37. Blanco, M.; Moreno-Torres, P.; Lafoz, M.; Ramírez, D. Design Parameter Analysis of Point Absorber WEC via an Evolutionary-Algorithm-Based Dimensioning Tool. *Energies* **2015**, *8*, 11203–11233. [CrossRef]

38. Crabb, J. Synthesis of a directional wave climate. In *Power from Sea Waves*; Academic Press: London, UK, 2000; pp. 41–74.

39. Rusu, E. Evaluation of the Wave Energy Conversion Efficiency in Various Coastal Environments. *Energies* **2014**, *7*, 4002–4018. [CrossRef]

40. Hong, Y.; Eriksson, M.; Boström, C.; Waters, R. Impact of generator stroke length on energy production of a direct drive wave energy converter. *Energies* **2016**, *9*, 730. [CrossRef]

41. Belmont, M.R. Increases in the average power output of wave energy converters using quiescent period predictive control. *Renew. Energy* **2010**, *35*, 2812–2820. [CrossRef]

42. Ringwood, J.V.; Bacelli, G.; Fusco, F. Energy-maximizing control of wave-energy converters: The development of control system technology to optimize their operation. *IEEE Contr. Syst. Mag.* **2014**, *34*, 30–55. [CrossRef]

43. Hultman, E.; Ekergård, B.; Tobias, K.; Salar, D.; Leijon, M. Preparing the Uppsala University Wave Energy Converter Generator for Large-Scale Production. In Proceedings of the 5th International Conference on Ocean Energy, Halifax, NS, Canada, 4–6 November 2014; pp. 1–12.

44. Myhr, A.; Bjerkseter, C.; Ågotnes, A.; Nygaard, T.A. Levelised cost of energy for offshore wind turbines in a life cycle perspective. *Renew. Energy* **2014**, *66*, 714 728. [CrossRef]

45. Stansby, P.K.; Carpintero, E.; Moreno, T.; Stallard, T. Large capacity multi-float configurations for the wave energy converter M4 using a time-domain linear diffraction model. *Appl. Ocean Res.* **2017**, *68*, 53–64. [CrossRef]

46. Liberti, L.; Carillo, A.; Sannino, G. Wave energy resource assessment in the Mediterranean, the Italian perspective. *Renew. Energy* **2013**, *50*, 938–949. [CrossRef]

47. Det Norske Veritas. *Recommended Practice DNV-RP-C205 Environmental Conditions and Environmental Loads*; Det Norske Veritas: Hamburg, Germany, 2010.

48. Danielson, O.; Erikkson, M.; Leijon, M. Study of a longitudinal flux permanent magnet linear generator for wave energy converter. *Int. J. Energy Res.* **2006**, *30*, 1130–1145. [CrossRef]

49. Previsic, M. *The Future Potential of Wave Power in the United States*; United States Department of Energy: Washington, DC, USA, 2012.

50. Falcão, A.F.O. Modelling and control of oscillating-body wave energy converters with hydraulic power take-off and gas accumulator. *Ocean Eng.* **2007**, *34*, 2021–2032. [CrossRef]

Article

Cost Optimization of Mooring Solutions for Large Floating Wave Energy Converters

Jonas Bjerg Thomsen [1,*], Francesco Ferri [1], Jens Peter Kofoed [1] and Kevin Black [2]

[1] Department of Civil Engineering, Aalborg University, Thomas Manns Vej 23, 9220 Aalborg Øst , Denmark;
ff@civil.aau.dk (F.F.); jpk@civil.aau.dk (J.P.K.)

[2] Tension Technology International Ltd., 69 Parkway, Eastbourne, East Sussex BN20 9DZ, UK;
black@tensiontech.com

* Correspondence: jbt@civil.aau.dk; Tel.: +45-99-40-8552

Received: 11 November 2017; Accepted: 3 January 2018; Published: 9 January 2018

Abstract: The increasing desire for using renewable energy sources throughout the world has resulted in a considerable amount of research into and development of concepts for wave energy converters. By now, many different concepts exist, but still, the wave energy sector is not at a stage that is considered commercial yet, primarily due to the relatively high cost of energy. A considerable amount of the wave energy converters are floating structures, which consequently need mooring systems in order to ensure station keeping. Despite being a well-known concept, mooring in wave energy application has proven to be expensive and has a high rate of failure. Therefore, there is a need for further improvement, investigation into new concepts and sophistication of design procedures. This study uses four Danish wave energy converters, all considered as large floating structures, to investigate a methodology in order to find an inexpensive and reliable mooring solution for each device. The study uses a surrogate-based optimization routine in order to find a feasible solution in only a limited number of evaluations and a constructed cost database for determination of the mooring cost. Based on the outcome, the mooring parameters influencing the cost are identified and the optimum solution determined.

Keywords: mooring; station keeping; wave energy; optimization; meta-model; surrogate model; cost; wave energy converters (WEC)

---

1. Introduction

The rising demand for sustainable and renewable energy in the world has led to increasing research into and development of alternative energy resources. By now, energy from, e.g., wind and solar is well developed and an active part of the energy mix in industrialised countries worldwide. Despite a comparatively large energy potential, one resource that still is not a part of the energy mix is wave energy. During the last few decades, the amount of research in wave energy absorption has been significant, resulting in a considerable amount of concepts for new wave energy converter (WECs), with varying levels of development. Despite the effort, the wave energy sector is not yet at a commercial stage, and further improvement must take place before wave energy can contribute to the energy mix.

The work in [1–3] list the levelized cost of energy (LCOE) for a range of energy resources, indicating the high price of wave energy compared to oil and gas (O&G) and even other renewable resources. As a result, there is an urgent demand to decrease the cost in order for wave energy to evolve from the current pre-commercial stage. According to [4,5], several parameters can be improved and take a considerable part in the cost reduction of wave energy. Despite different evaluations of the importance, station keeping moorings are listed as a driver towards lower cost, as they are estimated by [6,7] to compose 20–30% of the total structural cost of a WEC. In [8], the mooring is estimated to take up 8% of the CAPEX cost.

In addition to the cost, by now, mooring has also taken part in the failure of several WECs due to insufficient durability of the mooring system [7,9,10]. Consequently, a Danish research project entitled "Mooring Solutions for Large Wave Energy Converters" (MSLWEC) was initiated in 2014, which aimed at reducing the cost of the system, improving the applied design procedure and increasing the durability of the systems. Figure 1 illustrates a diagram of the project work and presents how the previous tasks have provided the basis and experience for this paper.

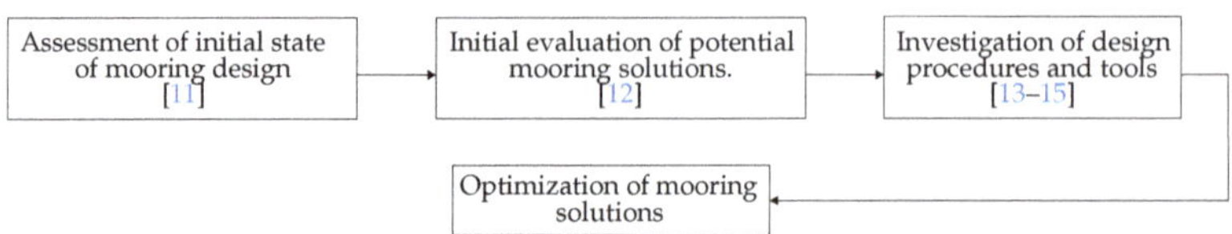

Figure 1. Flow diagram of the previous and related work [11–15] in the Mooring Solutions for Large Wave Energy Converters (MSLWEC) project.

The project took its point of departure in the Danish wave energy sector and four WECs, all considered to be large floating structures with passive moorings, meaning that the mooring does not take an active part in the energy conversion. The four WECs are the Floating Power Plant [16], KNSwing [17,18], LEANCON Wave Energy [19] and Wave Dragon [20], cf. Figure 2. In the early research works, the initial layout and design procedures were investigated (cf. [11]), and they highlighted a significant need for a more thorough and detailed design and investigation of the applied mooring systems. This conclusion was based on the fact that all environmental conditions were not fully included, and in most cases, a quasi-static approach was used. Finally, the project also concluded that there was a common tendency of applying traditions from the O&G sector in, e.g., using heavy mooring chains for the system. The following task [12] evaluated the initially-applied systems, identified the use of mooring chains as an inefficient solution and instead highlighted compliant, synthetic ropes as a potentially inexpensive and useful solution and, furthermore, identified a single anchor leg mooring (SALM) system as a strong potential for WEC mooring. Other studied like, e.g., Ref. [21,22] similarly identified synthetic ropes as an appealing solution.

Figure 2. The four WECs considered in the mooring optimization assessment. From left to right: Floating Power Plant, KNSwing , LEANCONWave Energy and Wave Dragon.

The following study [13] produced a large database of experimental data, used it to validate the initially-considered quasi-static design approach, and it identified a clear underestimation of mooring loads in extreme seas. Consequently, there was a need for a full dynamic analysis, as well as for further optimization of the initial mooring solutions since there was a clear possibility that these would not be capable of surviving in design storm conditions with extreme wind, wave and current.

Many different software packages are available for mooring analysis and have the capability to analyse the motion response and tensions under environmental load exposure according to limit states defined in design standards as in [23–26]. A number of tools was investigated in the project (cf. [14]) and has also been listed in other publications like [27]. A selected software package was validated against the experimental data in order to gain knowledge about the applicability of the tools on initial design, and the software proved its ability to model line tensions with acceptable overestimation of the tension without major tuning of the numerical model.

As presented in [13,27,28], the mooring characteristics are highly dependent on the site specification, mooring layout, materials, etc., and the response is, therefore, highly affected by the choices. The mooring design procedure is iterative, and it can be extremely time consuming to cover the full design space in order to find a solution that fulfils all defined requirements and, at the same time, provides a low cost. Based on the gained experience from the previous work on selecting tools, building hydrodynamic models and designing mooring, an optimization procedure must be utilized in order to find a cost-optimized solution, which introduces more reliable and fully-designed solutions compared to the initial layout.

In several studies like, e.g., [29–31], the energy absorption has been the objective of optimization with additional investigation of mooring line loads. The studies vary in the level of detail, the number of investigations of mooring configurations and the applied methodology, but they generally focus on operational conditions and the aim of improving the energy absorption as much as possible. This type of mooring is consequently reactive or active, and the actual cost of the moorings is not the objective of the studies. Naturally, optimizing mooring loads might reduce the needed strength of, e.g., lines and thereby reduce cost, but no actual cost investigation was done. Other optimization studies like, e.g., Ref. [32] treat the WEC farm layout in order to achieve the most feasible layouts for energy harvesting.

A passive mooring does not take an active part in the power take-off (PTO), and the cost is mostly determined by the extreme sea states during which survivability must be ensured. The present study focusses on the large floating WECs with passive moorings and uses an optimization procedure to reduce the mooring cost while securing that a reliable solution is found. The study continues the already presented work in the project "Mooring Solutions for Large Wave Energy Converters" and uses the four Danish WECs in Figure 2 as case studies.

The paper is structured with four sections including this Introduction. Section 2 describes the applied method and the design variables for the four cases, while Section 3 presents the results from each case. In Section 4, the work is summarized and discussed.

2. Method

This section presents the methodology used for the optimization study and describes the four cases by means of environmental conditions, design limits and choice of optimization parameters.

2.1. Mooring Cases

The presented WECs are planned for deployment at different locations and have differences in their design requirement. Because of this, four cases are defined, one for each WEC and its mooring system, which is designed and optimized for the relevant deployment site. In previous publications like [12], the potential of different mooring solutions was assessed, and based on this, three different mooring solutions are considered for the four WECs. These are illustrated in Figure 3 and cover: (a) a single anchor leg mooring (SALM) system with submerged buoys, a deformable tether and a nylon hawser; (b) a taut turret system with nylon lines; and (c) a single point mooring (SPM) system with nylon lines and hawser.

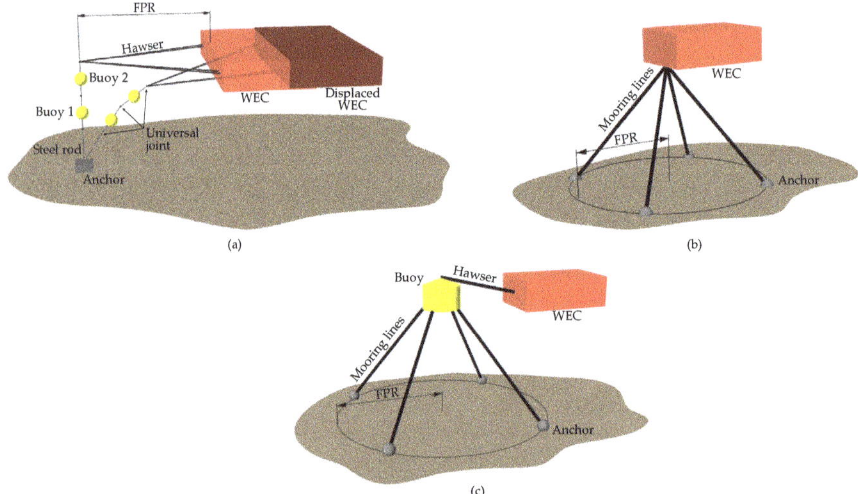

Figure 3. General illustration of the mooring layouts considered in this study. (**a**) A single anchor leg mooring (SALM) system with the illustration of the displaced layout; (**b**) a taut turret system; (**c**) a synthetic and taut single point mooring (SPM) system. The figure defines the footprint radius (FPR) for each layout.

All lines are composed of a chain part at the seabed connection, a synthetic line and a chain part at the fairlead. This is necessary for re-tensioning, to account for creep and to account for installation tolerances. A 2 m-long piece of chain is located at the seabed, while a chain length corresponding to 6% of the total line length is located at the WEC connection. Table 1 lists the system applied to each WEC.

Table 1. Definition of design cases for each of the WECs in Figure 2. The table list all considered environmental conditions together with the defined restraints on surge and pitch. ✗ denotes no limit.

WEC	Case 1	Case 2	Case 3	Case 4
	Floating Power Plant	**KNSwing**	**LEANCON**	**Wave Dragon**
Mooring solutions	Taut turret	Taut turret	SALM	Taut SPM
	Figure 3b	Figure 3b	Figure 3a	Figure 3c
Water depth, h	30 m	40 m	25 m	25 m
Significant wave height, H_s	6.6 m	9.9 m	8.3 m	8.3 m
Peak wave period, T_p	9.3 s	11.4 s	10.5 s	10.5 s
Relative depth, h/L_p	0.14	0.12	0.1	0.1
Current velocity, v_c	1.3 m/s	1.0 m/s	1.5 m/s	1.5 m/s
Wind velocity, v_w	33.0 m/s	39.9 m/s	34.0 m/s	34.0 m/s
Surge design limit	±29 m	±44 m	±30 m	±27 m
Pitch design limit	±15°	✗	✗	✗

2.1.1. Environmental Conditions

Each WEC is planned for deployment at a specific site either at the DanWEC test facility in Denmark, the Danish part of the North Sea or at the Belgian coast. Prior to the optimization, the environmental conditions for each site were assessed, and the 100-year extreme conditions were specified and can be read from Table 1. It is assumed that all waves are long-crested (2D), irregular and distributed in a JONSWAP spectrum with a peak enhancement factor $\gamma = 3.3$.

The current is assumed steady over time, while the wind is modelled as a steady component and a time-varying gust component. The latter is described by an NPD wind spectrum according to [33]. Both wind and current are assumed to be varying in the vertical direction and are modelled with a power law profile.

The wind speed defined in Table 1 corresponds to the 1-hour mean value at a height of 10 m, while the current velocity is at the still water level (SWL). All environmental loads are assumed to be aligned; hence, a full 2D problem is analysed.

2.1.2. Design Criteria

Different design standards are available and can be used in the design of mooring systems, e.g., [23–26]. These standards define the necessary requirements and ensure survivability in operational and extreme conditions by considering design in the ultimate, fatigue and accidental limit state (ULS, FLS and ALS). For a passive mooring system, the mooring should not affect the PTO, and [34] shows that ULS has the most dominant influence on the cost. The present study, therefore, only considers the ULS, which is the reason why the 100-year extremes are listed in Table 1. In several publications like, e.g., [35,36], the safety levels defined by the design standards have been discussed. A relaxation of safety levels for WECs compared to O&G structures has been suggested, due to the lower consequence if a failure occurs. In [35], the safety levels for API-RP-2SK and DNV-OS-E301 are compared, and it is found that the API standard provides higher safety than the DNV. ISO and IEC provide similar safety factors as API. In other works like [37–40], the topic of reliability assessment of WECs is treated with detailed work on calibration of safety factors, estimation of extreme values and reliability assessment approaches.

In order to follow the suggestion from [35] of using relaxed safety, the DNV-OS-E301 standard and the most relaxed consequence class (CC1) are used in the present analysis.

The standard focuses primarily on ensuring sufficient strength of the mooring lines and anchors to withstand the induced tensions. According to DNV-OS-E301, the design tension is defined by Equation (1):

$$T_{C,mean}\gamma_{mean} + T_{C,dyn}\gamma_{dyn} \leq S_C \tag{1}$$

where S_C is the characteristic strength corresponding to 95% of the minimum breaking strength T_{MBS} and $\gamma_{mean} = 1.10$ and $\gamma_{dyn} = 1.5$ are the safety factors for respectively the mean and dynamic part of the line tension. $T_{C,mean}$ is the mean tension, while $T_{C,dyn}$ is the dynamic part of the tension and is defined by Equation (2):

$$T_{C,dyn} = T_{MPM} - T_{C,mean} \tag{2}$$

where T_{MPM} is the most probable maximum with a 63% probability of exceedance when the extreme peaks tend to follow a Gumbel distribution. By applying the safety factors for the given consequence class, a target annual probability of failure of 10^{-4} is obtained.

According to DNV [41], the anchor should be designed for the same design tension as the lines, while the characteristic anchor resistance provided by the manufacturer is reduced by a safety factor $\gamma_m = 1.3$.

DNV-OS-E301 only considers survivability in ULS, and hence, the excursion is not specified in the standard as a design criterion. However, the WECs are equipped with umbilicals, which puts a limit on the allowable excursion, because tensions in these must be prohibited [27]. The design of the umbilical is, therefore, often a part of the mooring design and a compromise between the cost of umbilical and mooring. In the present study, the excursion limit in Case 3 was defined by the developer, while it was approximated for the remaining cases by assuming a lazy-S layout for the umbilical; cf. Figure 4. By accounting for the minimum bending radius of a suitable cable, clearance between seabed and sea surface and for water level variations (high water level (HWL) and low water level (LWL)), it is possible to calculate a cable length and an allowable excursion. This limit is defined in Table 1 and illustrated in Figure 4. Naturally, this is only a pragmatic approach to obtain an estimation of the excursion limit, but in the final design, a more detailed investigation and design of the interaction of the umbilical on the WEC must be included in the mooring design.

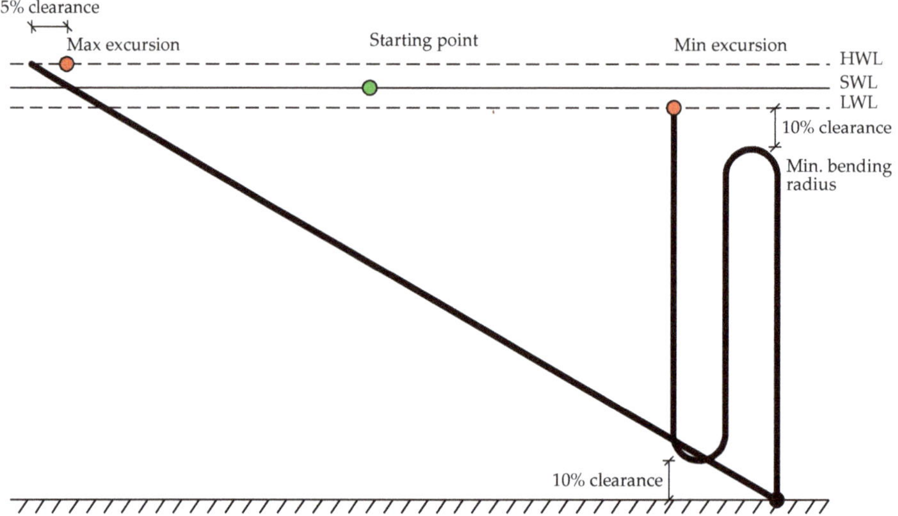

Figure 4. The considered umbilical configuration and definition of the allowable excursion. HWL, high water level; SWL, still water level; LWL, low water level.

The WEC in Case 1 is equipped with a wind turbine, which puts additional restraints on the motions. In operational conditions, the power production of the turbine must be considered, while it is merely vital to secure stability in the ULS. For the given turbine, a pitch limit of $\pm 15°$ was defined by the developer; cf. Table 1.

2.1.3. Optimization Parameters

Considering the layout of the mooring solutions in Figure 3, several parameters can be varied for each solution and will change the characteristics of the system. Table 2 lists the optimization parameters for each case together with the maximum and minimum values considered for each parameter. As seen, the mooring and type of optimization parameters for Cases 1 and 2 are identical, while they vary for Cases 3 and 4.

Table 2. Definition of the considered optimization parameters for each case and the applied value ranges. ✗ denotes that the parameter is not considered.

Parameter	Case 1		Case 2		Case 3		Case 4	
	Min.	Max.	Min.	Max.	Min.	Max.	Min.	Max.
Mooring line diameter (mm)	40	192	40	192	✗	✗	40	192
No. of mooring lines (-)	4	10	4	10	✗	✗	4	10
Hawser line diameter (mm)	✗	✗	✗	✗	40	192	✗	✗
Footprint radius (m)	30	40	80	250	40	75	25	100
Buoy 1 diameter (m)	✗	✗	✗	✗	1.5	6.0	3.5	15.0
Buoy 2 diameter (m)	✗	✗	✗	✗	1.5	6.0	✗	✗
No. of optimization parameters	3		3		4		4	

For all cases, the Bridon Superline Nylon [42] is considered with the range of diameters listed in the table. The lines are modelled with a non-linear stiffness curve and structural parameters according to [42].

The definition of the footprint radius (FPR) is illustrated in Figure 3 and varies significantly for each case. In Cases 1 and 4, the maximum FPR is chosen in order to ensure that there is no interaction between the WEC and the lines; cf. Figure 5. For Case 2, this problem is not present, and the upper value is chosen based on the length of the device. For Case 3, it is based on the umbilical.

Figure 5. Illustration of the WEC in Case 1 and the necessary limit on footprint radius (FPR) in order to avoid interaction between the line and structure.

The unstretched mooring line length is not considered as a direct optimization parameter, even though it is varied for each configuration. For the WECs in Cases 1, 2 and 4, a vertical pretension is specified and fixed for all simulations, and therefore, the line length is varied dependent on the FPR and line stiffness in order to achieve this pretension. In Case 4, the vertical pretension is calculated to ensure vertical equilibrium between line tension and buoyancy of the surface buoy in calm seas. In Case 3, the line length is varied directly because of the FPR. In this case, the steel rods are not a part of the optimization, but are dimensioned based on the obtained design tension in the rods. The maximum dimensions of the buoys are decided based on the length of the tether. In Case 4, the value is based on the investigation of the available buoys from commercial manufacturers.

In Case 4, the hawser length is not varied since the zero-position of the buoy is fixed, and there is no pretension in the line. In early stages of the work, it was intended to use a nylon hawser, but a suitable solution was not found, so a rigid bar is used instead.

The anchor type is determined prior to the optimization as drag embedded anchors with vertical strength for Cases 1, 2 and 4, while a gravity-based anchor is used for Case 3. The anchors are designed based on the achieved tensions in the lines at the anchor point, and hence, the necessary anchor size is determined after each simulation is completed and is not an optimization parameter.

As seen from Table 2, the optimization considers between 4 and 10 lines in the layouts. In some applications (cf., e.g., [11]), configurations with three lines are considered. In this study, however, this number is considered insufficient in order to increase durability and redundancy, which is considered a vital parameter for a mooring system [27]. This is mainly a problem in the ALS, but must be taken into account early in the process. Figure 6 presents the influence from the number of mooring lines on the static behaviour of a mooring. Considering the first graph, it is indicated that the horizontal mooring stiffness, as expected, increases with the number of lines. By having only three lines, it is possible to achieve around 40% of the stiffness with 10 lines. Considering the second graph, the advantage of having more than three lines is clear. The figure presents the relationship between intact stiffness and the stiffness when one line is broken. With only three lines, almost all stiffness is lost, and even though two lines are remaining, the durability of the system can be expected to be much less. With 4–10 lines, 60–80% of the stiffness remains after one line fails. By also considering the third graph, it is seen that the static position of the device at failure of a line is significantly worse when only having three lines. The displacement is approximately 55% larger than with 10 lines with one broken, while it decreases to 12% with 4 lines. The displacement is illustrated in Figure 6 with examples for 3 and 6 lines, where the displacement in the latter is almost negligible and hardly visible.

A traditional SALM system does not provide the same amount of redundancy as only one line connects the buoy to the seabed. For the given system, the hawser is composed of four lines, meaning that there is some redundancy if one nylon line breaks. However, in general, the system is much more vulnerable to failure. As a result, DNV-OS-E301 requires an additional safety factor of 1.2 in tension.

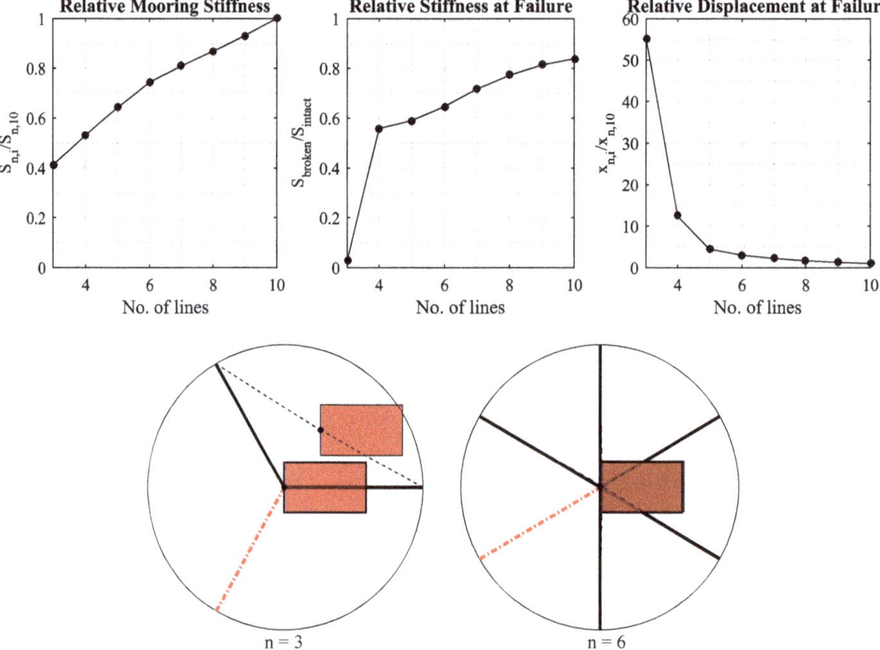

Figure 6. Influence of mooring line number on intact horizontal stiffness, stiffness after failure of one line and the static position of the WEC after failure. The figure also illustrates two examples of the WEC position after a line failure. The red line indicates the broken line.

In a final design, it is necessary to ensure survivability in both ALS and ULS. As previously stated, this paper focusses on the ULS, but the above considerations and the fact that a solution with three lines has been discarded account for a simple and coarse evaluation of the ALS.

2.2. Mooring Analysis Procedure

There are different methods available for the evaluation of the wave-WEC interaction, generally divided into experimental and numerical methods. Experiments provide highly reliable solutions, but are time consuming and often expensive, and it is difficult to include all environmental loads without introducing significant sources of errors, especially in small-scale tests. Sophisticated non-linear numerical models can be established through, e.g., computational fluid dynamics (CFD) or smoothed particle hydrodynamics (SPH), but the computational demands for such calculations are immense, which is why the methods are not suitable for initial mooring design where many iterations are needed and the simulation time is long. Often, the boundary element method (BEM) is used, which assumes inviscid, irrotational fluids and uses linear potential flow theory with its assumption of small steepness waves and small amplitude structure motion. In the ULS, this theory is stressed significantly, due to the extreme wave condition and compliant mooring, which allows large motions, but this work builds on the experience gained in previous work [15] where the applicability was validated for initial design and analysis.

In this study, the open-source BEM code NEMOH [43] is used, which provides the frequency-dependent added mass, radiation damping, wave excitation and Kochin coefficients. As explained in [44], the latter can be used for the calculation of the mean wave drift coefficients using the far-field formulation.

The output of NEMOH is coupled to the commercial time domain mooring solver OrcaFlex [45], which utilizes the Cummins equation to determine the coupled response of the WEC and mooring lines, by including all first and second order wave load contributions together with the wind and current loads. The output of the software package is a time series of line tension and WEC motions. The construction of the numerical model follows the procedure presented in [15].

Using linear theory on Case 4 is not without problems, as the device is an overtopping WEC where a safety system increases the draught in extreme seas to increase overtopping and decrease the loads. The BEM cannot directly account for overtopping, and by increasing the structure volume below the SWL, the loads are actually increased. In later research, this problem will be addressed and an improved model will be used. At present, the model is used with this inaccuracy, which especially is causing a significant problem with second order drift. Consequently, this first optimization of Case 4 has only been for 1st order wave loading, while both 1st and 2nd order effects are considered for the remaining cases.

2.3. Cost Database

For the calculation of the cost of each mooring system configuration, a database was constructed where the cost was calculated based on specification of the mooring components. The database included a high level of detail and considered the cost of both CAPEX and OPEX for a lifetime of 20 years. The database was based on the authors' experience and knowledge from other marine projects, manufacturers and providers.

The following subsection provides some of the assumptions behind the numbers, but it is not possible to list costs directly in this paper.

2.3.1. CAPEX

Planning, design, survey and engineering comprise a significant part of the mooring cost and were included as a fixed price based on the type of mooring system. The cost included assumption of the entire design process from determination of environmental conditions, site and seabed surveys (including vessel and labour), analysis and detailed engineering, together with estimation of the cost for review and certification and planning of the installation. All cost was found based on the assumption of the duration of each task and the expected day-rates. Final inspection and project management were also included.

The manufacturing and procurement cost is primarily based on prices provided by manufacturers of the given components. For the nylon line, the cost is dependent on the minimum breaking strength (MBS) and includes the protective jackets, a chain part at the connection point between line and WEC and the cost of connections. The anchor cost is based on the type of anchor and necessary weight to ensure survivability and includes the chain at the anchor point.

The cost of buoys and stainless steel rods is based on a fixed cost per unit weight and with the estimation of the cost for the universal joint, top-swivels and connection hardware. The cost of fairleads with tensioners was included and dependent on the number of lines.

Estimating the cost of a turret system is highly complicated and is a significant part of the overall cost. For the given cases where turrets are included, the cost is a fixed price based on experience from comparable applications.

All component cost includes certification and delivery to a North European port.

The installation and hook-up cost is mostly dependent on the cost of the needed vessels and the amount of labour needed. For each type of mooring system, the types of vessels, cranes, etc., were decided, and the current day-rates for these were collected. Generally, it was assumed that one line could be installed per day and a total of two days were needed for tensioning of all lines. For each vessel day, four person-days were needed. The installation cost included cost for preparation in the port and estimation of waiting time for a suitable weather window.

The decommissioning cost was based on a dismantling cost corresponding to 110% of the installation and hook-up, while material disposal was estimated to be 1 % of the CAPEX. A fixed value for seabed clean-up was found and dependent on the type of mooring system, as some systems have more seabed interaction than others.

The CAPEX is based on existing experience and current cost of vessels and labour. Naturally, this introduces some uncertainties in the calculations, particularly because presently, the cost rates are at the lowest level in a number of years due to the decline in the oil industry. Similarly, installation cost, etc., has been made with the assumption of waiting time, which is highly dependent on the site and season. For final cost evaluations, more specific input can be given to the cost database to decrease uncertainties, but presently, it provides generic and realistic estimations on the cost.

2.3.2. OPEX

The OPEX cost is based on a lifetime of 20 years and includes insurance, inspection and maintenance. The insurance is estimated as 1%/year of the total CAPEX, while 4 vessel days and 16 person-days are assumed for inspection and maintenance. This corresponds to checking and adjustment of tension after one month and after 1, 5 and 10 years. Replacements throughout the lifetime are assumed to correspond to 1.5% of the CAPEX based on experience.

Similar to the CAPEX, the OPEX cost introduces some uncertainties mainly resulting from the fact that some level of novelty is seen in the concepts. This means that no historical evidence is available to support the reliability for a 20-year deployment. Additionally, the risk of external damage to the system is difficult to take into account in the cost evaluation. Finally, the insurance cost takes up a considerable part of the OPEX, but in most industries, there is an increasing trend in this cost. Any type of failures due to mooring will expectedly lead to a significant increase in the insurance cost, and at present, several years of deployment are needed to build a reliability record in the industry and more knowledge about OPEX cost.

2.4. Optimization Procedure

According to the problem defined in Section 1, the objective of the current work is to minimize the cost of the applied mooring system while securing that the found system satisfies the ULS. This results in the optimization problem in Equation (3):

$$\min_{x \in \mathcal{D}} f(x), \tag{3}$$

where x is the variable vector, $f(x)$ is the objective function and $\mathcal{D}$ is the design space; cf. Table 2. Since the cost and response are directly related to the combination of variables in the design space, a full simulation is required for each mooring configuration. The objective function, therefore, evaluates the cost of the mooring based on a dynamic simulation of the complete system under extreme conditions. As the complete design space has a significant size, it is not feasible to evaluate the complete space to find the minimum solution, and instead, a methodology should be applied aiding in the search for minimums with only a limited number of function evaluations. For the objective function used in this study, no derivative information is available for identifying the minimum as no analytical description of the response surface can be constructed. Therefore, a derivative-free surrogate-based (also denoted meta-model) optimization algorithm is used, which has the clear advantage of only requiring a relatively limited number of function evaluations compared to, e.g., evolutionary algorithms [32]. In the surrogate-based optimization, the response surface is described by a surrogate model, constructed from the results of a limited number of function evaluations and used to identify the minimums. Several variations of derivative-free algorithms have been developed and presented in, e.g., [46,47], and this paper utilizes the surrogate-based algorithm in the MATLAB Surrogate Model Toolbox (MATSuMoTo) [48]; the model will be further explained in later sections. Figure 7 illustrates the steps in the algorithm as defined in [29,48].

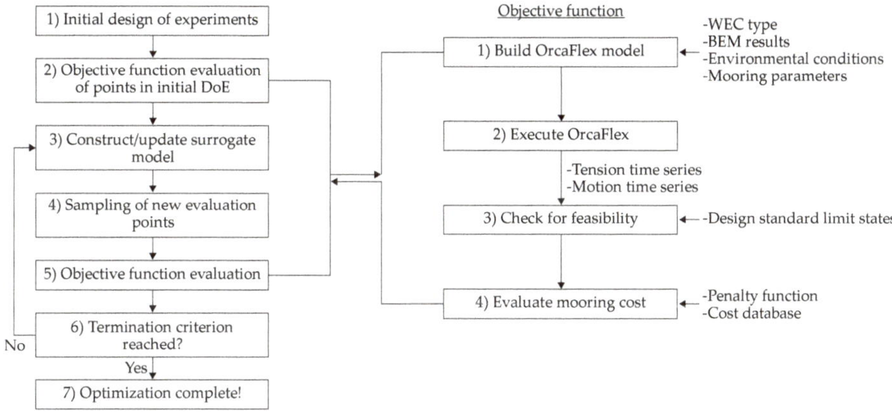

Figure 7. Flowchart of the applied optimization procedure. BEM, boundary element method.

As illustrated in Figure 7, the procedure is initiated by evaluating the objective function in a number of tests (Steps 1–2) used to construct a surrogate model (Step 3). A new number of points are selected (Step 4), and the objective function is evaluated at these points (Step 5) and used to update the model (back to Step 2). This procedure is continued until a chosen termination criterion is met. The following sections will provide an explanation of each step in the procedure.

2.4.1. Initial Design of Experiments

In order to make the surrogate model, it is essential to have initial knowledge of the design surface by conducting a number of simulations. In order to ensure that sufficient data are obtained and an efficient surrogate model can be constructed, the sampling points must be chosen to provide as much information on the response surface as possible. A design of experiments (DoE) strategy is initiated, which covers several different concepts [49,50]:

- Latin hypercube
- Corner
- Latin square
- Full factorial
- Fractional factorial

It is not possible to determine prior to the evaluation which strategy is the best choice, but according to [49], Latin hypercube is often suitable. The present study utilizes this method. A sufficient number of evaluations in the DoE must be chosen, and in the given study, the numbers were based on the following Equation (4):

$$n_{\mathrm{DoE}} = 2(d+1) \tag{4}$$

where n_{DoE} is the number of samples in the DoE and d is the number of optimization parameters [50].

2.4.2. Surrogate Model

In general terms, the surrogate model can be expressed by Equation (5):

$$f(x) = s(x) + \epsilon(x) \tag{5}$$

where $f(x)$ is the output of the objective function at point x, $s(x)$ is the output from the surrogate model and $\epsilon(x)$ is the difference between them [48]. Many different types of surrogate models are available and presented in [46,51] and are generally either interpolating (radial basis functions (RBF) and kriging) or non-interpolating (polynomial regression models and multivariate adaptive regression splines (MARS)) [51].

Different surrogate models suit different problems, and it is not possible to determine beforehand which model to choose without testing each of them. Considering that the purpose of the optimization is to find the optimum in a short time, testing each surrogate model to find the best is not feasible. In order to limit the influence of a bad surrogate model choice, model ensembles (or mixture models) can be used, which utilize weighted combinations of two or more models and emphasize the models that perform well (low error and high correlation coefficients) and restrict the influence from the poor models (large error and low correlation coefficients). The mixture models are represented as Equations (6) and (7) [51]:

$$s_{mix}(x) = \sum_{r=1}^{N} w_r s_r(x), \tag{6}$$

$$\sum_{r=1}^{N} w_r = 1, \tag{7}$$

where $s_{mix}(x)$ is the output of the mixture model at point x, N is the number of surrogate models in the mix, s_r is the output of the r-th model and w_r is the weight of it. MATSuMoTo uses the Dempster–Shafer theory to combine the models and takes advantage of correlation coefficients, maximum absolute error, median absolute deviation and root mean square error to calculate the weight w_r for each model, based on the performed objective function evaluations in the procedure [52].

Naturally, the number of surrogate models in the mix results in a larger number of models that need to be updated in each loop in the procedure and puts higher demands to the computational effort. On the other hand, the chance of selecting a poor model is minimized, and a much better description of the response surface is obtained [51].

Several studies like [46,51] have compared and investigated which models and model ensembles perform best, and [51] found that the use of RBF, either alone or in combination with other models, generally provided a reliable solution. Consequently, this study uses a model ensemble of a cubic RBF and a MARS model.

2.4.3. Sampling Technique

Different methods can be considered for choosing the sampling points in each loop. In general, either a randomized sampling technique or the constructed response surface can be used. In this study, the randomized method is used.

Within the randomized method, two different strategies can be used. A local search can be considered where the current optimum points are perturbed. This method is most suitable for problems where only a single minimum is present in the response surface, as the solver will tend to search towards one minimum and will not necessarily find more. The other strategy is a global search where the solver still perturbs the best points so far, but also selects a number of uniformly-distributed points in the whole design space. As the optimization progresses, the perturbation is decreased in order to find the best solution. When MATSuMoTo is no longer improving the output over a consecutive number of trials, the algorithm will restart with a new DoE and construct a new model based on the new evaluations in order to aim the search toward other areas of the design space [50].

When selecting the best candidate points, two criteria are used: the distance and response surface criteria. The distance criterion is based on the distance to already evaluated points, while the response surface criterion is based on the value predicted by the surrogate model. A score is assigned to each point as a weighted sum of these two. In order to select a point close to the expected minimum, a large weight is put on the response surface criterion, while a large weight must be put on the distance criterion to select points in unexplored areas.

The optimization can either be considered as an integer or mixed-integer problem dependent on the variables. The different problems affect the sampling strategy as presented in [48,53,54], and for the mooring cost optimization, the definition of parameters are listed in Table 3. Most parameters are considered integer, such as line number, line diameters and footprint radius. Considering the installation process and allowable tolerances during installation, it is not relevant to consider the footprint radius as a continuous parameter. The buoys can be produced in custom sizes and are, therefore, continuous parameters in the optimization procedure.

Table 3. Definition of integer or continuous parameters.

Parameter	Integer	Continuous
Mooring line diameter (mm)	✓	✗
No. of mooring lines (-)	✓	✗
Hawser line diameter (mm)	✓	✗
Footprint radius (FPR) (m)	✓	✗
Buoy 1 volume (m^3)	✗	✓
Buoy 2 volume (m^3)	✗	✓

2.4.4. Penalty Function

Naturally, the least expensive mooring solution will be the one consisting of the least amount of materials, because cost for components, installation, etc., will be small, but this solution might not fulfil the design requirements defined in Section 2.1.2. In order to ensure that the solver accounts for this and to search for the least expensive solution that also fulfils the requirements, a cost penalty is applied to the inadequate solutions.

Different types of penalty functions can be applied when the criteria are exceeded, e.g., letting the cost be a fixed and high value or adding a fixed value to the actual cost. The first solution is not applicable as the result will provide a plateau on the response surface, and it will be more difficult for the solver to detect which solutions perform the best. The latter function will not provide a plateau, but will not be dependent on the performance of the system; hence, if the least expensive solution results in the largest exceedance of the requirements, it will still appear less expensive than a more expensive solution, which performs better and is closer to satisfying the requirements. Consequently, the following penalty function is applied through the following Equations (8) and (9):

$$p_i = \frac{X_i - X_{i,C}}{X_{i,C}} \tag{8}$$

$$\text{Penalty} = \sum_{i=1}^{N} p_i \cdot scale + Fp \tag{9}$$

where p_i is the penalty associated with the violation of one design criterion, N is the number of design criteria, X_i is the simulated motion or tension and $X_{i,C}$ is the associated design criterion. The *scale*-factor is defined as a fixed value in € and ensures that the mooring solutions that perform worst have the largest prizes, so that MATSuMoTo will diverge from them. The Fp-factor is defined as a fixed penalty in €, which is used to ensure that none of the insufficient solutions are less expensive than the most expensive and adequate solution. The *scale*- and Fp-factors must be determined beforehand by assessing the extreme cost difference between the possible solutions.

2.4.5. Termination Criterion

The optimization is finished when the solutions are converged; hence, the minimum mooring cost is found. As stated in [48,51], MATSuMoTo is asymptotically complete, indicating that if an indefinite number of calculations is performed, the global minimum will be found with a probability of one. However, the termination criterion is listed as a maximum number of evaluations, after which the minimum cost is identified.

OrcaFlex is generally computationally effective and can run more simulations simultaneously. It is possible to run a potentially large number of simulations, and 300 evaluations were selected.

Considering Figure 8, the progress of the optimization for each of the cases is plotted. It is clearly observed how the optimization procedures manage to identify less expensive solutions until reaching a value where the solution converges. In the analysis of each case, the routine had one restart as mentioned previously, which means that the optimization is not based on one model constructed from 300 evaluations, but two models with less evaluations. Often, an approach is used where the routine is not restarted in order to check the convergence with a higher number of evaluations, but this has not been done in this research, and it is not expected to affect the result. From the figure, it is concluded that even less evaluations would have been sufficient for identifying an optimum. For Case 4, it is seen how the solutions in the DoE already provided a solution close to the optimum, but still manages to make further improvement.

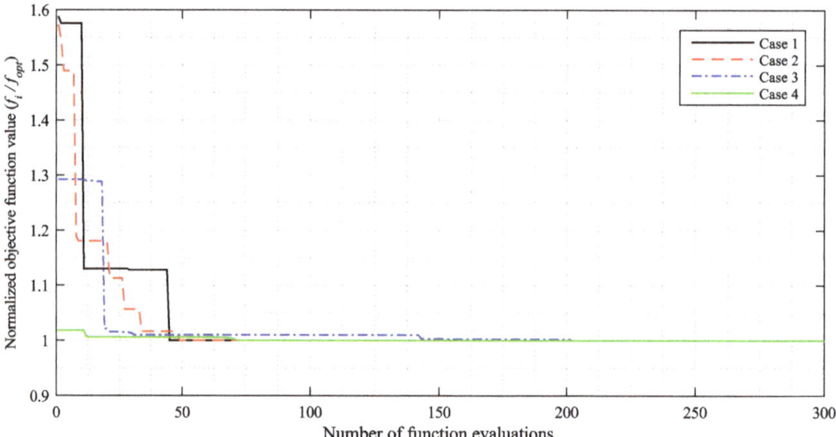

Figure 8. Progress of the found objective function values for the four optimization cases. The function values f_i have been normalized according to the optimum value f_{opt}.

3. Results

This section presents the outcome of the optimization routine when using the method as described in the previous section. Each case is presented separately and followed by a common discussion in the next section.

3.1. Case 1

The mooring solution in Case 1 consists of a synthetic turret system, and based on the optimization, the optimal layout has parameters as listed in Table 4. The mooring cost has been normalized according to the total CAPEX and OPEX of the WEC.

Table 4. Results of the optimization of the mooring system for Case 1.

Parameter	Optimum Value
No. of mooring lines (-)	6
Mooring line diameter (mm)	192
Footprint radius (m)	40
Horizontal stiffness (kN/m)	954
Normalized mooring system cost $\left(\frac{c_{mooring}}{c_{total}}\right)$	0.13

Figure 9a illustrates the evaluated points, with the optimum as a red marker. Figure 9b presents a contour plot for the cost of the mooring when normalized according to the total WEC cost. The figure presents systems with and without cost penalty, but the satisfying solutions are found in the normalized cost range 0.13–0.15. In this plot, it is possible to detect the influence of each parameter. In the top right plot, the footprint radius is plotted against the number of lines. The line diameter is kept constant in this plot and corresponds to the optimum. It is clearly seen that the footprint radius only provides a small influence on the cost, while an approximately 15% cost reduction can be achieved by varying the number of lines. Naturally, a large number of lines increases the cost due to installation, the amount of anchors and line materials, etc., but it also highly influences the mooring stiffness, corresponding motions and line tensions. Decreasing the line number from six to four decreases the mooring stiffness by 32% and results in larger excursions. Fewer lines also need to each take up a larger part of the load, and the tension with four lines is 10% larger than with six lines, resulting in insufficient strength. Increasing the line number to 10 lines increases the stiffness by 63%, but the larger number of lines means that each line only takes 75% of the tension experienced in the system with six lines. This means that the anchor size can be decreased. The installation cost is primarily dependent on the number of lines and is, therefore, largest for the system with 10 lines (80% larger than the cost for four lines); but since the motion limit is exceeded with four lines, the cost penalty has been applied to the system, and the difference appears smaller in Figure 9. Clearly, the optimum value is found as a balance between finding the lowest number of lines, where low installation cost is present, and finding a high enough number of lines to ensure small tensions in the lines with a corresponding need for smaller anchors and certainty on the line strength.

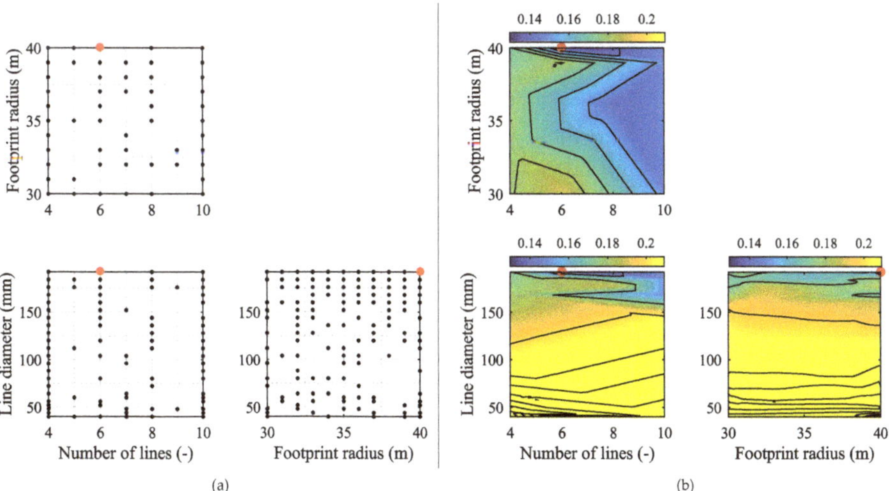

Figure 9. (a) Sample point from the optimization of Case 1; (b) cost contour plot of the optimization of Case 1. The mooring cost has been normalized according to the total WEC cost.

Considering the number of lines against the line diameter, it is clear that the most dominating parameter is the diameter. Having a small line results in a very compliant system, which in many cases causes a violation of the surge limit, while the line also provides less and insufficient strength. It is necessary to use some of the largest considered diameters in order to find a suitable solution, and the least expensive is found by using the strongest lines and reducing the line number.

In the final plot, this tendency is also observed as the footprint radius only provides minor importance while the line diameter is paramount to optimize. Considering all the graphs by Figure 9, it is noticeable that a large part of the evaluation provides insufficient solutions, and it would merely have been necessary to consider a few of the largest line diameters.

3.2. Case 2

As presented in Table 2, Case 2 considers an overall mooring design similar to the system in Case 1. However, this structure is located at a greater water depth and has less restriction on surge and none on pitch; cf. Table 1. By also considering Table 2, it is observed that this case allows for a larger footprint radius, which provides a larger range to find an optimum solution. Table 5 presents the results from the optimization.

Table 5. Results of the optimization of the mooring system for Case 2.

Parameter	Optimum Value
No. of mooring lines (-)	6
Mooring line diameter (mm)	192
Footprint radius (m)	100
Horizontal stiffness (kN/m)	548
Normalized mooring system cost $\left(\frac{c_{mooring}}{c_{total}} \right)$	0.17

Figure 10a presents the sample points. It is clearly indicated that the solver identified the need for large line diameters and aimed the search at these diameters. Similarly, it is indicated that smaller footprint radii were sufficient.

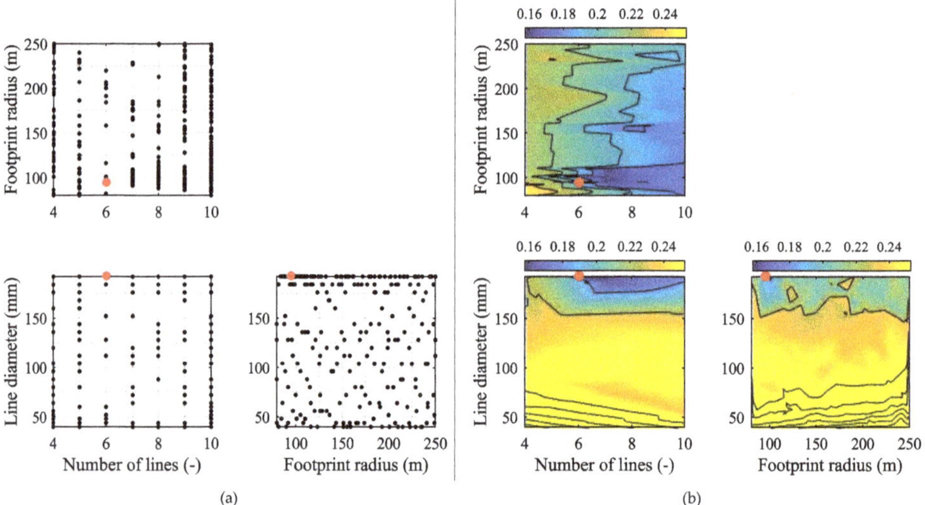

Figure 10. (a) Sample point from the optimization of Case 2; (b) cost contour plot of the optimization of Case 2. The mooring cost has been normalized according to the total WEC cost.

Figure 10b presents the cost contour for the problem where the working solutions are found in the normalized cost range 0.15–0.19. Similar to the previous case, one parameter (the optimum) is kept constant in each diagram, and the influence from the remaining can be identified.

When considering the footprint radius against the number of lines, a similar tendency as in Case 1 is observed. Variation in the number of lines provides a cost difference of approximately 20%. This difference is primarily caused by the influence on anchor and line loads. A large number of anchors might cause a high cost, but having too few lines causes large loads and insufficient strength. The cost is less independent of the footprint radius.

In the plot of line diameter against the number of lines, the influence from the former is seen to be crucial for the cost. Decreasing the line diameter leads to a paramount cost increase, because of the penalty function, as the line strength tends to become insufficient or the compliance so large that the excursion limit is exceeded. The number of lines plays a less important role, but it is seen how too few lines cause a higher cost.

In the final plot, the footprint radius and line diameter are presented with the number of lines kept constant. Similar conclusions can be made from this, which indicates that the main parameter is the line diameter. The footprint radius has a minor influence, but by decreasing it, the line material is decreased and results in some minor cost reduction.

3.3. Case 3

Case 3 consists of a larger number of optimization parameters compared to the first two cases. Table 6 presents the result of the optimization.

Table 6. Results of the optimization of the mooring system for Case 3.

Parameter	Optimum Value
Hawser line diameter (mm)	176
Footprint radius (m)	50
Buoy 1 diameter (m)	3.7
Buoy 2 diameter (m)	6
Horizontal stiffness (kN/m)	46
Normalized mooring system cost $\left(\frac{C_{mooring}}{C_{total}} \right)$	0.48

When considering Figure 11a, it is obvious that the code identified the minimum and concentrated the evaluation around this point. In Figure 11b, where the working range of the normalized cost is 0.48–0.52, it is apparent that a significant parameter for the cost is the line diameter. Similar to Cases 1 and 2, only the largest line diameters provide sufficient strength, but in this system, a larger range of diameters is adequate. Considering the very low stiffness presented in Table 6, it will be expected to obtain smaller loads. In the top graph, the influence from the two buoys can be identified. Buoy 1, which is the bottom buoy, provides the lowest influence on the cost, while the top buoy provides the most influence in the system and hence determines the cost. The best solution is found by having a large buoy at the top, while the other can be relatively small. The footprint radius (and thereby, the hawser line length) also provides an important influence on the system, as it highly determines the overall stiffness. Having a large radius provides large compliance and mainly causes problems with exceedance of the surge restraint. A small radius still ensures satisfaction of all requirements, but results in larger loads, hence the need for a larger anchor. Compared to Cases 1 and 2, the mooring cost of this system is taking up a larger part of the total cost. This is partly because the device itself is less expensive, but also because the gravity type anchor in this system is expensive and other types should be investigated.

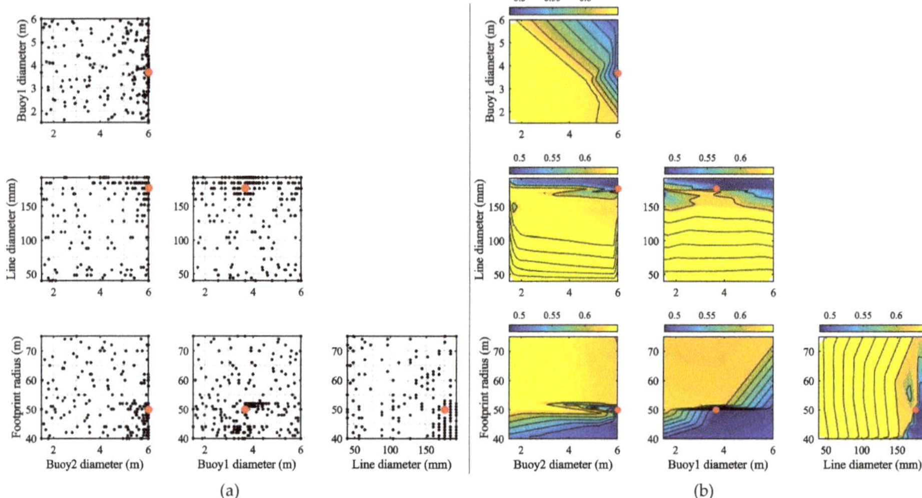

Figure 11. (**a**) Sample point from the optimization of Case 3; (**b**) cost contour plot of the optimization of Case 3.

3.4. Case 4

As described in Section 2.2, the hydrodynamic model of the WEC in Case 4 provides a significant inaccuracy, which also affects the optimization procedure. Table 7 presents the result for this case.

Table 7. Results of the optimization of the mooring system for Case 4.

Parameter	Optimum Value
No. of mooring lines (-)	10
Mooring line diameter (mm)	192
Footprint radius (m)	91
Buoy diameter (m)	3.5
Horizontal stiffness (kN/m)	373
Normalized mooring system cost $\left(\frac{C_{mooring}}{C_{total}} \right)$	0.22

Due to the implications of using linear theory on an overtopping device, very large loads are seen and cause difficulty in finding an adequate solution. Consequently, the strongest line is chosen and with the highest number of lines. Figure 12 clearly identifies the problem of finding a solution since a significant part of the design space gives a high cost due to the penalty function. The normalized cost of the working systems is in the range 0.22–0.24. The number of lines provides an influence as it helps distribute the loads into more lines and thereby secures sufficient strength in each and also lighter anchors. Similar, the footprint radius can be used for modifying the cost through its effect on the stiffness and load on anchors. Clearly, a large radius is desired as the longer lines introduce more compliance.

The buoy size also plays an import part in the system response and cost. By having a large buoy with high buoyancy, the stiffness of the system is primarily an effect of the line stiffness, while also a large pretension in obtained. A smaller buoy results in more vertical compliance and reduces the line tensions. The optimizer identifies a small buoy as the most feasible solution and instead requires strong lines and a larger footprint radius.

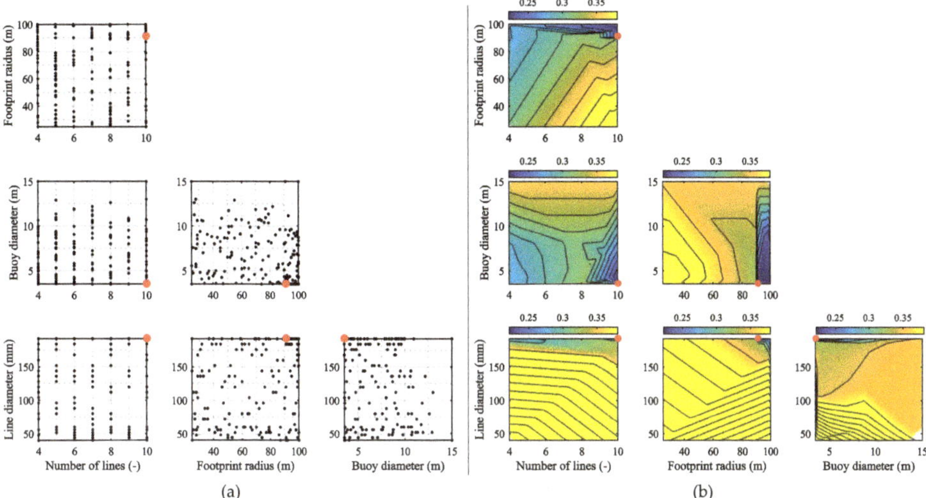

Figure 12. (**a**) Sample point from the optimization of Case 4; (**b**) cost contour plot of the optimization of Case 4.

4. Discussion and Conclusions

This paper used a surrogate-based optimization model to find the most suitable mooring configuration for four large floating WECs, considering design in the ULS and aiming to find the least costly solutions. Based on the presented environmental conditions and design constraints for each device, numerical models were constructed in the BEM code NEMOH and time domain model OrcaFlex. In this connection, a database of costs was constructed and used to calculate the total lifetime cost for each solution.

Based on the optimization routine, four working solutions were found. Even though it is not possible to detect from the normalized values in Figures 9–12, it was found that the cost for Cases 1, 2 and 3 approached comparable values despite some differences in the mooring layouts as Case 3 is equipped with a buoy and hawser, while Case 1 and 2 are turret systems. The cost for these two become high due to the turret system, while Case 4 becomes expensive due to larger loads and more anchors. Case 3 provides a clear low value when compared to the other cases and can both be explained by the fact that the mooring layout is much different, but also by the fact that the WEC is extremely light compared to the other devices and with a much smaller draught, which induces smaller loads on the WEC. As the devices are different in layout, it is not possible to directly compare the cost between the cases. In [8], the cost of mooring for a single point absorber buoy is listed to take up 8% of the total CAPEX, which for the current four cases is in the range 8–25%. The larger structures, hence, result in relatively more expensive moorings due to increased loads.

For all cases, the line diameter provided the largest impact on the cost of the mooring, as a relatively large line diameter was needed in the layouts in order to ensure sufficient strength to avoid line failure and to provide enough stiffness to avoid undesirably large motions. In the SALM system, some cost was saved by adjusting primarily the size of the top buoy as it highly influences the stiffness; cf. Figure 13. When having a small buoy, the stiffness is low for small motions, but the risk increases of fully stretching the system where the stiffness curve becomes steep and large tensions can occur (also seen in Figure 13). Similarly, the footprint radius was important to restrain in order to avoid large motions. In addition to the line diameter, the SPM system was influenced by most parameters. In this case, very high loads were obtained due to the problems of using linear theory on the device, but it was shown how the number of lines and footprint radius could affect the stiffness and thereby loads and cost; cf. Figure 13. The surface buoy provided a great impact on the cost as it is a paramount influencer on the stiffness. Having a large buoy provides a much stiffer system and thereby introduces large line loads, also because much more pretension is in the lines due to higher buoyancy. The study showed that decreasing the size as much as possible provided the least expensive solution. Similar to the SALM system, the use of a small buoy in this solution increases the risk of fully stretching the system, with risk of high loads. In this case, however, it was still found more feasible to use a small buoy; cf. Figure 13. For this particular case, there might be more cost savings, possibly by first improving the model to account for the energy dissipation by overtopping. Afterwards, it is highly relevant to consider a synthetic hawser, as well. This will introduce additional compliance to the system, but needs to be balanced with the stiffness from the buoy. Having a large buoy implies that much tension will be put on the hawser, and in this case, a small buoy is expected.

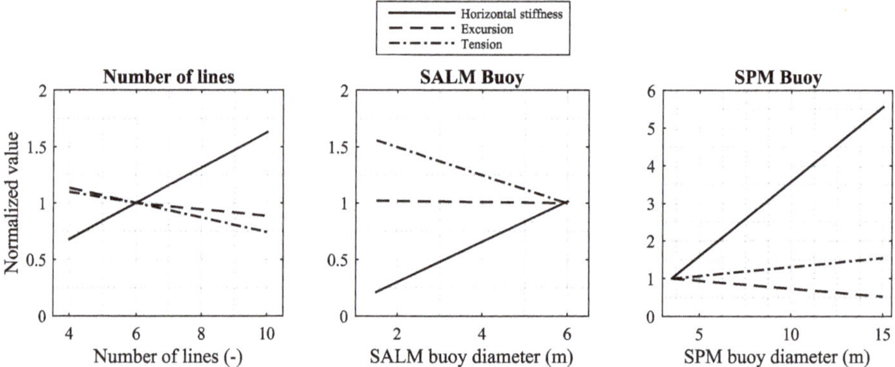

Figure 13. The influence of number of lines and buoy sizes on the mooring stiffness, tension and excursion.

When considering Figures 9–12, it is clear that some of the optimum solutions are found in minimums on the response surface where the gradient is large. Even minor changes to the system layout in these areas can result in significant changes of the cost, meaning that the solution is very sensitive to input parameters and their uncertainties. In some applications, it would be reasonable to search for areas where also the gradient is low in order to find solutions where the cost estimate is more reliable. In addition, the study also showed that some parameters had a minor effect on the mooring cost. From a safety point of view, it would be beneficial, e.g., to use more lines if the effect on cost is only minor in order to apply more redundancy and safety. These considerations are not included in the optimization routine, but a very strong benefit of the surrogate-based model is that information of the entire response surface is achieved and allows for additional and manual evaluation of the surface and potential use of other solutions than the global minimum.

The paper showed that it was possible to use a surrogate-based optimization routine to determine an optimum mooring solution with only a limited number of evaluations. The total computational time for each case was in the range of 25–30 h, which is reasonably low for a design process. For each case, a solution fulfilling all specified design criteria and ensuring survivability was found, and the parameters affecting the mooring cost for each layout were identified. In future studies, it would be natural to improve the hydrodynamic models further and investigate their further potential cost reduction. This study only considered a surrogate-based optimization procedure with a limited number of models, while many others exist, both surrogate models and other types of optimization routines. The advantage of the present method is considerable where only a limited number of function evaluations is needed and can be used for this type of problem.

Acknowledgments: The work in this study has been funded by the Energy Technology Development and Demonstration Program (EUDP) through the project "Mooring Solutions for Large Wave Energy Converters" (Grant Number 64014-0139). The authors wish to acknowledge all project partners, particularly Floating Power Plant, KNSwing, LEANCON Wave Energy and Wave Dragon, for input to the project, data on the WECs and review of this paper.

Author Contributions: Jonas Bjerg Thomsen, Francesco Ferri and Jens Peter Kofoed defined the overall outline of the study. Jonas Bjerg Thomsen produced the numerical models and implemented the optimization routine. Kevin Black collected cost data and produced the cost database, with input from Jonas Bjerg Thomsen, Francesco Ferri and Jens Peter Kofoed. Jonas Bjerg Thomsen ran the routine, analysed the data and made the outline and first draft of the paper. Francesco Ferri, Jens Peter Kofoed and Kevin Black provided the review and input to the finalization of the paper.

Conflicts of Interest: The authors declare no conflicts of interest.

References

1. World Energy Council. Bloomberg new energy finance. In *World Energy Perspective—Cost of Energy Technologies*; World Energy Council: London, UK, 2013.

2. Ocean Energy Systems (OES); International Energy Agency (IEA). *International Levelized Cost of Energy for Ocean Energy Technologies*; OES: Portland, OR, USA, 2017.

3. International Renewable Energy Agency (IRENA). *REthinking Energy 2017: Accelerating the Global Energy Transformation*; International Renewale Energy Agency: Abu Dhabi, UAE, 2017.

4. Carbon Trust. *Accelerating Marine Energy: The Potential for Cost Reduction—Insights from the Carbon Trust Marine Energy Accelerator*; Carbon Trust: London, UK, 2011.

5. Low Carbon Innovation Coordination Group. Technology Innovation Needs Assessment (TINA), Marine Energy Summary Report. Available online: https://www.carbontrust.com/media/168547/tina-marine-energy-summary-report.pdf (accessed on 5 November 2017).

6. Fitzgerald, J. Position Mooring of Wave Energy Converters. Ph.D. Thesis, Chalmers Univerisity of Technology, Goteborg, Sweden, 2009.

7. Martinelli, L.; Ruol, P.; Cortellazzo, G. On mooring design of wave energy converters: The Seabreath application. *Coast. Eng. Proc.* **2012**, *1*, doi:10.9753/icce.v33.structures.3.

8. Neary, V.S.; Lawson, M.; Previsic, M.; Copping, A.; Hallett, K.C.; LaBonte, A.; Rieks, J.; Murray, D. *Methodology for Design and Economic Analysis of Marine Energy Conversion (MEC) Technologies*; Sandia National Laboratories: Albuquerque, NM, USA, 2014.

9. Christensen, L.; Friis-Madsen, E.; Kofoed, J. The Wave Energy Challenge: The Wave Dragon case. In Proceedings of the POWER-GEN 2005 Europe Conference, Milan, Italy, 28–30 June 2005.

10. Holmberg, P.; Andersson, M.; Bolund, B.; Strandanger, K. Wave power - Surveillance study of the development. *Elforsk Rapp.* **2011**, *11*, 47.

11. Thomsen, J.B.; Ferri, F.; Kofoed, J.P. Assessment of Current State of Mooring Design in the Danish Wave Energy Sector. In Proceedings of the 11th European Wave and Tidal Energy Conference, Nantes, France, 6–11 September 2015.

12. Thomsen, J.; Kofoed, J.; Delaney, M.; Banfield, S. Initial Assessment of Mooring Solutions for Floating Wave Energy Converters. In Proceedings of the Twenty-Sixth (2016) International Ocean and Polar Engineering Conference, Rhodes, Greece, 26 June–2 July 2016; Chung, J., Muskulus, M., Kokkinis, T., Wang, A., Eds.; International Society of Offshore & Polar Engineers: Cupertino, CA, USA, 2016; Volume 1, pp. 590–596.

13. Thomsen, J.; Ferri, F.; Kofoed, J. Experimental testing of moorings for large floating wave energy converters. In *Progress in Renewable Energies Offshore*; Soares, C., Ed.; CRC Press LLC: Boca Raton, FL, USA, 2016; pp. 703–710.

14. Thomsen, J.B.; Ferri, F.; Kofoed, J.P. Screening of available tools for dynamic mooring analysis of wave energy converters. *Energies* **2017**, *10*, 853, doi:10.3390/en10070853.

15. Thomsen, J.; Ferri, F.; Kofoed, J. Validation of a tool for the initial dynamic design of Mooring systems for large floating wave energy converters. *J. Mar. Sci. Eng.* **2017**, *5*, 45, doi:10.3390/jmse5040045.

16. Floating Power Plant. Available online: http://www.floatingpowerplant.com/ (accessed on 5 November 2017).

17. Bingham, H.B.; Ducasse, D.; Nielsen, K.; Read, R. Hydrodynamic analysis of oscillating water column wave energy devices. *J. Ocean Eng. Mar. Energy* **2015**, *1*, 405–419.

18. Nielsen, K.; Bingham, H. MARINET experiment KNSWING testing an I-Beam OWC attenuator. *Int. J. Mar. Energy* **2015**, *12*, 21–34.
19. LEANCON Wave Energy. Available online: http://www.leancon.com/ (accessed on 5 November 2017).
20. Wave Dragon. Available online: http://www.wavedragon.net/ (accessed on 5 November 2017).
21. Ridge, I.; Banfield, S.; Mackay, J. Nylon fibre rope moorings for wave energy converters. In Proceedings of the IEEE OCEANS, Seattle, WA, USA, 20–23 September 2010; pp. 1–10.
22. Fitzgerald, J.; Bergdahl, L. Considering mooring cables for offshore wave energy converters. In Proceedings of the European Wave and Tidal Energy Conference, Porto, Portugal, 11–14 September 2007.
23. DNV. *DNV Offshore Standard DNV-OS-E301: Position Mooring*; DNV: Oslo, Norway, 2010.
24. American Petroleum Institute (API). *API-RP-2SK: Design and Analysis of Stationkeeping Systems for Floating Structures*; American Petroleum Institute: Washington, DC, USA, 2005.
25. International Organization for Standardization (ISO). *ISO 19901-7 2005: Stationkeeping Systems for Floating Offshore Structures and Mobile Offshore Units*; ISO: Geneva, Switzerland, 2013.
26. International Electrotechnical Commission (IEC). *IEC 62600-10: Assessment of Mooring System for Marine Energy Converters (MECs)*; IEC: Geneva, Switzerland, 2014.
27. Davidson, J.; Ringwood, J.V. Mathematical modelling of mooring systems for wave Energy converters—A review. *Energies* **2017**, *10*, 666, doi:10.3390/en10050666.
28. Pecher, A.F.S.; Kofoed, J.P. *Handbook of Ocean Wave*; Springer: Basel, Switzerland, 2017; Volume 7.
29. Ortiz, J.P.; Bailey, H.; Buckham, B.; Crawford, C. Surrogate based design of a mooring system for a self-reacting point absorber. In Proceedings of the The Twenty-fifth International Ocean and Polar Engineering Conference, Kona, HI, USA, 21–26 June 2015.
30. Fitzgerald, J.; Bergdahl, L. Including moorings in the assessment of a generic offshore wave energy converter: A frequency domain approach. *Mar. Struct.* **2008**, *21*, 23–46.
31. Vicente, P.C.; Falcão, A.D.O.; Justino, P.A. Optimization of mooring configuration parameters of floating wave energy converts. In Proceedings of the 30th International Conference on Ocean, Offshore and Artctic Engineering-OMAE2011, Rotterdam, The Netherlands, 19–24 June 2011.
32. Ferri, F. Computationally efficient optimisation algorithms for WECs arrays. In Proceedings of the 12th European Wave and Tidal Energy Conference, Cork, Ireland, 27 August–1 September 2017.
33. DNV. *DNV-RP-C205 Environmental Conditions and Environmental Loads*; DNV: Oslo, Norway, 2014.
34. Zanuttigh, B.; Martinelli, L.; Castagnetti, M. *Screening of Suitable Mooring Systems*; Aalborg University: Aalborg, Denmark, 2012; Volume 2.
35. Paredes, G.M.; Bergdahl, L.; Palm, J.; Eskilsson, C.; Pinto, F.T. Station keeping design for floating wave energy devices compared to floating offshore oil and gas platforms. In Proceedings of the 10th European Wave and Tidal Energy Conference, Aalborg, Denmark, 2–5 September 2013; Volume 10.
36. Johanning, L.; Smith, G.; Wolfram, J. Towards design standards for WEC moorings. In Proceedings of the 6th European Wave and Tidal Energy Conference, Glasgow, UK, 29 August–2 September 2005; Volume 29, pp. 223–229.
37. Ambühl, S.; Kramer, M.; Sørensen, J. Different reliability assessment approaches for wave energy converters. In Proceedings of the 11th European Wave and Tidal Energy Conference, Nantes, France, 6–11 September 2015.
38. Ambühl, S.; Kramer, M.; Sørensen, J. Reliability-based Calibration of Partial Safety Factors for Wave Energy Converters. In Proceedings of the 12th International Conference on Applications of Statistics and Probability in Civil Engineering, Vancouver, BC, Canada, 12–15 July 2015.
39. Ambühl, S.; Kramer, M.; Kofoed, J.; Sørensen, J.; Ferreira, C. Reliability assessment of wave Energy devices. In *Safety, Reliability, Risk and Life-Cycle Performance of Structures and Infrastructures*; Deodatis, G., Ellingwood, B., Frangopol, D., Eds.; CRC Press LLC: Boca Raton, FL, USA, 2014; pp. 5195–5202.
40. Ambühl, S.; Kofoed, J.; Sørensen, J. Stochastic modeling of long-term and extreme value estimation of wind and sea conditions for probabilistic reliability assessments of wave energy devices. *Ocean Eng.* **2014**, *89*, 243–255.
41. DNV. *DNV Offshore Standard DNV-RP-E301: Design and Installation of Fluke Anchors*; DNV: Oslo, Norway, 2012.
42. Bridon. Wire and Fibre Rope Solutions. 2016. Available online: http://www.bridon.com/uk/ (accessed on 5 November 2017).

43. Babarit, A.; Delhommeau, G. Theoretical and numerical aspects of the open source BEM solver NEMOH. In Proceedings of the 11th European Wave and Tidal Energy Conference (EWTEC2015), Nantes, France, 6–11 September 2015.

44. Newman, J.T. The drift force and moment on ships in waves. *J. Ship Res.* **1967**, *11*, 51–60.

45. Orcina Ltd. *Orcaflex User Manual*; Orcina Ltd.: Ulverston, UK, 2013.

46. Rios, L.M.; Sahinidis, N.V. Derivative-free optimization: A review of algorithms and comparison of software implementations. *J. Glob. Optim.* **2013**, *56*, 1247–1293.

47. Conn, A.R.; Scheinberg, K.; Vicente, L.N. *Introduction to Derivative-Free Optimization*; SIAM: Philadelphia, PA, USA, 2009.

48. Müller, J. *Matsumoto: The Matlab Surrogate Model Toolbox for Computationally Expensive Black-Box Global Optimization Problems*; Cornell University: New York, NY, USA, 2014.

49. Cavazzuti, M. *Optimization Methods: From Theory to Design Scientific and Technological Aspects in Mechanics*; Springer Science & Business Media: Dordrecht, The Netherlands, 2012.

50. Müller, J. MATSuMoTo Code Documentation. 2014. Available online: https://ccse.lbl.gov/people/julianem/Manual_MATSuMoTo_matlab.pdf (accessed on 5 November 2017).

51. Müller, J.; Shoemaker, C.A. Influence of ensemble surrogate models and sampling strategy on the solution quality of algorithms for computationally expensive black-box global optimization problems. *J. Glob. Optim.* **2014**, *60*, 123–144.

52. Müller, J.; Piché, R. Mixture surrogate models based on Dempster-Shafer theory for global optimization problems. *J. Glob. Optim.* **2011**, *51*, 79–104.

53. Müller, J.; Shoemaker, C.A.; Piché, R. SO-MI: A surrogate model algorithm for computationally expensive nonlinear mixed-integer black-box global optimization problems. *Comput. Oper. Res.* **2013**, *40*, 1383–1400.

54. Müller, J.; Shoemaker, C.A.; Piché, R. SO-I: A surrogate model algorithm for expensive nonlinear integer programming problems including global optimization applications. *J. Glob. Optim.* **2014**, *59*, 865–889.

Article

Hydrodynamic Investigation of a Concentric Cylindrical OWC Wave Energy Converter

Yu Zhou [1,2], Chongwei Zhang [1,2] and Dezhi Ning [1,2,*]

1   State Key Laboratory of Coastal and Offshore Engineering, Dalian University of Technology,
    Dalian 116024, China; yuzhouzhouyu@mail.dlut.edu.cn (Y.Z.); chongweizhang@dlut.edu.cn (C.Z.)
2   Offshore Renewable Energy Research Center, Dalian University of Technology, Dalian 116024, China
*   Correspondence: dzning@dlut.edu.cn

Received: 22 March 2018; Accepted: 16 April 2018; Published: 18 April 2018

Abstract: A fixed, concentric, cylindrical oscillating water column (OWC) wave energy converter (WEC) is proposed for shallow offshore sites. Compared with the existing shoreline OWC device, this wave energy device is not restricted by the wave directions and coastline geography conditions. Analytical solutions are derived based on the linear potential-flow theory and eigen-function expansion technique to investigate hydrodynamic properties of the device. Three typical free-surface oscillation modes in the chamber are discussed, of which the piston-type mode makes the main contribution to the energy conversion. The effects of the geometrical parameters on the hydrodynamic properties are further investigated. The resonance frequency of the chamber, the power extraction efficiency, and the effective frequency bandwidth of the device is discussed, amongst other topics. It is found that the proposed OWC-WEC device with a lower draft and wider chamber breadth has better power extraction ability.

Keywords: oscillating water column; wave energy; wave diffraction; eigen-function expansion; potential flow theory; air chamber

---

1. Introduction

Sustainable energy from ocean waves has drawn people's attention for over 40 years. So far, there have been a large number of wave energy converters proposed [1,2]. The energy extraction from waves mainly relies on the hinged mechanical motions [3], individual body oscillations [4], water overtopping to reservoirs [5], or trapped air pockets in oscillating water column (OWC) chambers [6]. Compared with those mechanical oscillators, the OWC wave energy converter has no moving parts in the water, which is an attractive feature in terms of survivability and maintenance convenience [7].

Up to now, the OWCs have been the most widely used wave energy converters in practice. The first OWC device deployed into the sea on a large scale was in Japan [8]. After that, a number of OWC devices have been proposed and deployed. For example, a breakwater equipped with OWC plant was finished with a 40 kW Wells turbine in the Sea of Japan [9]. A small shoreline OWC (75 kW) was built for academic studying at the island of Islay in Scotland. In Guangdong Province of China, a shore-fixed OWC device (100 kW) was completed in 2001 [10]. Recently, a new U-type OWC has been being installed in breakwaters, which is expected to show dramatically improved absorption width for fixed installations [11]. However, OWCs inserted in breakwaters can be installed in zones with low available energy [12]. It should be noticed that the above-mentioned OWC-WECs are often based on shoreline or near-shore regions, which not only have the potential to harm the coastline environment but also are limited in terms of the deployment scale. To reduce the environmental impact and increase the scale of deployment, the wave energy exploitation tends to progress to offshore sites [13]. Typical examples are the 110 kW Mighty Whale [14] and the OE Buoy in Galway Bay [15].

In this study, a concentric cylindrical OWC-WEC is innovatively designed for offshore sites with multiple wave directions. From economic and risk considerations, a fixed OWC structure is preferred. As shown in Figure 1, the structure can be considered as an upside-down cylindrical bucket supported by a cylinder. A concentric cylindrical chamber is formed. A beam is used to connect the chamber and the cylinder as shown in Figure 1b. The energy is transformed from the wave to the air by an oscillating water column moving up and down in the chamber. As the bidirectional airflow flows through the orifice on the top of the chamber, the turbine is driven to be an electricity generator. Compared with the shoreline OWC devices that can only capture the wave energy from one direction, the application of the present-type wave energy device is not restricted by the wave direction and coastline geography condition.

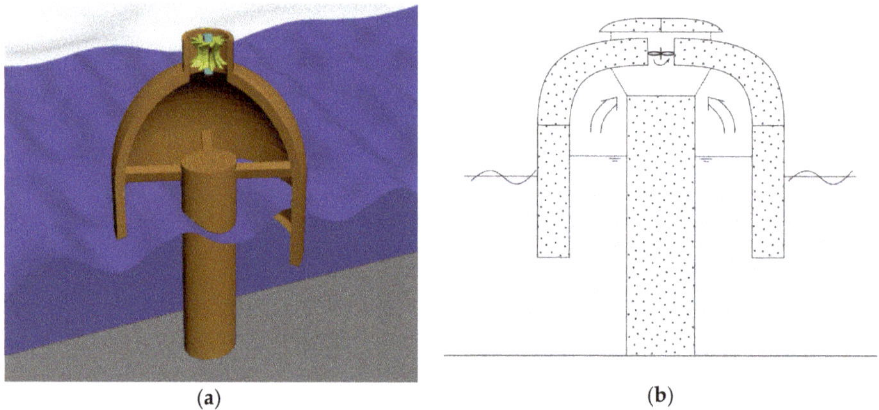

(a)	(b)

Figure 1. Concept of the concentric cylindrical OWC-WEC, (a) 3D effect picture; (b) the profile of the chamber.

At the preliminary study stage, the efficiency and hydrodynamic properties of the structure must be identified before the optimized dimension parameters can be given. The analytic method has been used for the preliminary research of the OWC-WECs efficiently by scholars. For example, Garrett [16] derived the linear diffraction solution of a fistulous suspended cylinder in shallow water. Sarmento and Falcão [17] derived a two dimensional analytical model to investigate the effects of linear and nonlinear power take-off system on hydrodynamic properties of OWC-WEC. Zhu and Mitchell [18] presented a diffraction analytical solution of waves around a fistulous cylinder without thickness. Martin-Rivas and Mei [19] presented a linear theory of an OWC which is installed on a straight coast for studying the effects of the coastline on wave power conversion. Martin-Rivas and Mei [20] simulated a thin cylindrical OWC-WEC, which was installed in a breakwater. It is concluded that air compressibility can broaden the effective bandwidth of extraction efficiency with the specific chamber volume. Konispoliatis [21] developed an analytical model to study the hydrodynamics of an OWC-WECs array. Three-unit arrays are discussed, which can be used as a floating wind-turbine foundation. In the present study, an analytical solution based on linear potential-flow theory is derived for the present OWC-WEC.

The present paper is organized as follows. Section 2 describes the power take-off model and analytical solutions of the hydrodynamic problem. In Section 3, the accuracy and convergence of the analytical model are validated. Then, effects of environmental and geometrical parameters on the resonant wave modes and hydrodynamic properties are studied systematically. Conclusions are summarized in Section 4.

2. Mathematical Problem

The proposed OWC-WEC structure is fixed and rigid in the sea. The submerged part can be considered as a concentric cylindrical geometry as in Figure 2. In this study, the draft of the chamber is d. The water depth is h. The bottom-mounted cylinder and the internal and external surfaces of the chamber have the radius R_1, R_2, and R_3, respectively. A Cartesian coordinate system O-xyz is defined with the origin O on the undisturbed free surface, z-axis pointing vertically upwards, and x-axis coinciding with the direction of incident waves. The structure is axisymmetric about the Oz axis.

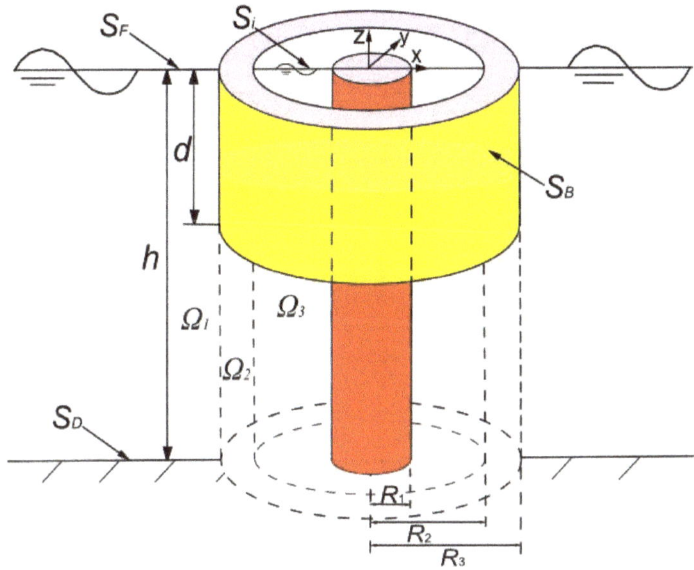

Figure 2. Schematic of the proposed OWC-WEC foundation.

2.1. Power Take-Off Model

In present study, the air in the chamber of the OWC-WEC is considered to be compressible and motion-isentropic. Additionally, all time-dependent variables in the problem are assumed to be harmonic. Then, the time-independent air pressure is introduced:

$$P_a = \mathrm{Re}[P_0 e^{-i\omega t}] \tag{1}$$

in which ω is the wave angular frequency, t is the time, Re is the real part of a complex variable, P_0 is the complex amplitude of the pressure, and $i = \sqrt{-1}$. A Wells turbine, which is installed at the top of the air chamber, is forced to rotate by oscillating air flow. The mass flux rate of the air across the Wells turbine is proportional to the air pressure in the chamber [17,22], and the relationship between the air mass flux and the turbine characteristics can be expressed as [18]

$$q_0 = \left(\frac{KD}{N\rho_a} - \frac{i\omega V_0}{c^2 \rho_a} \right) P_0 \tag{2}$$

in which q_0 is the amplitude of the air volume flux, V_0 is the mean chamber volume, ρ_a is the air density, N is the turbine rotational speed in r.p.m., D is diameter of the turbine rotor, c is the sound velocity in air, and K is an empirical coefficient depending on the design of the turbines.

Based on the linear theory, the volume flux in the chamber is the sum of volume fluxes due to wave radiation and diffraction. The wave radiation occurs in the situation when the wave motion is purely caused by the oscillating air pressure of the chamber. The wave diffraction is due to scattering of incident waves when the air pressures inside and outside the chamber are identical. Thus, complex amplitude of the volume flux has two components

$$q_0 = q_D + P_0 q_R \tag{3}$$

in which q_R and q_D are connected with the radiation and diffraction problems, respectively. For a radiation problem, an added mass coefficient $\tilde{C}$ and a damping coefficient $\tilde{B}$ can be obtained [23]. Complex amplitude of the volume flux q_R, which is caused by the unit air pressure, can be expressed as

$$q_R = \left(-\tilde{B} + i\tilde{C} \right) \tag{4}$$

By substituting Equations (3) and (4) into Equation (2), the air pressure in the chamber can be expressed as follows

$$P_0 = \frac{q_D}{\left[\left(\frac{KD}{N\rho_a} + \tilde{B} \right) - i \left(\tilde{C} + \frac{\omega V_0}{c^2 \rho_a} \right) \right]} \tag{5}$$

in which q_D is solved from the wave diffraction problem, and $\tilde{B}$ and $\tilde{C}$ can be obtained by solving the radiation problem with unit forced pressure.

The time-averaged value of the power captured by the turbine is [18]

$$P_{out} = \frac{KD}{2N\rho_a} |P_0|^2 \tag{6}$$

The power extraction efficiency ζ can be calculated as [18]

$$\zeta = \frac{kP_{out}}{\rho g A C_g / 2} = \frac{gkR_2}{\omega C_g} \frac{\chi |Q_D|^2}{(\chi + B)^2 + (\beta + C)^2} \tag{7}$$

in which $Q_D = \omega q_D / (AR_2 g)$, $(B, C) = \left(\omega \rho \tilde{B}, \omega \rho \tilde{C} \right) / R_2$, $\chi = \rho KD\omega / (N\rho_a R_2)$, $\beta = V_0 \omega^2 \rho / \left(c_a^2 R_2 \rho_a \right)$ are nondimensional parameters. χ characterizes the turbine, β represents the air compressibility. C_g is the group velocity of the incident wave, g is the gravitational acceleration, ρ denotes the water density, and k is the incident wave number.

2.2. Solution of Boundary Value Problem

Under the assumption that the fluid is incompressible, inviscid, and flow-irrotational, the proposed problem is considered. The complex velocity potential $\varphi(x, y, z, t)$ around the structure is introduced to describe the properties of the fluid. For small amplitude wave, the potential φ satisfies the following linearized boundary value problem

$$\nabla^2 \varphi = 0, \text{ in } \Omega \tag{8}$$

$$\frac{\partial \varphi}{\partial n} = 0, S_B \text{ and } S_D \tag{9}$$

$$\frac{\partial \varphi}{\partial t} + g \frac{\partial \varphi}{\partial z} = \begin{cases} -\sigma_R P/\rho, & \text{on } S_i \\ 0, & \text{on } S_F \end{cases} \tag{10}$$

in which Ω, S_D, S_i, S_F, and S_B represent the mean fluid domain, water bottom, internal free surface, external free surface, and wet body surface, respectively. The operator $\partial/\partial n$ denotes the normal derivative of a variable on the impenetrable solid surface. For completeness, a radiation condition is also required on the open boundary. For the time-harmonic, velocity potential with angular frequency ω can be expressed as

$$\varphi(x,y,z,t) = \mathrm{Re}\left[\phi(x,y,z)e^{-i\omega t}\right] \tag{11}$$

in which ϕ is a spatial function. It should be highlighted that σ_R is used as a switch between the radiation and diffraction problem. For the radiation problem, $\sigma_R = 1$ is set and the pressure amplitude in chamber is $P_0 = 1$ in Equation (1). For the diffraction problem, $\sigma_R = 0$ is set.

Based on the axisymmetric geometry of the OWC-WEC, the radiation and diffraction problem can be solved in the cylindrical polar coordinates (r, θ, z) with

$$x = r\cos\theta \text{ and } y = r\sin\theta \tag{12}$$

The fluid domain can be divided into three subdomains, as shown in Figure 2. The external subdomain is Ω_1 with $r \geq R_3$ and $-h \leq z \leq 0$, the lower subdomain Ω_2 with $R_2 \leq r \leq R_3$ and $-h \leq z \leq -d$, and the internal subdomain Ω_3 with $R_1 \leq r \leq R_2$ and $-h \leq z \leq 0$. Correspondingly, the potentials in Ω_1, Ω_2, and Ω_3 are expressed by $\phi^{(1)}$, $\phi^{(2)}$, and $\phi^{(3)}$, respectively.

The boundary value problem with respect to $\phi^{(1)}$ can be founded as follows

$$\nabla^2\phi^{(1)} = 0, \text{in } \Omega_1 \tag{13}$$

$$\frac{\partial\phi^{(1)}}{\partial z} = 0 \text{ for } r \geq R_3 \text{ and } z = -h \tag{14}$$

$$\frac{\partial\phi^{(1)}}{\partial z} = \frac{\omega^2}{g}\phi^{(1)}, \text{ for } r \geq R_3 \text{ and } z = 0 \tag{15}$$

$$\left\{ \begin{array}{l} \phi^{(1)} = \phi^{(2)} \\ \partial\phi^{(1)}/\partial r = \partial\phi^{(2)}/\partial r \end{array} \right., \text{ for } r = R_3 \text{ and } -h \leq z \leq -d \tag{16}$$

$$\partial\phi^{(1)}/\partial r = 0, \text{ for } r = R_3 \text{ and } -d \leq z \leq 0 \tag{17}$$

The radiation condition is also needed for $r \to \infty$.

For $\phi^{(2)}$, the boundary value problem corresponds to the following

$$\nabla^2\phi^{(2)} = 0, \text{ in } \Omega_2 \tag{18}$$

$$\frac{\partial\phi^{(2)}}{\partial z} = 0, \text{ for } R_2 \leq r \leq R_3 \text{ and } z = -d \tag{19}$$

$$\frac{\partial\phi^{(2)}}{\partial z} = 0, \text{ for } R_2 \leq r \leq R_3 \text{ and } z = -h \tag{20}$$

$$\left\{ \begin{array}{l} \phi^{(2)} = \phi^{(1)} \\ \partial\phi^{(2)}/\partial r = \partial\phi^{(1)}/\partial r \end{array} \right., \text{ for } r = R_3 \text{ and } -h \leq z \leq -d \tag{21}$$

$$\left\{ \begin{array}{l} \phi^{(2)} = \phi^{(3)} \\ \partial\phi^{(2)}/\partial r = \partial\phi^{(3)}/\partial r \end{array} \right., \text{ for } r = R_2, \text{ and } -h \leq z \leq -d \tag{22}$$

The $\phi^{(3)}$ could be obtained from the following boundary value problem

$$\nabla^2\phi^{(3)} = 0, \text{ in } \Omega_2 \tag{23}$$

$$\frac{\partial\phi^{(3)}}{\partial z} = 0, \text{ for } R_1 \leq r \leq R_2 \text{ and } z = -h \tag{24}$$

$$\frac{\partial \phi^{(3)}}{\partial z} - \frac{\omega^2}{g}\phi^{(3)} = \frac{i\omega\sigma_R P_0}{\rho g}, \text{ for } R_1 \leq r \leq R2, \text{ at } z = 0 \tag{25}$$

$$\begin{cases} \phi^{(3)} = \phi^{(2)} \\ \partial\phi^{(3)}/\partial r = \partial\phi^{(2)}/\partial r \end{cases}, \text{ for } r = R_2, \text{ and } -h \leq z \leq 0 \tag{26}$$

$$\partial\phi^{(3)}/\partial r = 0, \text{ for } r = R_2, \text{ and } -d \leq z \leq 0 \tag{27}$$

$$\partial\phi^{(3)}/\partial r = 0, \text{ for } r = R_1, \text{ and } -h \leq z \leq 0 \tag{28}$$

In Ω_3, P_0 is constant, and then the linear transformation can be made

$$\phi^{(3)} = \phi^* - i\sigma_R P_0/\rho\omega \tag{29}$$

in which ϕ^* satisfies the original system (23)–(28) in Ω_3, and then the inhomogeneous free surface condition (25) can be rewritten as

$$\frac{\partial\phi^*}{\partial z} = \frac{\omega^2}{g}\phi^*, \text{ for } R_1 \leq r \leq R_2 \text{ and } z = 0 \tag{30}$$

2.3. Mathematical Solution

In cylindrical polar coordinate system, the Laplace's equation of ϕ could be written as

$$\nabla^2\phi(r,\theta,z) = \frac{1}{r^2}\frac{\partial^2\phi}{\partial\theta^2} + \frac{1}{r}\frac{\partial}{\partial r}\left(r\frac{\partial\phi}{\partial r}\right) + \frac{\partial^2\phi}{\partial z^2} = 0 \tag{31}$$

Based on the method of separation of variables, ϕ may be assumed to be $\phi(r,\theta,z) = R(r)\Theta(\theta)Z(z)$. Then, Equation (31) can be rewritten by three ordinary differential equations as

$$\frac{d^2 Z}{dz^2} + \alpha^2 Z = 0 \tag{32}$$

$$\frac{d^2\Theta}{d\theta^2} + \beta^2\Theta = 0 \tag{33}$$

$$r\frac{d}{dr}\left(r\frac{dR}{dr}\right) - \left(\alpha^2 r^2 + \beta^2\right)R = 0 \tag{34}$$

Both α^2 and β^2 are constants here.

In subdomain Ω_1, the solution of Equation (32) has the form:

$$Z(z) = \sum_{n=0}^{\infty} B_n Z_n^{(I)} \tag{35}$$

in which B_n denote constant coefficients and $Z_n^{(I)}$ are vertical eigen-functions. Further, taking the boundary conditions (i.e., Equations (14), (15), (19), (20), (24) and (30)) into account leads to the expression $Z_n^{(I)}$ for each subdomain as

$$Z_n^{(1)}(z) = Z_n^{(3)}(z) = \begin{cases} \cosh k(z+h), & n = 0 \\ \cos k_n(z+h), & n \geq 1 \end{cases} \tag{36}$$

$$Z_n^{(2)}(z) = \begin{cases} 1, & n = 0 \\ \cos k_n^{(2)}(z+h), & n \geq 1 \end{cases} \tag{37}$$

in which $\{k_n; n = 1, 2, \cdots\}$ and k are positive real roots of the dispersion equations

$$\omega^2/g - k\tanh kh = 0 \tag{38}$$

$$\omega^2/g + k_n \tan k_n h = 0 \tag{39}$$

and $\left\{k_n^{(2)}; n = 1, 2, \cdots\right\}$ can be obtained from

$$k_n^{(2)} = n\pi/(h - d) \tag{40}$$

In each subdomain, the set of vertical eigen-functions is complete and satisfies the orthogonality relations as

$$\int_{-h}^{0} Z_m^{(1)} Z_n^{(1)} dz = 0, \text{ and } \int_{-h}^{-d} Z_m^{(2)} Z_n^{(2)} dz = 0 \text{ for } m \neq n \tag{41}$$

For Θ, with the continuity condition (i.e., $\Theta(\theta) = \Theta(\theta + 2\pi)$), the solution of Equation (33) can be written as

$$\Theta(\theta) = \sum_{m=0}^{\infty} B_m \Theta_m \tag{42}$$

in which B_m denote constant coefficients and Θ_m are circumferential eigen-functions as

$$\Theta_m(\theta) = \cos(m\theta), m = 0, 1, 2, \cdots \tag{43}$$

Equation (34) is the modified Bessel's equation

$$r\frac{d}{dr}\left(r\frac{dR}{dr}\right) - \left(k_n^2 r^2 + m^2\right)R = 0, \text{ for } m, n = 0, 1, 2, \cdots \tag{44}$$

with $k_0 = -ik$. The solution can be expressed as

$$R^{(1)}(r) = \sum_{m=0}^{\infty} \sum_{n=0}^{\infty} R_{mn}^{(1)}, \text{ for } m, n = 0, 1, 2, \cdots \tag{45}$$

In Ω_1, the potential of the incident wave can be written as

$$\phi_I(r, \theta, z) = \underbrace{\left(-\frac{igA}{\omega \cosh kh}\right)}_{E} \left[e^{ikr\cos\theta}\right] Z_0^{(1)}(z) = E\left[\sum_{m=0}^{\infty} \varepsilon_m i^m J_m(kr)\cos(m\theta)\right] Z_0^{(1)}(z) \tag{46}$$

in which A is the wave amplitude, $\varepsilon_m = \{1, \text{ for } m = 0; 2, \text{ for } m \geq 1\}$ is the Neumann symbol, and J_m is the first-kind of Bessel function of order m. The radiation condition is also satisfied to guarantee that the radiated waves propagate away from the structure. Then, the R_{mn} term can be written as

$$R_{mn}^{(1)}(r) = \begin{cases} B_{mn} H_m^{(1)}(kr), n = 0 \\ B_{mn} K_m(k_n r), n \geq 1 \end{cases}, \text{ for } m = 0, 1, 2, \cdots \tag{47}$$

in which B_{mn} is a constant coefficient, and $H_m^{(1)}$ and K_m are the first-kind of Hankel function and the second-kind of modified Bessel function, respectively, both of order m.

In Ω_2, $R_{mn}^{(2)}$ has the following expression

$$R_{mn}^{(2)} = \begin{cases} C_{00} + D_{00} \ln r, m = 0, n = 0 \\ C_{m0} r^m + D_{m0} r^{-m}, m \neq 0, n = 0 \\ C_{mn} I_m\left(k_n^{(2)} r\right) + D_{mn} K_m\left(k_n^{(2)} r\right), n \geq 1 \end{cases} \tag{48}$$

in which C_{mn}, D_{mn} also denote constant coefficients, and I_m is the first-kind of modified Bessel functions of order m. Note that the following relationship is satisfied for the case of $n = 0$

$$I_m(k_0 r) = (-1)^m J_m(kr) \text{ and } K_m(k_0 r) = \frac{\pi}{2} i^{m+1} H_m^{(1)}(kr) \tag{49}$$

In Ω_3, it has

$$R_{mn}^{(3)} = \begin{cases} A_{mn} J_m(kr) + E_{mn} Y_m(kr), n = 0 \\ A_{mn} I_m(k_n r) + E_{mn} K_m(k_n r), n \geq 1 \end{cases} \tag{50}$$

in which A_{mn}, E_{mn} are constant coefficients, and Y_m is the second kind of Bessel functions of order m. Further, with the boundary condition at $r = R_1$, the following relationship is further satisfied

$$A_{mn} = S_{mn} E_{mn} \text{ with } S_{mn} = \begin{cases} -\frac{dJ_m(ka)}{dr} \Big/ \frac{dY_m(ka)}{dr}, n = 0 \\ -\frac{dI_m(k_n a)}{dr} \Big/ \frac{dK_m(k_n a)}{dr}, n \geq 1 \end{cases} \tag{51}$$

Then, the general solution of ϕ in each subdomain can be expressed as

$$\begin{aligned} \phi^{(1)}(r, \theta, z) &= \sigma_D E \sum_{m=0}^{\infty} \varepsilon_m i^m \cos(m\theta) J_m(kr) Z_0^{(1)}(z) \\ &+ E \sum_{m=0}^{\infty} \varepsilon_m i^m \cos(m\theta) \left[\sum_{n=0}^{\infty} B_{mn} P_{mn}^{(1)}(r) Z_n^{(1)}(z) \right] \end{aligned} \tag{52}$$

$$\phi^{(2)}(r, \theta, z) = \sum_{m=0}^{\infty} \cos(m\theta) \sum_{n=0}^{\infty} \left[C_{mn} P_{mn}^{(2)}(r) + D_{mn} Q_{mn}^{(2)}(r) \right] Z_n^{(2)}(z) \tag{53}$$

$$\phi^{(3)}(r, \theta, z) = \sum_{m=0}^{\infty} \cos(m\theta) \sum_{n=0}^{\infty} \left[A_{mn} P_{mn}^{(3)}(r) Z_n^{(3)}(z) \right] - \frac{i\sigma_R P_0}{\rho\omega} \tag{54}$$

with

$$P_{mn}^{(1)}(r) = \left\{ H_m^{(1)}(kr), n = 0; K_m(k_n r), n \geq 1 \right\} \tag{55}$$

$$P_{mn}^{(2)}(r) = \left\{ 1, m = 0, n = 0; r^m, m \neq 0, n = 0; I_m\left(k_n^{(2)} r\right), n \geq 1 \right\} \tag{56}$$

$$Q_{mn}^{(2)}(r) = \left\{ \ln r, m = 0, n = 0; r^{-m}, m \neq 0, n = 0; K_m\left(k_n^{(2)} r\right), n \geq 1 \right\} \tag{57}$$

$$P_{mn}^{(3)}(r) = \begin{cases} J_m(kr) + S_{mn} H_m^{(1)}(kr), n = 0 \\ I_m(k_n r) + S_{mn} K_m(k_n r), n \geq 1 \end{cases} \tag{58}$$

Equations (52)–(54) have to satisfy the boundary conditions at $r = R_2$ and $r = R_3$. Additionally, $\sigma_D = 1 - \sigma_R$ is a switch to the diffraction problem. In order to take advantage of the orthogonality relations, the boundary conditions can be rewritten as

$$\int_{-h}^{-d} \phi^{(1)}(R_3, \theta, z) Z_n^{(2)}(z) dz = \int_{-h}^{-d} \phi^{(2)}(R_3, \theta, z) Z_n^{(2)}(z) dz \tag{59}$$

$$\int_{-h}^{-d} \phi^{(3)}(R_2, \theta, z) Z_n^{(2)}(z) dz = \int_{-h}^{-d} \phi^{(2)}(R_2, \theta, z) Z_n^{(2)}(z) dz \tag{60}$$

$$\int_{-h}^{0} \frac{d\phi^{(1)}(R_3, \theta, z)}{dr} Z_n^{(1)}(z) dz = \int_{-h}^{-d} \frac{d\phi^{(2)}(R_3, \theta, z)}{dr} Z_n^{(1)}(z) dz \tag{61}$$

$$\int_{-h}^{0} \frac{d\phi^{(3)}(R_2, \theta, z)}{dr} Z_n^{(3)}(z) dz = \int_{-h}^{-d} \frac{d\phi^{(2)}(R_2, \theta, z)}{dr} Z_n^{(3)}(z) dz \tag{62}$$

By substituting the general solutions of ϕ into these conditions and truncating each infinite series to M terms, it can lead to the following linear system of equations

$$\sum_{n=0}^{M} B_{mn}\left[E\varepsilon_m i^m P_{mn}^{(1)}(R_3)a_{nl}\right] - \sum_{n=0}^{M} C_{mn}\left[P_{mn}^{(2)}(R_3)b_{nl}\right] - \sum_{n=0}^{M} D_{mn}\left[Q_{mn}^{(2)}(R_3)b_{nl}\right]$$
$$= \left[-\sigma_D E\varepsilon_m i^m J_m(kR_3)a_{0l}\right] \tag{63}$$

$$\sum_{n=0}^{M} A_{mn}\left[P_{mn}^{(3)}(R_2)a_{nl}\right] - \sum_{n=0}^{M} C_{mn}\left[P_{mn}^{(2)}(R_2)b_{nl}\right] - \sum_{n=0}^{M} D_{mn}\left[Q_{mn}^{(2)}(R_2)b_{nl}\right] = e_{0l} \tag{64}$$

$$\sum_{n=0}^{M} B_{mn}\left[E\varepsilon_m i^m \frac{dP_{mn}^{(1)}(R_3)}{dr}d_{nl}\right] - \sum_{n=0}^{M} C_{mn}\left[\frac{dP_{mn}^{(2)}(R_3)}{dr}c_{nl}\right] - \sum_{n=0}^{M} D_{mn}\left[\frac{dQ_{mn}^{(2)}(R_3)}{dr}c_{nl}\right]$$
$$= -E\varepsilon_m i^m k\frac{dJ_m(kR_3)}{dr}d_{0l} \tag{65}$$

$$\sum_{n=0}^{M} A_{mn}\left[\frac{dP_{mn}^{(3)}(R_2)}{dr}d_{nl}\right] - \sum_{n=0}^{M} C_{mn}\left[\frac{dP_{mn}^{(2)}(R_2)}{dr}c_{nl}\right] - \sum_{n=0}^{M} D_{mn}\left[\frac{dQ_{mn}^{(2)}(R_2)}{dr}c_{nl}\right] = 0 \tag{66}$$

with

$$a_{nl} = \int_{-h}^{-d} Z_n^{(3)} Z_l^{(2)} dz = \int_{-h}^{-d} Z_n^{(1)} Z_l^{(2)} dz; b_{nl} = \int_{-h}^{-d} Z_n^{(1)} Z_l^{(1)} dz = \int_{-h}^{-d} Z_n^{(3)} Z_l^{(3)} dz;$$
$$c_{nl} = \int_{-h}^{-d} Z_n^{(2)} Z_l^{(3)} dz = \int_{-h}^{-d} Z_n^{(2)} Z_l^{(1)} dz; d_{nl} = \int_{-h}^{0} Z_n^{(3)} Z_l^{(3)} dz = \int_{-h}^{0} Z_n^{(1)} Z_l^{(1)} dz;$$
$$e_{0l} = \int_{-h}^{-d} i\sigma_R P_0/\rho\omega Z_l^{(2)} dz; \tag{67}$$

By letting $m = 0 \sim M$, $l = 0 \sim M$, and $n = 0 \sim M$ in the eigen-function expansions terms, $(M + 1)$ systems of equations can be obtained. For a specific m, the linear equations could be assembled as

$$[G(m)]_{[4\times(M+1)]\times[4\times(M+1)]}\mathbf{X} = \mathbf{T}, \tag{68}$$

with

$$\mathbf{X} = [A_{mn}, B_{mn}, C_{mn}, D_{mn}]^T \text{ for } n = 0, 1, \ldots, M \tag{69}$$

Once the coefficients A_{mn}, B_{mn}, C_{mn}, and D_{mn} are known, the velocity potential in each domain can be obtained based on Equations (52)–(54). The complex amplitude of volume flux can be calculated by the free-surface integration

$$q_D = \iint_{S_i} \frac{\partial\phi_3}{\partial z} ds \tag{70}$$

for the wave diffraction, and

$$q_R = \iint_{S_i} \frac{\partial\phi_3}{\partial z} ds \tag{71}$$

for the wave radiation.

Then, the wave loads on the OWC-WEC are given by:

$$f = \int_{S_B} pn dS \tag{72}$$

$$M_y = \int_{S_B} p[(z - z_0)n_x - (x - x_0)n_z]dS \tag{73}$$

in which $n = (n_x, n_y, n_z)$ is the unit normal vector on the body surface, (x_0, z_0) is the rotational center, $f = (f_x, f_y, f_z)$, and p can be calculated from the linearized Bernoulli's equation as

$$p = -i\omega\rho\phi \tag{74}$$

The wave loads on the beam shown in Figure 1 can be obtained if S_B is replaced with the surface chamber shell in Equations (72) and (73).

3. Results and Discussions

3.1. Convergence and Validation

The convergence of the results with respect to the truncation number M in Equations (63)–(66) is firstly tested. The water depth h, gravitational acceleration g, and water density ρ are used as the base of non-dimensionalization in the following sections. In the first case, the parameters are set as $R_1/h = 0.5$, $R_2/h = 1.0$, $R_3/h = 1.50$, $d/h = 0.5$, and $kh = 1.0$. Figure 3 shows the obtained non-dimensional value of added mass B, damping coefficient C, and volume flux Q_D with different M. It can be seen that the numerical convergence can be obtained as $M \geq 20$. Similar convergence can also be realized for other cases with the same value of M.

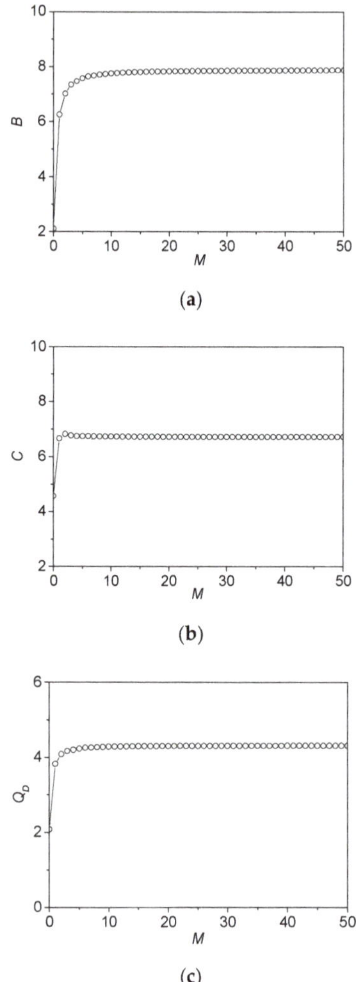

(a)

(b)

(c)

Figure 3. Convergence study with respect to parameter M, for (**a**) B, (**b**) C, and (**c**) Q_D.

Then, the validation of the present solution is examined. A higher-order BEM (boundary element method) [24] was applied to solve the diffraction problem with the same parameters as above. The rotation center is fixed at the bottom of the cylinder. Figure 4 shows the comparisons of wave forces in

x and z directions and moment about axis y obtained by the present method and BEM, respectively. From these comparisons, the numerical results given by two methods are in good agreement, which verifies the accuracy of the present analytical model.

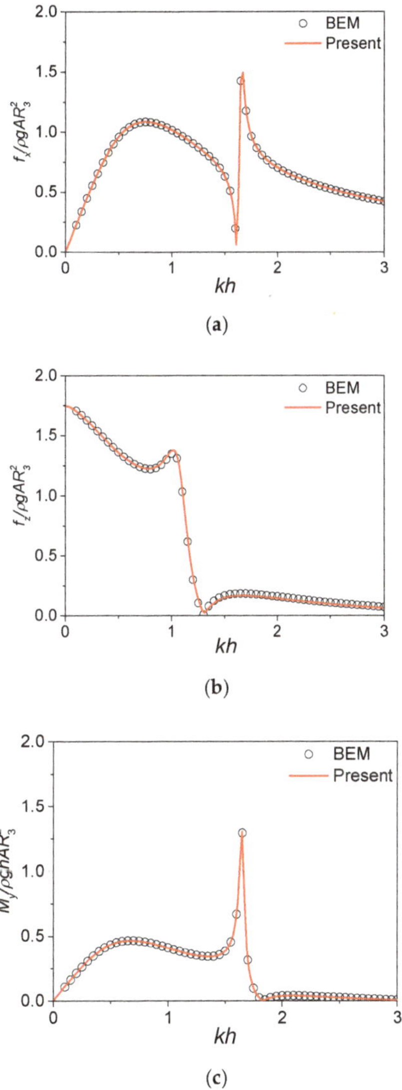

Figure 4. Comparison of hydrodynamic force amplitudes by the present model and the BEM: (**a**) wave force in x direction, (**b**) wave force in z direction, and (**c**) wave moment about axis y.

3.2. Determination of Turbine Rotate Speed

Effect of turbine rotate speed on the conversion efficiency is studied firstly. Martins-Rivas and Mei [18] simulated one OWC-WEC property in Pico Island, Azores, Portugal with N = 2000 r.p.m., K = 0.45 (for one turbine). As a preliminary study, the same sea depth and power take-off system were selected to simulate the present OWC-WEC. Figure 5 shows the efficiency distribution of the OWC-WEC, with different turbine rotate speeds N = 100 (r.m.p), N = 200 (r.m.p), N = 500 (r.m.p), N = 1000 (r.m.p), and N = 2000 (r.m.p). The remaining geometrical parameters are kept as R_1/h = 0.15, R_2/h = 0.50, R_3/h = 0.51, d/h = 0.2, K = 0.45, $V_0 = \pi R_2^2 h$ A/h = 0.1, and h = 10 m. It is found that there is one optimal conversion frequency for different turbine rotation speeds. Additionally, the optimal frequency shifts towards a lower frequency with the decrease of the turbine rotate speed. It is clear that the case of N = 200 (r.m.p) is consistent with the results obtained in Martins-Rivas and Mei's model. This is the result of the difference between the present and Martins-Rivas and Mei's models. Additionally, based on the semi-empirical Pierson-Moscowitz (P-M) energy density spectrum in 10 m sea depth, the wave frequency is in range of $1 \leq kh \leq 4$. Hence, the case of N = 200 (r.m.p) can be used to simulate the wave energy conversion in the present OWC-WEC.

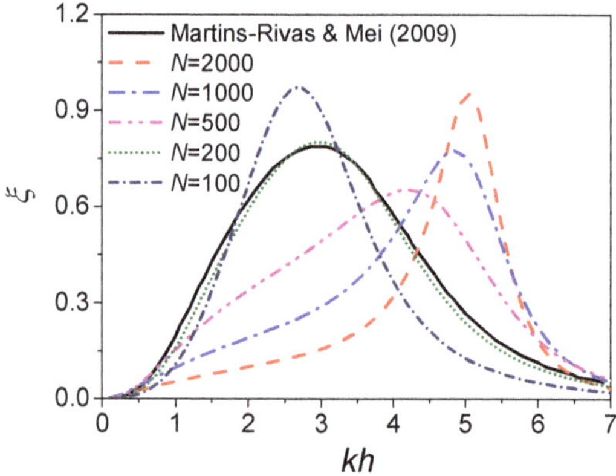

Figure 5. Distribution of hydrodynamic efficiency for different turbine rotational speed N on the efficiency ξ.

3.3. Wave Motion Inside the OWC Chamber

To study the wave phenomena inside the OWC chamber, an example is considered with the parameters of R_1/h = 0.15, R_2/h = 0.35, R_3/h = 0.4, d/h = 0.2, K = 0.45, N = 200 r.p.m, $V_0 = \pi R_2^2 h$, A/h = 0.1, and h = 10 m. The incident wave propagates in x-direction. Figure 6 shows the variation of the wave surface amplitude at two points $U_1(-0.20, 0.0)$ and $U_2(-0.30, 0.0)$ inside the chamber versus the dimensionless wave number kh. The vertical coordinate is normalized by the incident wave amplitude A. From the figure, it can be seen that both surface amplitudes at different positions vary in a similar trend with the wave frequency. In the low frequency region ($kh < 1.5$), the dimensionless surface amplitudes inside the chamber are near to 1, which is due to the fact that the chamber size is much smaller than the wave length, and the relating wave effects can be ignored. Then, three peaks occur at kh = 2.83, kh = 4.68, and kh = 8.15 with the increase of the wave frequency. W1, W2, and W3 are used to denote these three situations for convenience. It is also noticed that, compared with condition W1, the dimensionless surface amplitudes at the other two conditions show sharper peaks with singular feature.

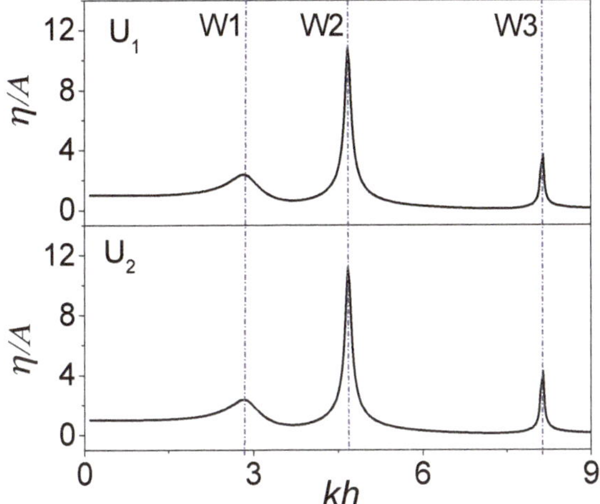

Figure 6. Variations of dimensionless surface amplitudes at two positions with *kh*.

Contours of surface oscillation amplitude inside the chamber are plotted in Figure 7 at situations W1, W2, and W3, respectively. Note that all amplitudes are normalized by the incident wave height *A*. From Figure 7, it can be seen that the wave amplitudes of the free surface are almost the same with each another at W1, which leads to a piston-type water oscillation in the chamber. However, the distributions of the free surface amplitudes at W2 and W3 are nearly symmetrical about $x = 0$ or $y = 0$. The phenomena in Figure 7b,c are in sloshing modes occurring at circumferential direction with order $m = 1$ and $m = 2$ in Equation (43), respectively.

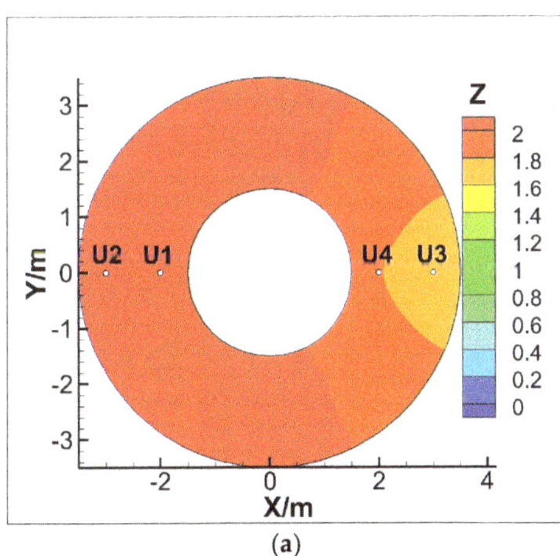

(a)

Figure 7. *Cont.*

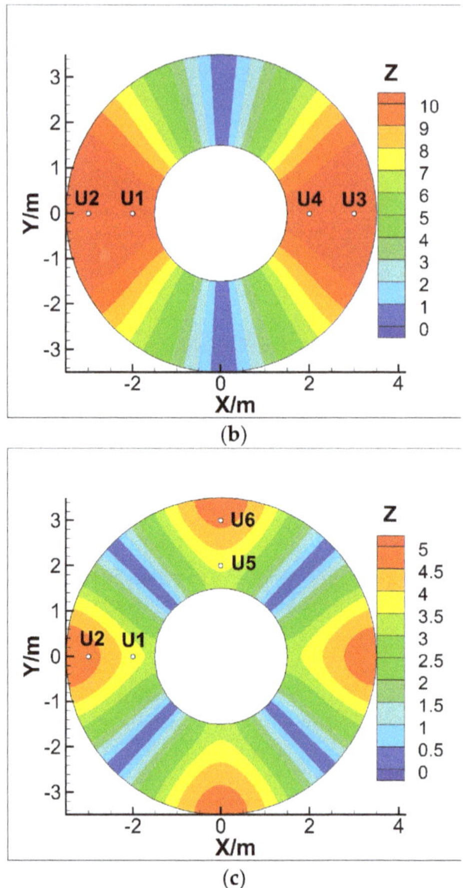

Figure 7. Contours of surface amplitudes in the chamber at (**a**) W1, (**b**) W2, and (**c**) W3 conditions.

To further analyze water motion inside the chamber, another four points $U_3(0.3,0)$, $U_4(0.2,0)$, $U_5(0,0.2)$, and $U_6(0,0.3)$ are considered as shown in Figure 7. Time series of the water elevations at these points are shown in Figure 8. From Figure 8a, it can be seen that the wave elevations at these four points in the chamber are always in phase and with the similar amplitude, as a feature of the piston-type oscillation. It further proves that the water inside the chamber at condition W1 oscillates as a piston, which can compress and expand the chamber air effectively. Strictly speaking, the wave amplitudes at these positions are not exactly same. This is due to the fact that a small-amplitude sloshing also occurs in the chamber at the same time, even though the piston motion is absolutely dominant at W1. For cases of W2 and W3, the free surface motions in the chamber are either centrosymmetric or antisymmetric about the plane $x = 0$ in Figure 8b,c. This means that, except for W1, there is no evident variation of the total air volume in the chamber. Figure 9 presents the distribution of power extraction efficiency ζ with dimensionless wave number kh. It can be seen that the extraction efficiency ζ reaches the peak at $kh = 2.83$, which just coincides with condition W1, i.e., the piston-type mode. Hence, it is the piston-type mode that makes the main contribution to the wave energy conversion.

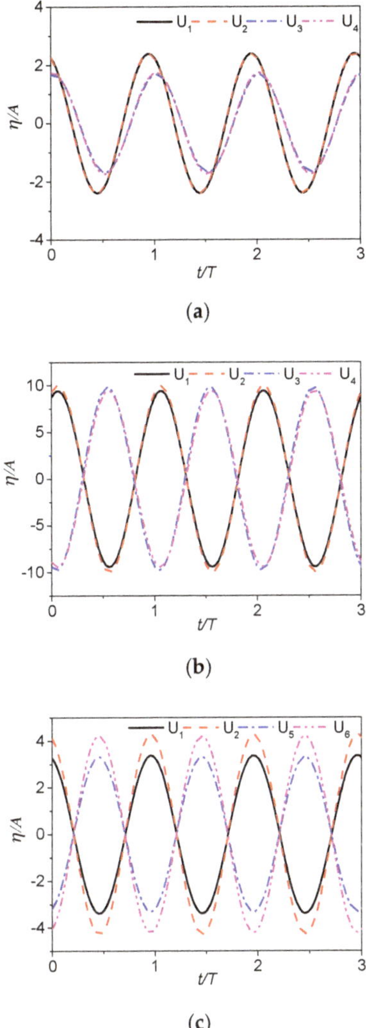

Figure 8. Time series of the free-surface elevation at test positions in the chamber at conditions (**a**) W1, (**b**) W2, and (**c**) W3.

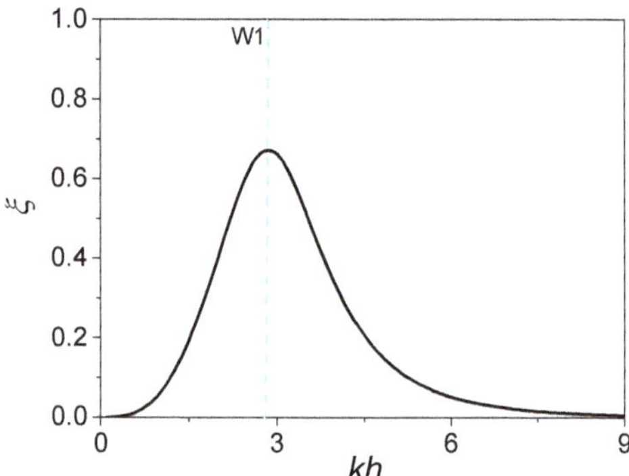

Figure 9. Distribution of extraction efficiency ζ with *kh*.

3.4. Effects of Geometry Parameters on Wave Loads and Extraction Efficiency

Both survivability and efficiency are important factors for a WEC. The effects of the OWC geometrical parameters on them are investigated in this section, which can be used as guidance for OWC optimization. To be specific, the chamber draft d, chamber breadth $b = R_2 - R_1$, and thickness of the chamber wall $d_{wall} = R_3 - R_2$ are taken into account. As the preliminary study, the main parameters are selected based on engineering practice to be with $D/h = 0.3$, $K = 0.45$, $N = 200$ r.p.m, $V_0 = \pi R_2^2 D$, $A/h = 0.1$, and $h = 10$ m.

Lateral wave force and bending moment are critical parameters for survivability and reliability in ocean engineering. Figure 10a,b shows the wave force f_x and bending moment M_y about sea bed on the whole OWC device with four chamber drafts, i.e., $d/h = 0.10$, $d/h = 0.20$, $d/h = 0.30$, and $d/h = 0.40$. The chamber breadth $b/h = 0.2$ and chamber wall thickness $d_{wall}/h = 0.05$ are fixed. Figure 10c shows the wave force $f_{x\text{beam}}$ of the chamber shell, and Figure 10d shows the chamber shell moment $M_{y\text{beam}}$ about the connection beam between the cylinder and the chamber. The beam is assumed above the water surface $D/h = 0.30$. In Figure 10a,b, two resonant peaks can be found within the considered range of $0 \leq kh \leq 6$ for both the wave force and moment. It is the first peak being of engineering interest. From the figure, it can be seen that as d/h change from 0.10 to 0.40, the first peaks of both the wave force f_x and moment M_y increase greatly due to the increase of the wet surface acting on the OWC-WEC. Hence, lower chamber draft is preferred to decrease the wave loads. In the high frequency domain, another resonant peak occurs. In Figure 10a, for the case of $d/h = 0.2$, the second peak occurs at $kh = 4.68$, which is exactly corresponding to condition W2 as shown in Figure 6. At $kh = 4.68$, the wave length $L/h = 0.67$, which is close to the chamber inner diameter $2R_2 = 0.70$. The incident waves are captured in $m = 1$ mode. Therefore, it can be concluded that the resonant sloshing mode at order $m = 1$ in Equation (43) induces an extreme wave load on the device at the second peak. The potential theory without considering the fluid viscidity leads to an over-prediction of the wave force at resonance. However, in real fluids, the wave loads at the W2 sloshing mode cannot be so large due to the viscous dissipation. Such peak wave force acting on the OWC-WEC increases sharply with the chamber draught. Additionally, the resonant sloshing frequency shifts towards a lower frequency. In other words, the sloshing resonance becomes much stronger with the increase of the chamber draught [16]. In Figure 10c,d, the beam force $f_{x\text{beam}}$ and bending moment $M_{y\text{beam}}$ vary with the same trend in Figure 10a,b. Compared with f_x and M_y, wave loads of the beam at the first peak decrease slightly. However, the resonant response at $m = 1$ mode is much higher than that in Figure 10a,b. Hence, in present OWC-WEC, lower chamber draught leads to higher survivability.

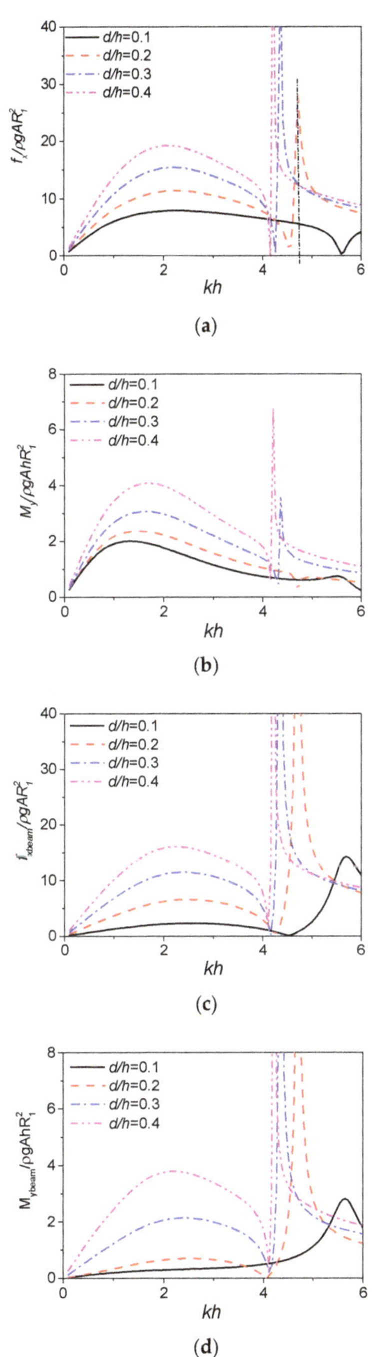

Figure 10. Effect of chamber draft on the hydrodynamic loads for $d/h = 0.10$, $d/h = 0.20$, $d/h = 0.30$, and $d/h = 0.40$. (**a**) Wave force f_x, (**b**) wave moment M_y, (**c**) wave force $f_{x\text{beam}}$, and (**d**) wave moment $M_{y\text{beam}}$.

Further, the effects of the chamber draught on power extraction efficiency are presented in Figure 11. A dash line at $\zeta = 0.3$ is marked in the figure to depict the effective frequency bandwidth (i.e., the part for $\zeta \geq 0.3$). The results indicate that the effective frequency bandwidth decreases largely with the reflected by a larger-draught. An explanation to such a phenomenon is that, in the high-frequency zone, the corresponding short waves with lower penetrability can be easily reflected by larger-draught chamber. From the figure, it also can be observed that the resonant frequency and maximal efficiency go up with the decrease of chamber draught. In other words, the proposed OWC device with a smaller chamber draft tends to possess a better capacity for wave energy conversion.

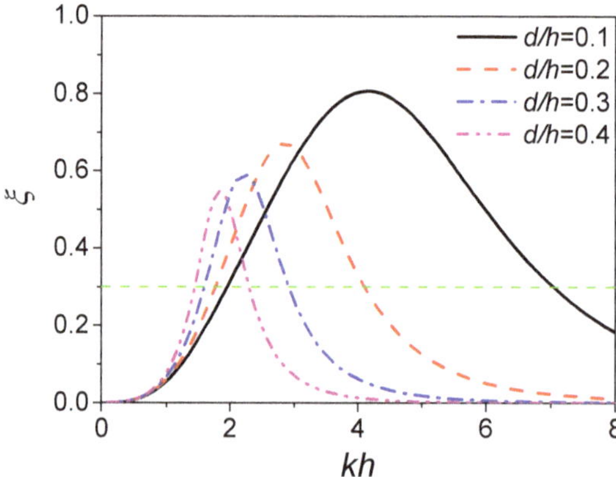

Figure 11. Effect of chamber draft on power extraction efficiency ζ for $d/h = 0.10$, $d/h = 0.20$, $d/h = 0.30$, and $d/h = 0.40$.

Figure 12 shows the effects of the chamber breadth on the hydrodynamic loads with four chamber breadths $b/h = 0.10$, $b/h = 0.20$, $b/h = 0.30$, and $b/h = 0.40$. The drafts and wall thickness of the chamber are set as constants $d/h = 0.20$ and $d_{wall}/h = 0.05$. In Figure 12a,b, there are also two resonant peaks shown in the considered range of kh for both the wave force and moment. As the chamber breadth increases, the wave moment M_y decreases greatly at the first resonant frequency. However, the wave force f_x on the device increases with the chamber breadth due to the increase of the chamber wall surface in domain Ω_1 leading to a larger wave force. In the high frequency domain, as b/h increases, the second resonant peak shifts towards a lower frequency and increases its magnitude slightly. For wider chamber breadth, the resonance of the sloshing model at order $m = 1$ occurs at the longer wave area. Figure 12c,d shows the wave loads of the connection beam with different chamber breadth. It also can be seen that the beam force $f_{x\text{beam}}$ and bending moment $M_{y\text{beam}}$ at the $m = 1$ mode are exceptionally high. Additionally, compared with f_x and M_y, the frequency bandwidth in W1 sloshing mode in Figure 12c,d increases with the chamber breadth; this is especially apparent for the bending moment $M_{y\text{beam}}$. Hence, for wider chamber breadth, the bending moment of the whole OWC-WEC decreases, even though the resonant loads of the connection beam increase greatly. Figure 13 shows the power extraction efficiency with different chamber breadths. It can be seen that the effective frequency bandwidth increases sharply with the chamber breadth for capturing longer waves. The reason for this is that the inertia of the water column in the chamber increases with chamber width. Thus, long wave energy captured can make a greater contribution to the piston-type oscillation in the chamber. It is also found that the resonant frequency decreases rapidly with chamber breadth. In other words, the proposed OWC with a wider chamber tends to possess better efficiency.

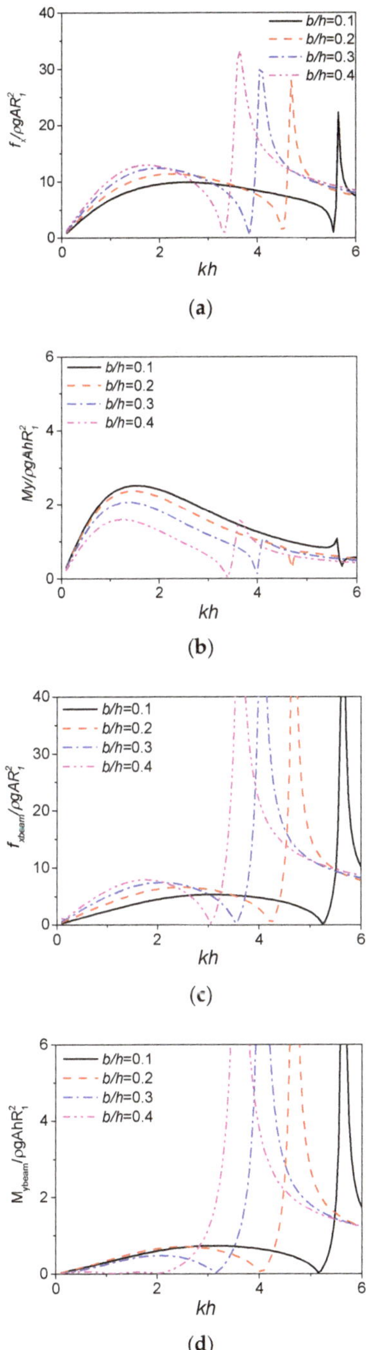

Figure 12. Effect of the chamber breadth on the wave loads for $b/h = 0.10$, $b/h = 0.20$, $b/h = 0.30$, and $b/h = 0.40$: (**a**) wave force f_x, (**b**) wave moment M_y, (**c**) wave force $f_{x\text{beam}}$, and (**d**) wave moment $M_{y\text{beam}}$.

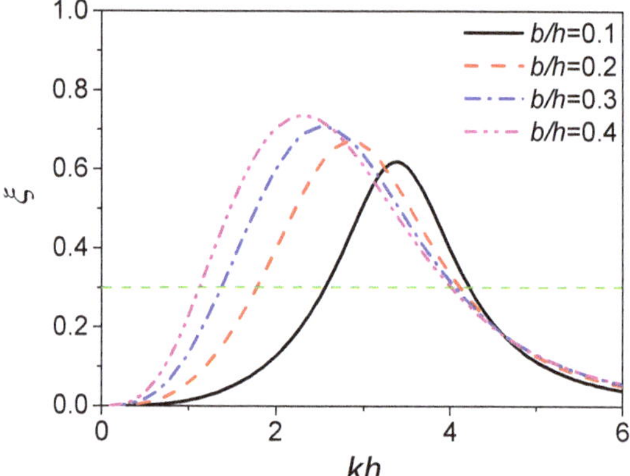

Figure 13. Effect of the chamber breadth on the energy extraction efficiency for $b/h = 0.10$, $b/h = 0.20$, $b/h = 0.30$, and $b/h = 0.40$.

Finally, the effects of the chamber wall thickness on the hydrodynamic loads and extraction efficiency are presented in Figures 14 and 15, respectively. Four different chamber wall thicknesses, i.e., $d_{wall}/h = 0.01$, $d_{wall}/h = 0.05$, $d_{wall}/h = 0.10$, and $d_{wall}/h = 0.15$, are considered. The other parameters are set as constant $d/h = b/h = 0.20$. It can be seen that the chamber wall thickness has little impact on the second resonant frequency, which is mainly decided by the chamber breadth. With the increase of the chamber wall thickness, at the first resonant peak, the wave force f_x increases, but the bending moment M_y decreases. An explanation to the latter is that, while increasing the chamber wall thickness, the wet surface of the proposed device in domain Ω_2 increases, which can reduce the bending moment M_y about the sea bed. Figure 15 shows the variation of extraction efficiency with four different thicknesses of the chamber wall. The results indicate that the extraction efficiency ζ and effective frequency bandwidth decrease as the wall thickness grows. However, compared with the results shown in Figures 11 and 13, the effect of the wall thickness on the resonant frequency is weaker. Therefore, for engineering considerations, the chamber wall thickness can be designed mainly according to extreme environmental loads and material strength.

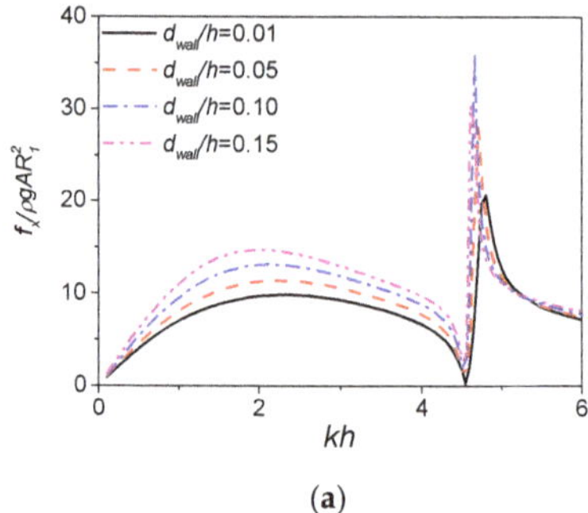

(a)

Figure 14. *Cont.*

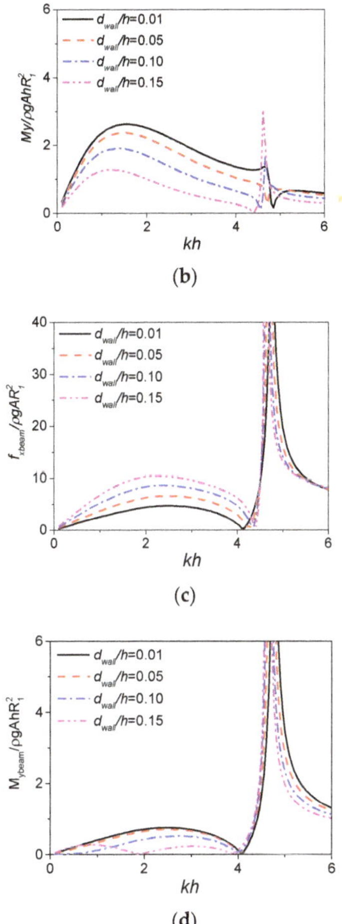

(b)

(c)

(d)

Figure 14. Effect of chamber wall thickness on the wave loads, for $d_{wall}/h = 0.01$, $d_{wall}/h = 0.05$, $d_{wall}/h = 0.10$, and $d_{wall}/h = 0.15$: (**a**) wave force f_x, (**b**) wave moment M_y, (**c**) wave force $f_{x\text{beam}}$, and (**d**) wave moment $M_{y\text{beam}}$.

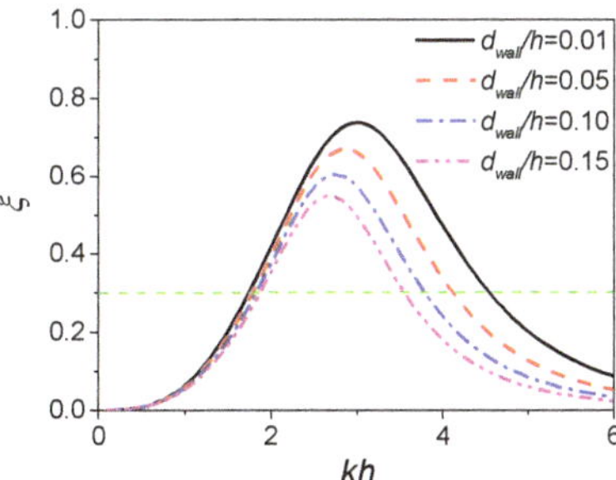

Figure 15. Effect of the chamber wall thickness on the energy conversion efficiency, for $d_{wall}/h = 0.01$, $d_{wall}/h = 0.05$, $d_{wall}/h = 0.10$, and $d_{wall}/h = 0.15$.

4. Conclusions

This research has provided an easy-to-deploy OWC-WEC design with good performance. This OWC-WEC is featured with an excellent adaption to offshore sites with multiple wave directions. The overall costs are expected to be reduced when a large-scale deployment is achieved. The structure has a mushroom appearance, which can be considered as an upside down cylindrical bucket supported by a column fixed on the seabed. A concentric cylindrical chamber is formed. The energy is transformed from the wave to the air by an oscillating water column moving up and down in the chamber. As the bidirectional airflow flows through the orifice on the top of the chamber, the turbine drives an electricity generator. Compared with the existing shoreline OWC devices, this wave energy device is not restricted by the wave directions and coastline geography conditions.

Then, analytical solutions are derived based on the linear potential-flow theory to investigate hydrodynamic properties of the wave energy converter. The wave motion inside the chamber and the effect of geometry parameters on the hydrodynamic loads and extraction efficiency are investigated. Three typical free-surface oscillation modes in the chambers are found and analyzed. The piston-type mode, which can compress the chamber air efficiently, makes the largest contribution to the energy conversion efficiency. Further, it is found that, as the chamber draft decreases, the resonant frequency and peak efficiency increase apparently, and the wave loads on the whole OWC-WEC also decrease greatly. Additionally, for the design of the connection beam, the wave loads at m = 1 mode should be considered, especially the OWC-WEC with wider chamber breadth. For wider effective frequency bandwidth, the OWC shell can be selected with lower draft and wider chamber breadth. Compared with the chamber draft and breadth, the chamber wall thickness has a relatively small impact on the wave energy conversion, which can be designed by environmental loads and material strength in engineering. In the next stage, a series of model experiments (e.g., similar to those in Viviano et al. [25]) will be carried out for further investigation based on the present analytical solutions.

Acknowledgments: This study is supported by the National Science Foundation of China (Grant Nos. 51679036, 51490672, and 51709038) and the Royal Academy of Engineering under the UK-China Industry Academia Partnership Programme (Grant No. UK-CIAPP\73).

Author Contributions: Yu Zhou, Chongwei Zhang and Dezhi Ning drafted the work, made substantial contributions to the design of the work, gave the final approval of the version to be published, and agreed to be accountable for all aspects of the work in ensuring that questions related to the accuracy or integrity of any part of the work are appropriately investigated and resolved.

Conflicts of Interest: The authors declare no conflict of interest.

References

1. Falcão, A.F.O. Wave energy utilization: A review of the technologies. *Renew. Sust. Energy Rev.* **2010**, *14*, 899–918. [CrossRef]
2. Borthwick, A.G. Marine Renewable Energy Seascape. *Engineering* **2016**, *2*, 69–78. [CrossRef]
3. Henderson, R. Design, simulation, and testing of a novel hydraulic power take-off system for the Pelamis wave energy converter. *Renew. Energy* **2006**, *31*, 271–283. [CrossRef]
4. He, F.; Huang, Z.; Law, W.K. An experimental study of a floating breakwater with asymmetric pneumatic chambers for wave energy extraction. *Appl. Energy* **2013**, *106*, 222–231. [CrossRef]
5. Monk, K.; Zou, Q.; Conley, D. An approximate solution for the wave energy shadow in the lee of an array of overtopping type wave energy converters. *Coast. Eng.* **2013**, *73*, 115–132. [CrossRef]
6. Falcão, A.F.O.; Justino, P.A.P. OWC wave energy devices with air flow control. *Ocean Eng.* **1999**, *26*, 1275–1295. [CrossRef]
7. Heath, T.V. A review of oscillating water columns. *Philos. Trans.* **2012**, *370*, 235–245. [CrossRef] [PubMed]
8. Falcão, A.F.O.; Henriques, J.C.C. Oscillating-water-column wave energy converters and air turbines: A review. *Renew. Energy* **2016**, *85*, 1391–1424. [CrossRef]
9. Masuda, Y.; Mccormick, M.E. Experiences in pneumatic wave energy conversion in Japan. In *Utilization of Ocean Waves—Wave to Energy Conversion*; ASCE Library: Reston, VA, USA, 2015; pp. 1–33.

10. Zhang, D.; Li, W.; Lin, Y. Wave energy in China: Current status and perspectives. *Renew. Energy* **2009**, *34*, 2089–2092. [CrossRef]
11. Boccotti, P. Comparison between a U-OWC and a conventional OWC. *Ocean Eng.* **2007**, *34*, 799–805. [CrossRef]
12. Naty, S.; Viviano, A.; Foti, E. Wave Energy Exploitation System Integrated in the Coastal Structure of a Mediterranean Port. *Sustainability* **2016**, *8*, 12. [CrossRef]
13. Bingham, H.B.; Ducasse, D.; Nielsen, K.; Read, R. Hydrodynamic analysis of oscillating water column wave energy devices. *J. Ocean Eng. Mar. Energy* **2015**, *1*, 405–419. [CrossRef]
14. Ogata, T.; Washio, Y.; Osawa, H.; Tsuritani, Y.; Yamashita, S.; Nagata, Y. The Open Sea Tests of the Offshore Floating Type Wave Power Device "Mighty Whale": Performance of the Prototype. *Oceans* **2002**, *4*, 1860–1866.
15. Alcorn, R.; Blavette, A.; Healy, M.; Lewis, A. FP7 EU funded CORES wave energy project: A coordinators' perspective on the Galway bay sea trials. *Underw. Technol.* **2016**, *32*, 51–59. [CrossRef]
16. Garrett, C.J.R. Bottomless harbours. *J. Fluid Mech.* **2006**, *43*, 433–449. [CrossRef]
17. Sarmento, A.J.N.A.; Falcão, A.F.O. Wave generation by an oscillating surface-pressure and its application in wave-energy extraction. *J. Fluid Mech.* **1985**, *150*, 467–485. [CrossRef]
18. Martins-Rivas, H.; Mei, C.C. Wave power extraction from an oscillating water column at the tip of a breakwater. *J. Fluid Mech.* **2009**, *626*, 395–414. [CrossRef]
19. Martins-Rivas, H.; Mei, C.C. Wave power extraction from an oscillating water column along a straight coast. *Ocean Eng.* **2009**, *36*, 426–433. [CrossRef]
20. Zhu, S.P.; Mitchell, L. Diffraction of ocean waves around a hollow cylindrical shell structure. *Wave Motion* **2009**, *46*, 78–88. [CrossRef]
21. Konispoliatis, D.N.; Mazarakos, T.P.; Mavrakos, S.A. Hydrodynamic analysis of three–unit arrays of floating annular oscillating–water–column wave energy converters. *Appl. Ocean Res.* **2016**, *61*, 42–64. [CrossRef]
22. Falcão, A.F.O. Wave-power absorption by a periodic linear array of oscillating water columns. *Ocean Eng.* **2002**, *29*, 1163–1186. [CrossRef]
23. Evans, D.V. Wave-power absorption by systems of oscillating surface pressure distributions. *J. Fluid Mech.* **1982**, *114*, 481–499. [CrossRef]
24. Teng, B.; Taylor, R.E. New higher-order boundary element methods for wave diffraction/radiation. *Appl. Ocean Res.* **1995**, *17*, 71–77. [CrossRef]
25. Viviano, A.; Naty, S.; Foti, E.; Bruce, T.; Allsop, W.; Vicinanza, D. Large-scale experiments on the behaviour of a generalised Oscillating Water Column under random waves. *Renew. Energy* **2016**, *99*, 875–887. [CrossRef]

Article

Arrays of Point-Absorbing Wave Energy Converters in Short-Crested Irregular Waves

Malin Göteman [1,*], Cameron McNatt [2], Marianna Giassi [1], Jens Engström [1] and Jan Isberg [1]

[1] Department of Engineering Sciences, Uppsala University, 75105 Uppsala, Sweden; marianna.giassi@angstrom.uu.se (M.G.); jens.engstrom@angstrom.uu.se (J.E.); jan.isberg@angstrom.uu.se (J.I.)

[2] Mocean Energy, Edinburgh EH9 3BF, UK; cameron.mcnatt@moceanenergy.com

* Correspondence: malin.goteman@angstrom.uu.se; Tel.: +46-18-471 5875

Received: 22 March 2018; Accepted: 13 April 2018; Published: 17 April 2018

Abstract: For most wave energy technology concepts, large-scale electricity production and cost-efficiency require that the devices are installed together in parks. The hydrodynamical interactions between the devices will affect the total performance of the park, and the optimization of the park layout and other park design parameters is a topic of active research. Most studies have considered wave energy parks in long-crested, unidirectional waves. However, real ocean waves can be short-crested, with waves propagating simultaneously in several directions, and some studies have indicated that the wave energy park performance might change in short-crested waves. Here, theory for short-crested waves is integrated in an analytical multiple scattering method, and used to evaluate wave energy park performance in irregular, short-crested waves with different number of wave directions and directional spreading parameters. The results show that the energy absorption is comparable to the situation in long-crested waves, but that the power fluctuations are significantly lower.

Keywords: wave energy; short-crested waves; multidirectional; arrays; parks; multiple scattering; power fluctuations

---

1. Introduction

Ocean waves provide a clean, renewable energy source with a large potential to contribute to the energy demand without negative environmental or climate impact. There is a large number of different technology approaches for conversion of wave energy to electricity, and very few have reached a commercial maturity. Common for many of the technologies is that a large-scale electricity production requires that many wave energy converters (WECs) are deployed together in arrays, or parks. In particular, this is true for the point-absorber concept considered in this study.

Since the devices in the park will interact hydrodynamically by scattered and radiated waves spreading throughout the park, it is of importance to determine the optimal park layout that achieves maximum electricity production with minimum power fluctuations and costs. Since the early works on wave energy, studying and optimizing wave energy array layouts have been main topics of interest [1–3], and remains an active area of research today [4–9].

Many parameters affect the performance of the park. The impact of increasing the number of devices in a park on both the energy absorption and power fluctuations was investigated in [10–14], and it was seen that increasing the number of WECs by around 30% may reduce the power fluctuations by roughly 7% and the average power of each device by 3% [14], and, in experiments, it was seen that, in an array with 24 WECs, up to 26% of the energy yield from an equivalent number of isolated WECs may be lost due to interference effects [13]. The effect of the individual device dimensions was studied in [14–16], and it was seen that the total performance of the park may increase if WECs of different dimensions are deployed together. The separation distance between devices and the layout of the park has been studied in a large number of works [11,13,17–21], and separation distances ranging from $4R$, where R is the radius of the float, up to a few hundred metres, have been established, above which the interaction effects between devices can be neglected. Obviously, the wave climate, the incident wave directions and the bathymetry are other factors that may affect the performance of the park, which has been studied in [14,22,23], among others. Most of the work has been based on numerical or analytical modelling, but a few examples of experimental studies exist, both in wave tank [13,24] and offshore [12].

Much of the mentioned works on optimal configurations of wave energy parks have considered only incident regular waves, and even if irregular waves have been considered, almost all have considered only long-crested, unidirectional waves. However, real ocean waves can be multidirectional, or short-crested, with irregular waves travelling in different directions simultaneously, and this is likely to affect the performance of the wave energy parks. An array of 12 oscillating wave surge converters was studied in short-crested waves in [7] and it was found that the average absorbed energy of the park was slightly lower than when it was operating in unidirectional waves. A similar conclusion was found in [25] for attenuator type WECs, and it was also found that the relative pitch motions were reduced in short-crested waves. Wave run-up on bottom-mounted cylinders was found to increase in short-crested waves studied with semi-analytical methods in [26], and the presence of near-trapped modes was reduced. The wake effect behind WEC arrays was studied in [17,27], and both papers found that the wake was reduced when directional wave spreading was taken into account. This effect and other wave energy array effects in short-crested waves was also investigated experimentally in [24].

From the abovementioned studies, it is clear that the performance of wave energy parks might change when operating in short-crested waves as opposed to in unidirectional, long-crested waves. In [28], a semi-analytical model for computing the hydrodynamical forces and interactions in a wave energy park of point-absorbing WECs was presented, based on the approximate method in [21]. An interaction distance cut-off was introduced, which enabled accurate and fast modelling of large parks with over 100 devices. The method was extended to enable point-absorber devices of different individual dimensions in [16], and has been coupled with a genetic algorithm for multiple parameter optimization of wave energy parks in [29]. The approach provides a fast and reliable method to assess and optimize all involved parameters in a park, including park layout and individual WEC dimensions. In the present paper, the method is extended to also describe short-crested waves, and used to study the performance of wave energy parks in more realistic seas.

The paper is organized as follows. The theory of multiple scattering with short-crested waves is described in Sections 2.1 and 2.2, with the notation and equations of motion for wave energy parks established in Section 2.1 and the multiple scattering method in Section 2.2. The resulting method is used to study short-crested waves and wave energy park performance and discussed in Sections 3.1–3.3. Conclusions from the study are presented in Section 4.

2. Method

2.1. Wave Energy Park Model

Consider an array of N point-absorbing wave energy converters, each consisting of a surface buoy with radius R^i and draft d^i connected to a linear generator at the seabed. The generator consists of a translator moving vertically within a stator, and is characterized by a generator damping Γ^i. The WEC model is based on the wave energy technology developed at Uppsala University, Sweden. The approach for the wave energy concept is based on simplicity to improve the life-expectancy of the device and reduce capital and maintenance costs—the generator consists of as few moving parts as possible, and is situated at the seabed to be protected from storms or other extreme wave conditions. By changing the buoy and generator dimensions, as well as the connection line length, the WEC can be adapted to different wave climates and deployment site conditions. Simulated and experimentally measured performance in different wave conditions was presented for earlier prototypes of the device in [30,31], and more information about the device can be found in [10,12].

The water depth is h and the density of the water is ρ. The coordinate system is chosen such that $z = 0$ at the still water surface and $z = -h$ at the seabed. Local cylindrical coordinate systems (r, θ, z) are defined at the origin (x^i, y^i, z) of each buoy. An illustration of the wave energy park is shown in Figure 1.

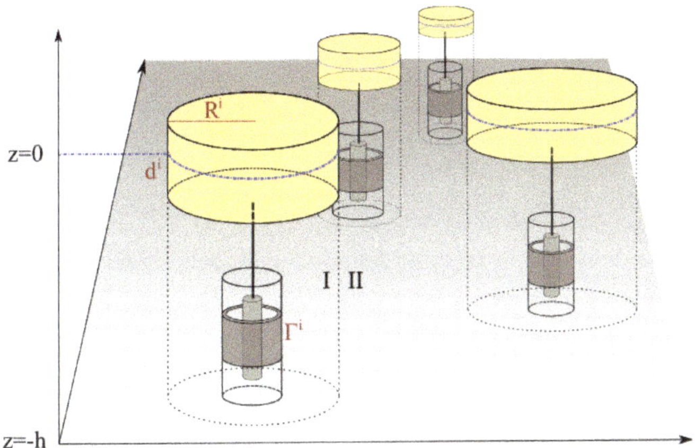

Figure 1. Illustration of the fluid domain of the wave energy park. Each wave energy converter is characterized by an individual radius R^i and draft d^i for the float and a power take-off damping Γ^i of the generator. The fluid domain is divided into interior domains I for each float and exterior domain II.

The motion of the buoys and the translator are given by the coupled equations of motion

$$m_b^i \ddot{\vec{x}}_b^i(t) = -\iint_{S^i} p(t)\vec{n}\,dS - \vec{F}_{\text{line}}^i(t) - m_b^i g\hat{z}, \tag{1}$$

$$m_t^i \ddot{z}_t^i(t) = \vec{F}_{\text{line}}^i(t) + \vec{F}_{\text{PTO}}^i(t) - m_t^i g\hat{z}, \tag{2}$$

where m_b^i and m_t^i are the mass of the buoy and translator, respectively, p contains both the dynamical and hydrostatical pressure, $\vec{F}_{\text{line}}^i$ is the line in the force connecting the buoy and the translator, and $\vec{F}_{\text{PTO}}^i$ is the damping force of the generator. The normal vector $\vec{n}$ is defined to point away from the body surface, into the fluid. The hydrodynamical forces will be solved using linear potential flow theory, described in more detail in Section 2.2.1. Under the constraints that the connection between the buoy and translator is stiff, and that the buoy is moving in heave only, a single equation of motion describes the dynamics of the system

$$m^i \ddot{z}^i(t) = F_{\text{exc}}^i(t) + F_{\text{rad}}^i(t) + F_{\text{stat}}^i(t) + F_{\text{PTO}}^i(t), \tag{3}$$

where $m^i = m_b^i + m_t^i$ is the total masses of the buoy and the translator, F_{exc}^i is the excitation force due to the incident and scattered waves, F_{rad}^i the radiation force due to the oscillations of the buoy, $F_{\text{stat}}^i = -\rho g \pi R^i z^i$ the hydrostatic restoring force and $F_{\text{PTO}}^i = -\Gamma^i \dot{z}^i(t)$ the power take-off damping of the generator.

The hydrodynamical forces will be computed in the frequency domain, where the equivalent equation of motion (3) reads

$$\left[-\omega^2(m + m_{\text{add}}) - i\omega(\Gamma + B) + \rho g \pi R\right]_j^i z^j(\omega) = F_{\text{exc}}^i(\omega), \tag{4}$$

where the radiation force was divided into an added mass term proportional to the acceleration and a radiation damping term proportional to the velocity, $F^i_{rad}(\omega) = [-i\omega m_{add} + B]^i_j i\omega z^j(\omega)$ and the excitation force can be divided into an excitation force coefficient vector and the incident waves, $F^i_{exc}(\omega) = f^i_{exc}(\omega)A^i(\omega)$. The incident waves will be irregular, short-crested waves, described in more detail in Section 2.2.2.

When the equations of motion have been solved in the frequency domain, the result is obtained in the time domain by the inverse Fourier transform, and the absorbed power of each WEC at each point in time is computed as

$$P^i(t) = \Gamma^i[\dot{z}(t)]^2, \tag{5}$$

and the total power of the park is the sum of the absorbed power of all devices in the park. The absorbed power will vary over time with the incident waves, and the resulting total power of the park will display power fluctuations. To connect the power from a park to the electrical grid, low power fluctuations are required. Large power peaks will require over-dimensional, expensive power electronics system, and instants with no power will create unwanted power outages. The power fluctuations can be quantified by the normalized variance, defined in terms of the standard deviation σ as

$$v = \frac{\sigma^2(P_{tot}(t))}{\bar{P}_{tot}}. \tag{6}$$

A common way to represent the interaction in a park is with the interaction factor, or q-factor [1], defined as the ratio between the time averaged power absorbed by the park, and the sum of all devices' individual power absorption if they were isolated,

$$q = \frac{\bar{P}_{tot}}{\sum_{i=1}^N \bar{P}^i_{isolated}}. \tag{7}$$

An individual q-factor for each WEC in the park can also be defined as the ratio between its averaged absorbed power and the average it would absorb in isolation,

$$q^i = \frac{\bar{P}^i}{\bar{P}^i_{isolated}}. \tag{8}$$

Although cases can be found where the interaction factor is larger than one, in realistic cases, the interaction effects will be destructive with $q < 1$, and optimization of the park interactions will aim to minimize destructive interactions and achieve an interaction factor as close to 1 as possible [3,22]. To allow for direct comparisons, in this paper, the individual q-factor is computed with the same isolated power absorption, i.e., the denominator $\bar{P}^i_{isolated}$ in Equation (8) is computed with the same irregular waves and corresponds to an isolated WEC situated at the origin $(x, y) = (0, 0)$. To avoid confusion, the computed values in Equations (7) and (8) will therefore be denoted normalized energy absorption.

2.2. Multiple Scattering Theory with Short-Crested Waves

2.2.1. Linear Potential Flow Theory

Consider a fluid that is incompressible and irrotational, implying that it can be described by potential flow theory using a fluid velocity potential satisfying the Laplace equation, $\nabla^2\Phi = 0$, where $\bar{u} = \nabla\Phi$ is the fluid velocity. Further assume that the viscosity of the fluid can be neglected and that the wave height is small compared to the wave length, so that the boundary constraints at the sea surface can be linearized. The fluid is then described by linear potential flow theory, and the details can be found in many text books such as [32].

The surface elevation of a wave with potential $\Phi(t)$ is given by

$$\eta(x,y,t) = -\frac{1}{g}\frac{\partial\Phi}{\partial t}\bigg|_{z=0},$$ (9)

where g is the gravitational acceleration constant. The surface elevation in the frequency domain is obtained by Fourier transform,

$$\hat{\eta}(x,y,\omega) = \int_{-\infty}^{\infty}\eta(x,y,t)e^{-i\omega t}\,dt.$$ (10)

Due to the linearity of the problem, the fluid potential can be decomposed into incident, scattered and radiated waves, and from the linear Bernoulli equation the dynamical pressure can be obtained from the fluid potential as $p = -\rho\frac{\partial\Phi}{\partial t}$, so that the hydrodynamical forces are obtained in the frequency domain as

$$\bar{F}_{exc}(\omega) = i\omega\rho\iint_S \phi_D\bar{n}dS,$$ (11)

$$\bar{F}_{rad}(\omega) = i\omega\rho\iint_S \phi_R\bar{n}dS,$$ (12)

where $\phi_D = \phi_I + \phi_S$ is the diffraction and ϕ_R the radiation potentials and the integration is taken over the wetted surface of the buoy. By solving the multiple scattering problem as described in Section 2.2.3, the fluid potentials can be solved for and the hydrodynamical forces computed according to (11) and (12). With the hydrodynamical forces obtained, the equations of motion in Equation (4) can be solved and the performance of the park evaluated.

2.2.2. Multidirectional Waves

Open ocean waves are assumed, i.e., no reflective structures exist so that there are no phase-locked waves—all the phases are randomly distributed, implying that the wave components are independent of each other. Thus, we can treat the irregular short-crested waves as a superposition of harmonic waves travelling in different directions.

A single harmonic wave travelling in direction χ away from the x-axis can be described by the surface elevation $\eta(x,y,t) = a\cos\left(\omega t - \bar{k}\cdot\bar{x} + \varphi\right)$, where φ the phase, and ω is the angular frequency related to the wave number vector $\bar{k} = (k\cos\chi, k\sin\chi, 0)$ by the dispersion relation for ocean waves. When the waves are composed of many waves travelling in independent directions, the surface elevation can be written as the superposition

$$\eta(x,y,t) = \sum_{n=-\infty}^{\infty}\sum_{m=1}^{M} a_{mn}e^{i(\omega_n t - \bar{k}_{mn}\cdot\bar{x})} = 2\sum_{n=0}^{\infty}\sum_{m=1}^{M}\left|a_{mn}\right|\cos(\omega_n t - \bar{k}_{mn}\cdot\bar{x} + \varphi_{mn}),$$ (13)

where a_{mn} is hermitian, i.e., $a_{mn}^* = a_{m-n}$, the phase is $\varphi_{mn} = \arg(a_{mn})$ and $\bar{k}_{mn}\cdot\bar{x} = k_n(x\cos\chi_m + y\sin\chi_m)$. The complex amplitude coefficients can be written in terms of the directional wave spectra as

$$2\left|a_{mn}\right|^2 = S(\omega_n,\chi_m)d\omega d\chi,$$ (14)

where $d\omega = \omega_{n+1} - \omega_n$ and $d\chi = \chi_{m+1} - \chi_m$. The expression for the surface elevation turns into an inverse Fourier integral when $d\omega$ becomes infinitesimal and we write $a_{mn} = A(\omega_n,\chi_m)d\omega d\chi$, such that

$$2\left|A(\omega,\chi)\right|^2 d\omega d\chi = S(\omega,\chi).$$ (15)

The energy in the waves is given by

$$E = \rho g \langle \eta^2 \rangle = \rho g \int_0^\infty d\omega \int_{-\pi}^{\pi} d\chi \, S(\omega, \chi). \tag{16}$$

The directional wave spectrum can be assumed to be decomposed into a direction independent spectrum and a directional spreading function (DSF)

$$S(\omega, \chi) = S(\omega) D(\omega, \chi), \tag{17}$$

where periodicity is required, $D(\omega, 2\pi) = D(\omega, 0)$, and conservation of energy requires that the directional spreading function is normalized

$$\int_{-\pi}^{\pi} D(\omega, \chi) d\chi = 1. \tag{18}$$

Several different directional spreading functions have been defined and studied in the literature. A common assumption is that the directional spreading function can be described by a unimodal model parametrized by the mean direction $\bar{\chi}$ and another parameter, such as the spreading parameter or a directional width, so that it is independent of the wave frequency. The default representation, still widely used, was defined in [33] and reads

$$D(\chi - \bar{\chi}) = F(s) \cos^{2s}\left(\frac{\chi - \bar{\chi}}{2}\right), \tag{19}$$

where $\bar{\chi}$ is the principal wave direction and $F(s)$ is defined such that the normalization constraint in Equation (18) is satisfied. In [34], the coefficient $F(s)$ was defined in terms of gamma functions as

$$F(s) = \frac{2^{2s-1}}{\pi} \frac{(\Gamma(s+1))^2}{\Gamma(2s+1)}, \tag{20}$$

which was later also used in [25], albeit presented differently, for the spreading parameter $s = 5, 15, 25$. Here, we will use the directional spreading function presented in [35] and used recently in [7],

$$D(\chi - \bar{\chi}) = \begin{cases} F(s) \cos^{2s}(\chi - \bar{\chi}), & |\chi - \bar{\chi}| < \frac{\pi}{2}, \\ 0, & \text{otherwise,} \end{cases} \tag{21}$$

with

$$F(s) = \frac{1}{\sqrt{\pi}} \frac{\Gamma(s+1)}{\Gamma(s+\frac{1}{2})} = \frac{1}{\pi} \frac{(2s)!!}{(2s-1)!!}, \tag{22}$$

which is simply 1 over the integral in Equation (18), hence the normalization constraint (18) is satisfied. For example, for the spreading parameter $s = 1$, the coefficient is $F(s) = 2/\pi$. Note that the argument in the cosine function of Equation (21) differs by a factor 2 from the convention defined in [33] and displayed in Equation (19). For discrete wave directions, the sum over all wave directions

$$\sum_{m=1}^{M} D(\chi_m - \bar{\chi}) d\chi \tag{23}$$

only converges to the integral value when $d\chi$ is infintesimally small. Hence, the coefficient will be defined as

$$F(s) = \frac{1}{\sum_{m=1}^{M} \cos^{2s}(\chi_m - \bar{\chi}) d\chi}, \tag{24}$$

which converges to the value in Equation (22) when $d\chi \to 0$.

The principal wave direction will be considered here as $\bar{\chi} = 0$, i.e., moving along the x-direction. The shape of the directional spreading function for different values of the spreading parameter s is shown in Figure 2. As can be seen from the figure, the higher the spreading parameter, the more energy in the waves is distributed along the principal wave direction.

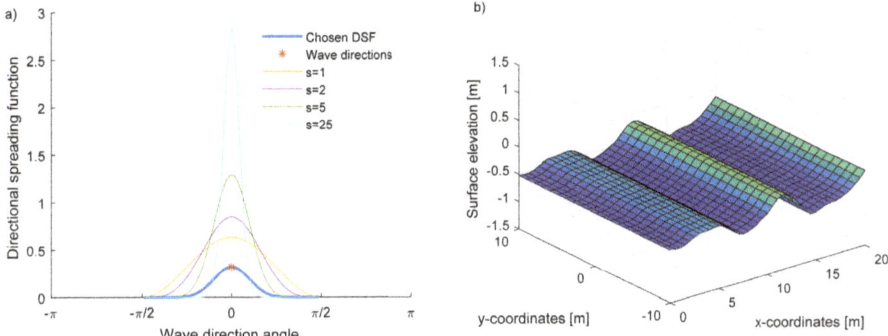

Figure 2. (**a**) Directional spreading function and (**b**) surface elevation of a long-crested, unidirectional wave, to be compared against the short-crested waves in Figures 3–5. For the case of only one wave direction $\chi = 0$, the value of the spreading parameter is irrelevant as $\cos^{2s}(0) = 1$ for any values of s. The factor $F(s)$ in Equation (24) equals $F(s) = 1/\pi$ in the case of a single wave direction, which guarantees normalization in Equation (18).

From Equations (15) and (17), the amplitude function can be also decomposed into a directional independent part and the directional dependent part. However, from Equation (15), only the modulus of the complex amplitude is known, and we add an unknown phase $\varphi(\omega, \chi)$, which is assumed to be uniformly distributed over $(-\pi, \pi)$,

$$
\begin{aligned}
A(\omega, \chi) &= |A(\omega, \chi)| e^{i\arg(A(\omega,\chi))} \\
&= \sqrt{\frac{S(\omega)D(\omega, \chi)}{2d\omega d\chi}} e^{i\arg(A(\omega,\chi))} \\
&= |A(\omega)| \sqrt{D(\omega, \chi)\frac{1}{d\chi}} e^{i\arg(A(\omega,\chi))} \\
&= A(\omega) \sqrt{D(\omega, \chi)\frac{1}{d\chi}} e^{i\varphi(\omega,\chi)},
\end{aligned}
\tag{25}
$$

where $D(\omega, \chi)$ is the directional spreading function defined in Equation (21). In this work, we will use incident irregular *unidirectional* waves for which the surface amplitude $A(\omega)$ is known, and compute the complex amplitudes for the short-crested waves according to the expression in Equation (25).

In the frequency domain, the surface elevation can be written as the product of the amplitude function and a transfer function $H_m(\omega)$,

$$
\hat{\eta}(x, y, \omega) = \sum_{m=1}^{M} A(\omega, \chi_m) e^{-ik(x\cos\chi_m + y\sin\chi_m)} d\chi = \sum_{m=1}^{M} \sqrt{D(\chi_m - \bar{\chi})d\chi} A(\omega) H_m(\omega),
\tag{26}
$$

where $H_m(\omega) = e^{-ik(x\cos\chi_m + y\sin\chi_m) + i\varphi(\omega,\chi_m)}$.

In the time domain, the inverse Fourier transform of the product becomes a convolution

$$
\eta(x, y, t) = \sum_{m=1}^{M} \sqrt{D(\chi_m - \bar{\chi})d\chi}\, \eta(0, 0, t) * h_m(t),
\tag{27}
$$

where $\eta(0,0,t) = \widehat{A(\omega)}$ is the surface elevation at point $(x,y) = (0,0)$ and $h_m(t)$ is the inverse Fourier transform of the transfer function $H_m(\omega)$.

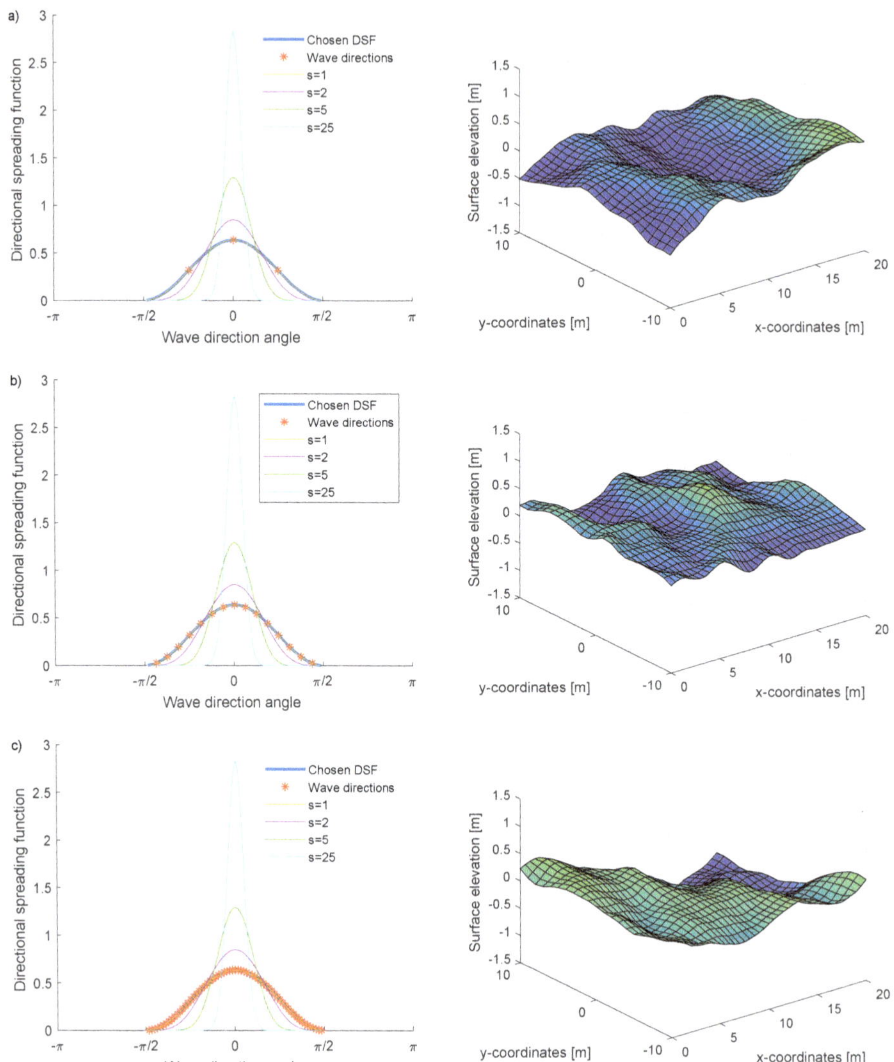

Figure 3. Directional spreading function $D(\chi_m - \bar{\chi})$ in Equation (21) for $s = 1$ and increasing wave directions from 3, 15 and 63, as shown in the directional spreading function plots. A time instant of the surface elevation is shown to the right. (**a**) three wave directions $\chi_m = \pm m\pi/4$ with $m = 0,1$; (**b**) 15 wave directions $\chi_m = \pm m\pi/16$ with $m = 0,\ldots,7$; (**c**) 63 wave directions $\chi_m = \pm m\pi/64$ with $m = 0,\ldots,31$.

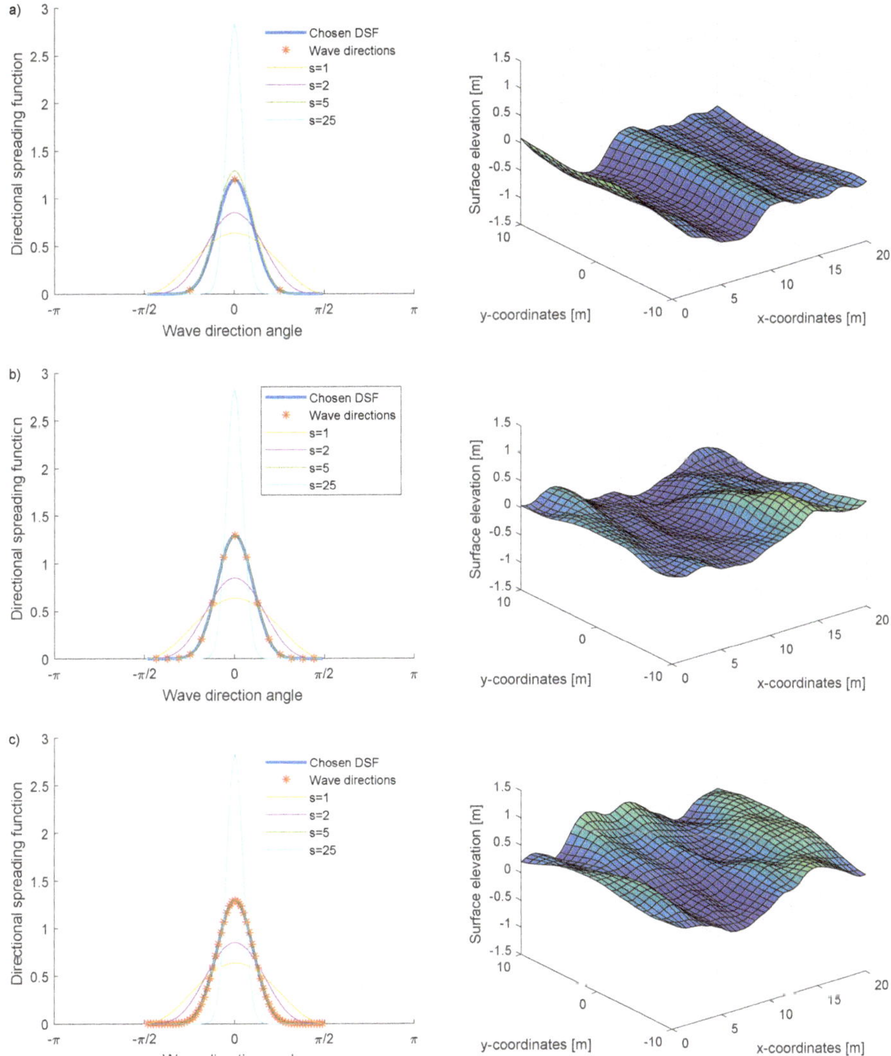

Figure 4. Directional spreading function $D(\chi_m - \bar{\chi})$ in Equation (21) for spreading parameter $s = 5$ increasing wave directions from 3, 15 and 63, as shown in the directional spreading function plots. A time instant of the surface elevation is shown to the right. (**a**) three wave directions $\chi_m = \pm m\pi/4$ with $m = 0,1$; (**b**) 15 wave directions $\chi_m = \pm m\pi/16$ with $m = 0,\ldots,7$; (**c**) 63 wave directions $\chi_m = \pm m\pi/64$ with $m = 0,\ldots,31$.

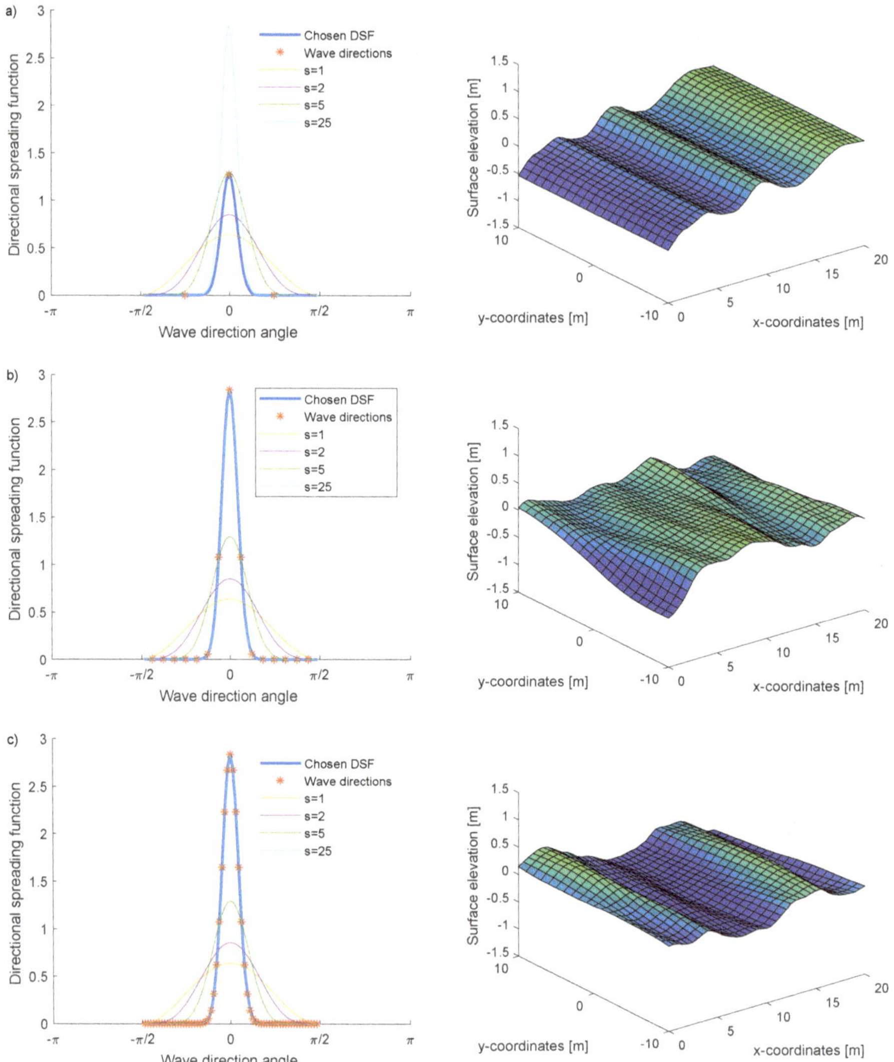

Figure 5. Directional spreading function $D(\chi_m - \bar{\chi})$ in Equation (21) for spreading parameter $s = 25$ increasing wave directions from 3, 15 and 63, as shown in the directional spreading function plots. A time instant of the surface elevation is shown to the right. Note that, due to the few wave directions, the magnitude of the directional spreading function defined by Equation (24) does not completely converge to the integral value in (22), but takes a smaller value. (**a**) three wave directions $\chi_m = \pm m\pi/4$ with $m = 0, 1$; (**b**) 15 wave directions $\chi_m = \pm m\pi/16$ with $m = 0, \ldots, 7$; (**c**) 63 wave directions $\chi_m = \pm m\pi/64$ with $m = 0, \ldots, 31$.

The fluid potential of the incident wave in Equation (10) can be written in the frequency domain as

$$\phi_0(x, y) = \sum_{m=1}^{M} \frac{ig}{\omega} A(\omega, \chi_m) \psi_0(z) e^{i[-k(x\cos\chi_m + y\sin\chi_m)]} \, d\chi, \tag{28}$$

where $\psi_0(z)$ is the vertical eigenfunction defined as

$$\psi_0(z) = \frac{\cosh(k(z+h))}{\cosh(kh)}. \tag{29}$$

Consider now a buoy located at position (x^i, y^i) with local cylindrical coordinates (r, θ, z). Due to properties of Bessel functions, the incident fluid potential in Equation (28) can be rewritten in the local coordinate system as

$$
\begin{aligned}
\phi_0(r,\theta,z) &= \sum_{m=1}^{M} \frac{ig}{\omega} A(\omega, \chi_m) \psi_0(z) e^{i[-k(x^i \cos \chi_m + y^i \sin \chi_m)]} e^{-ikr \cos(\theta - \chi_m)} \, d\chi \\
&= \sum_{m=1}^{M} \frac{ig}{\omega} A(\omega, \chi_m) \psi_0(z) e^{-ik(x^i \cos \chi_m + y^i \sin \chi_m)} \, d\chi \sum_{n=-\infty}^{\infty} (-i)^n J_n(kr) e^{in(\theta - \chi_m)} \\
&= \sum_{n=-\infty}^{\infty} Z_0(z) \underbrace{\left[\sum_{m=1}^{M} \frac{ig}{\omega} A(\omega, \chi_m) e^{-ik(x^i \cos \chi_m + y^i \sin \chi_m)} \frac{(-i)^n}{Z_0(0)} e^{-in\chi_m} \, d\chi \right]}_{A_{0n}^i} J_n(kr) e^{in\theta} \\
&= \sum_{n=-\infty}^{\infty} Z_0(z) A_{0n}^i J_n(kr) e^{in\theta},
\end{aligned}
\tag{30}
$$

where $Z_0(z)$ is the normalized vertical eigenfunction satisfying $\psi_0(z) = Z_0(z)/Z_0(0)$ and $J_n(kr)$ are oscillating Bessel functions.

2.2.3. Multiple Scattering

Any point in the global coordinate system can be written in terms of the local cylindrical coordinate system defined at the origin (x^i, y^i) at each buoy as $(x, y, z) = (x^i + r \cos \theta, y^i + r \sin \theta, z)$. At each buoy, the fluid domain is divided into an interior region I beneath the buoy and an exterior region II outside $r > R^i$ (see Figure 1). By separation of variables, a solution to the Laplace equation and the linear boundary constraints at the free surface, the seabed and at the buoy surfaces can be found for the two fluid domains on the general form

$$
\phi^{i(\mathrm{I})} = \frac{V^i}{2L^i}\left((z+h)^2 - \frac{r^2}{2}\right) + \sum_{n=-\infty}^{\infty}\left[\gamma_{m0}^i \left(\frac{r}{R^i}\right)^{|n|} + \sum_{m=1}^{\infty} \gamma_{mn}^i \cos(\lambda_m^i(z+h)) \frac{I_n(\lambda_m^i r)}{I_n(\lambda_m^i R^i)}\right] e^{in\theta} \tag{31}
$$

$$
\phi^{i(\mathrm{II})} = \sum_{n=-\infty}^{\infty}\left[\alpha_{0n}^i Z_0(z) \frac{H_n(kr)}{H_n(kR^i)} + \sum_{m=1}^{\infty} \alpha_{mn}^i Z_m(z) \frac{K_n(k_m r)}{K_n(k_m R^i)}\right] e^{in\theta}, \tag{32}
$$

where I/II define the interior/exterior regions and $L^i = h - d^i$ where h is the water depth and d^i the individual drafts of the buoys. The wave number $k_0 = -ik$ is a solution to the dispersion relation $\omega^2 = gk \tanh(kh)$, and the Hankel functions $H_n(kr)$ correspond to propagating modes. The wave numbers k_m, $m > 0$ are consecutive roots to the dispersion relation $\omega^2 = -gk_m \tan(k_m h)$, and the Bessel functions $K_n(k_m r)$ correspond to evanescent modes. $I_n(\lambda_m r)$ are modified Bessel functions with argument $\lambda_m^i = m\pi/L^i$.

The term proportional to the velocity V^i in the potential in the interior domain is required to satisfy the inhomogeneous boundary constraint at the body surface of the oscillating buoy. If the buoy is stationary, its velocity vanishes $V^i = 0$, and the potential in the interior region only contains its second, homogeneous term.

Solving the multiple scattering problem is essentially a matter of finding the unknown coefficients in expressions (31) and (32) by requiring continuity of the potentials and their derivatives at the domain boundaries $r = R^i$. The problem can be solved with incident waves and oscillating buoys occurring simultaneously, but for clarity will here be solved as two separate problems, where, in the scattering problem, the buoys are held fixed and there are incident waves, and in the radiation problem the buoys are free to oscillate, but there are no incident waves. The derivation follows the same procedure as shown in [16,28], and the reader is referred to those references for further details.

Scattering Problem

Consider incident short-crested waves as described by the fluid potential in Equation (30) and let the buoys be fixed, with zero velocity $V^i = 0$. The potential in the exterior domain of any buoy will be a superposition of incident wave ϕ_0^i, scattered waves ϕ_S^i, and incoming waves that are scattered off the other buoys $\phi_S^j, j \neq i$,

$$\phi^{i,\mathrm{II}} = \phi_0^i + \phi_S^i + \sum_{j \neq i} \phi_S^j \Big|_i. \tag{33}$$

By using Graf's addition theorems for Bessel functions, the outgoing waves from one buoy can be written as an incoming wave in the local coordinates of another cylinder as

$$\phi_S^j \Big|_i = \sum_{n=-\infty}^{\infty} \left[Z_0(z) J_n(kr) \sum_{l=-\infty}^{\infty} T_{0ln}^{ij} \alpha_{0l}^j + \sum_{m=1}^{\infty} Z_m(z) \frac{I_n(k_m r)}{I_n(k_m R)} \sum_{l=-\infty}^{\infty} T_{mln}^{ij} \alpha_{ml}^j \right] e^{in\theta_i}, \tag{34}$$

where the expressions for $T^{ij} = T_{mln}^{ij}$ are given in the Appendix. Continuity between the interior (31) and exterior solutions (32) and their derivatives along the boundaries $r = R^i$ between the interior and exterior domains implies the infinite system of equations

$$\sum_{m=0}^{\infty} D_{smn}^i \alpha_{mn}^i = - \sum_{m=0}^{\infty} \widetilde{D}_{smn}^i \left(A_{0n}^i \delta_{m0} + \sum_{j \neq i} \sum_{l=-\infty}^{\infty} T_{mln}^{ij} \alpha_{ml}^j \right), \tag{35}$$

together with an expression for the coefficients γ in the interior solution in terms of the coefficients α, given in Equation (A5). The matrices D, $\widetilde{D}$ and their components are defined in the appendix.

To solve for the unknown coefficients α^i from Equation (35), we truncate the infinite sums at the vertical cut-off Λ_z for the vertical sums $\sum_m Z_m(z)$ and the angular cut-off Λ_θ for the angular sums $\sum_n e^{in\theta}$. The system of equations take the form

$$\begin{bmatrix} 1 & -B^1 T^{12} & \cdots & -B^1 T^{1N} \\ -B^2 T^{21} & 1 & \cdots & -B^2 T^{2N} \\ \vdots & \vdots & \ddots & \vdots \\ -B^N T^{N1} & -B^N T^{N2} & \cdots & 1 \end{bmatrix} \begin{bmatrix} \alpha^1 \\ \alpha^2 \\ \vdots \\ \alpha^N \end{bmatrix} = \begin{bmatrix} B^1 A_{0n}^1 \\ B^2 A_{0n}^2 \\ \vdots \\ B^N A_{0n}^N \end{bmatrix}, \tag{36}$$

where $B^i = -[D^i]^{-1} \widetilde{D}^i$ is the single-body diffraction matrix and A_{0n}^i is defined for the incident, short-crested waves as in Equation (30). The matrix on the left-hand side is called the diffraction matrix. The diagonal entries are identity matrices, and the non-diagonal entries $B^i T^{ij}$ account for the hydrodynamical multiple scattering interactions. Neglecting the multiple scattering would result in the diffraction matrix being the identity matrix. Solving the equation in (36) is the computationally most extensive part of the computation, as the diffraction matrix is a large, quadratic matrix of size $N(\Lambda_z + 1)(2\Lambda_\theta + 1)$.

When the scattering coefficients α in the diffraction potential in the exterior domain have been solved from Equation (36), the coefficients γ in the potential in the interior domain can be found from Equation (A5), and the heave excitation force can be computed from Equation (11) as

$$F_{\mathrm{exc}}^i(\omega) = 2\pi i \omega \rho \left[\gamma_{00}^i \frac{(R^i)^2}{2} + 2R^i \sum_{m=1}^{\Lambda_z} \gamma_{m0}^i \frac{(-1)^m}{\lambda_m^i} \frac{I_1(\lambda_m^i R^i)}{I_0(\lambda_m^i R^i)} \right]. \tag{37}$$

Radiation Problem

The radiation problem follows the same procedure as the scattering problem, but solves the problem when there are no incident waves and the buoys are free to oscillate with independent velocities V^i.

The potential in the exterior domain of any buoy will be a superposition of radiated and scattered waves of the own body, and incoming waves that are scattered and radiated off the other buoys,

$$\phi^{i,\mathrm{II}} = \phi_R^i + \phi_S^i + \sum_{j\neq i} \left(\phi_R^j + \phi_S^j \right)\Big|_i . \tag{38}$$

By using Graf's addition theorems for Bessel functions, the outgoing waves from one buoy can be written as an incoming wave in the local coordinates of another cylinder as in (34), and again continuity between the fluid domains implies the infinite system of equations

$$\sum_{m=0}^{\infty} D_{smn}^i \alpha_{mn}^i = -\sum_{m=0}^{\infty} \widetilde{D}_{smn}^i \left(\sum_{j\neq i} \sum_{l=-\infty}^{\infty} T_{mln}^{ij} \alpha_{ml}^j \right) - V^i R_s^i \delta_{0n}, \tag{39}$$

together with Equation (A5), where R_s^i is the radiation vector defined in the appendix and α contains coefficients for both the scattered and radiated waves in (38). As in the scattering problem, the infinite sums are truncated to obtain a finite system of equations

$$\begin{bmatrix} 1 & -B^1 T^{12} & \cdots & -B^1 T^{1N} \\ -B^2 T^{21} & 1 & \cdots & -B^2 T^{2N} \\ \vdots & \vdots & \ddots & \vdots \\ -B^N T^{N1} & -B^N T^{N2} & \cdots & 1 \end{bmatrix} \begin{bmatrix} \alpha^1 \\ \alpha^2 \\ \vdots \\ \alpha^N \end{bmatrix} = \begin{bmatrix} V^1 B_R^1 \\ V^2 B_R^2 \\ \vdots \\ V^N B_R^N \end{bmatrix}, \tag{40}$$

where $B_R^i = [D^i]^{-1} R_s^i$. When the scattering and radiation coefficients α have been solved from Equation (40), the coefficients γ in the potential in the interior domain can be found from Equation (A5), and the heave radiation force can be computed from Equation (12) as

$$F_{\mathrm{rad}}^i(\omega) = 2\pi i \omega \rho \left[\frac{V^i L^i}{4} \left((R^i)^2 - \frac{(R^i)^4}{4(L^i)^2} \right) + \gamma_{00}^i \frac{(R^i)^2}{2} + 2R^i \sum_{m=1}^{\Lambda_z} \gamma_{m0}^i \frac{(-1)^m}{\lambda_m^i} \frac{I_1(\lambda_m^i R^i)}{I_0(\lambda_m^i R^i)} \right]. \tag{41}$$

There is an implicit velocity dependence in the coefficients γ that can be factored out, and the radiation force can be written in terms of added mass and radiation damping as

$$F_{\mathrm{rad}}^i(\omega) = \sum_{i=1}^{N} [i\omega m_{\mathrm{rad}} - B]_j^i V^j. \tag{42}$$

2.3. Numerical Implementation

To transform Equation (35) to a finite system of linear equations, the vertical and angular cut-offs have been chosen as $\Lambda_z = 20$ (index $s, m = 1, \ldots, 20$) and $\Lambda_\theta = 3$ (index $l, n = -3, -2, \ldots, 2, 3$). In [28], these cut-off values were shown to produce results with a high accuracy as compared to computations performed with the state-of-art software WAMIT (version 7.062, Massachusetts Institute of Technology, MA, USA). The theory and equations described in Sections 2.1 and 2.2 have been implemented and solved in a MATLAB (version R2017a, MathWorks Inc., MA, USA) script.

Although the method allows the use of an interaction distance cut-off [28] to speed up the computations with attained high accuracy, in this paper, that option has not been used, and full hydrodynamical interaction is computed between all devices.

The water depth has been chosen to constant $h = 25$ m, based on the depth at the test site Lysekil on the west coast of Sweden. In this paper, the dimensions of the WECs in the park have been chosen constant and equal for all WECs. The buoy radius has been chosen to $R = 3$ m, the draft to $d = 0.5$ m and the power take-off damping to $\Gamma = 200$ kNs/m.

The incident short-crested waves are computed as described in Section 2.2.2, where the incident long-crested waves with amplitude $A(\omega)$ is the Fourier transform of the incident long-crested waves at the origin, $\eta(0,0,t)$. The time-series of incident waves $\eta(0,0,t)$ is measured by a Datawell Waverider buoy installed at the Lysekil test site, Sweden. One hour of incident wave data has been used, which was measured on 1 January 2015 with a sampling frequency of 2.56 Hz. The sea state is characterized by a significant wave height of $H_s = 1.88$ m and an energy period of $T_e = 5.98$ s.

3. Results and Discussion

3.1. Irregular, Short-Crested Waves

The surface elevation in time domain is plotted at the same instant in time for different values of the spreading parameter s and different number of wave directions in Figures 2–5, together with the corresponding directional spreading function. As is expected from the form of the directional spreading function and clear from the figures, a larger spreading parameter s implies that more energy in the waves is propagating along the principal wave direction, with the asymptotic state being a unidirectional wave. Hence, the short-crested wave in Figure 5a with $s = 25$ takes a similar form as the long-crested, unidirectional wave in Figure 2. In Figures 3–5, the spreading parameter is kept constant to $s = 1$, $s = 5$ and $s = 25$, respectively, while the number of wave directions is increased from three in subfigures (a), 15 in subfigures (b) and 63 in subfigures (c).

Hence, the upper rows, i.e., Figures 3a, 4a and 5a all have three wave directions but increasing spreading parameter from $s = 1$ to $s = 25$, and equivalently the Figures 3b, 4b and 5b share the same 15 wave directions but increasing spreading parameter, and Figures 3c, 4c and 5c all have 63 wave directions but increasing spreading parameter. As can be seen from the figures, not only the spreading parameter affects the surface elevation, but also the number of wave directions. The fewer wave directions, the more the waves resemble a long-crested, unidirectional wave, which should be expected.

To analyse the effect of wave directions and spreading parameter even further, the surface elevations for all cases is shown in Figure 6. Here, the spreading parameter increases with the columns, with $s = 1$ in the left column, $s = 5$ in the middle and $s = 25$ in the right column, and the number of wave directions increases with the rows, with three wave directions in the top row, 15 in the middle and 63 wave directions in the bottom row. In the upper left figure, the three wave directions $\{-\pi/4, 0, \pi/4\}$ are clearly distinguishable, whereas this is more vague in the other cases. As can be seen in the figures, the fewer wave directions (upper row) and the higher spreading parameter (right column), the more the resulting waves resemble long-crested waves.

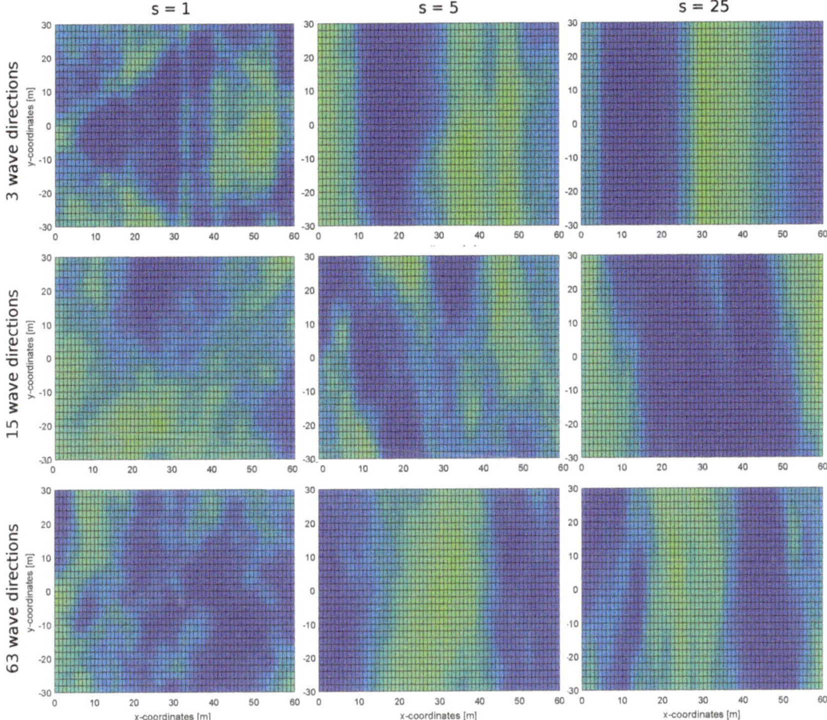

Figure 6. Surface elevations corresponding to the surfaces in Figures 3–5, but here shown in two dimensions and over a larger ocean surface. The crests/throughs are shown in green/blue color according to the three-dimensional plots in Figures 3–5. The spreading parameter increases with the columns, with $s = 1$ in the left column, $s = 5$ in the middle and $s = 25$ in the right column. The number of wave directions increases with the rows, with three wave directions in the top row, 15 in the middle and 63 wave directions in the bottom row.

3.2. Wave Energy Arrays in Short-Crested Waves

Due to the random phase in Equation (25), different short-crested waves will be generated in each simulation. The random phase is a function of both the wave direction and the frequency, hence all wave direction and wave frequency components of the composed waves will receive independent phases each time. For an evaluation of the wave energy park performance in short-crested waves, the simulations should be carried out a number of times with different random phases, and the average taken.

In Figure 7a,b, two different simulations of park performance in short-crested waves are shown. The park consists of 16 WECs in a gridded layout with 20 m separation distance. In the figure, the color of each WEC shows the normalized energy absorption q^i of Equation (8), i.e., value $q^i > 1$ implies that the WEC absorbs more wave power than it would in isolation at the origin. As can be seen from Figure 7a,b, the result differs for two different random phases in the short-crested waves; there is no clear pattern of wave shadowing or constructive/destructive interactions. In Figure 7c, an average of 50 simulations such as the ones shown in Figure 7a,b is shown. When an average is taken, a pattern emerges, and it is clear that wave shadowing occurs: the first row at $x = 0$ m absorbs more power than the second row at $x = 20$ m, and so on. The case of long-crested, unidirectional waves is shown in Figure 7d, where as expected the absorbed power by each WEC decreases strictly with the number of rows perpendicular to the incident wave direction. One can observe that when an average over a multiple of short-crested waves with different random phases is taken, the result converges to the long-crested wave case.

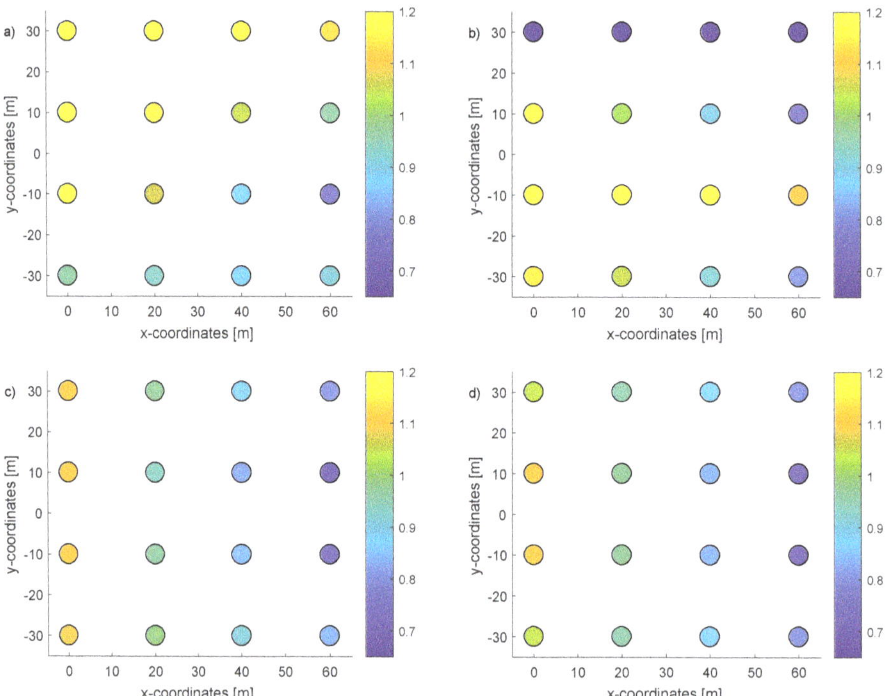

Figure 7. Time averaged energy absorption in a park consisting of 16 Wind Energy Conversion Systems (WECs.) The absorbed energy is divided by the absorbed energy of a single WEC in isolation to give the normalized energy absorption of Equation (8); hence, a value above 1 shows that the device absorbs more energy than an isolated device at the origin would, whereas a value below 1 shows that the absorption decreases. All short-crested wave simulations are run with 15 wave directions and spreading parameter $s = 5$. (**a**,**b**) two simulations in short-crested waves with different random phases; (**c**) average over 50 runs of short-crested waves with different random phases; (**d**) park in long-crested waves propagating along the x-axis.

The convergence of the normalized energy absorption is shown in more detail in Figure 8. Here, for each simulation number S, the average of the ratio of the individual normalized energy absorption in short-crested and long-crested waves is computed, and the average is taken over all the simulations $s = 1, \ldots, S$,

$$\frac{1}{S} \sum_{s=1}^{S} \left(\frac{1}{N} \sum_{i=1}^{N} \frac{q^i_{\text{short},s}}{q^i_{\text{long},s}} \right). \tag{43}$$

As can be seen from the figure, the power absorption for the parks in short-crested waves converges to the unidirectional case; after 30 simulations with random phases, the difference is less than 5% and, after 50 simulations (shown in Figure 7c), the difference is 1.6%.

In Figure 9a, the total instant power of a park with 16 wave energy devices (with park layout shown in Figure 7) is shown both in long-crested and short-crested waves. The power has been divided by the average power of a single device in isolation times the number of devices in a park. As is clear from the figure, the power output of the park in these short-crested waves have smaller power peaks than in the long-crested waves. This is a desired behaviour as connection to the electricity grid requires a good electricity quality with small power fluctuations. This is also an expected result, since the WECs in the same row perpendicular to the unidirectional wave direction will be excited simultaneously by the incident long-crested waves, whereas this does not happen in short-crested waves. Nevertheless, although it is an expected result, it is still important to quantify the reduction of the power fluctuations in short-crested waves, so that realistic values are used as design parameters for the electrical system of a park. A useful measure of the power fluctuations in a park is given by the variance (6). However, low fluctuations is only one measure of good park performance; in addition, large energy absorption is required, which can be quantified in terms of the normalized energy absorption (7). An optimal park will have a high energy absorption and low power fluctuations, simultaneously.

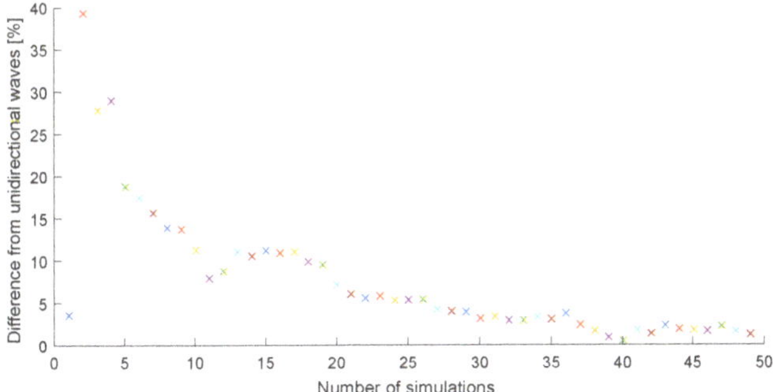

Figure 8. Convergence of the normalized energy absorption with number of simulations. For each simulation S, the average over the 1:Sth simulations is shown for the average of the energy absorption as in Equation (43), corresponding to the park of 16 WECs in Figure 7. Hence, Figure 7c, which is the average over all 50 simulations, corresponds to the last point in this figure.

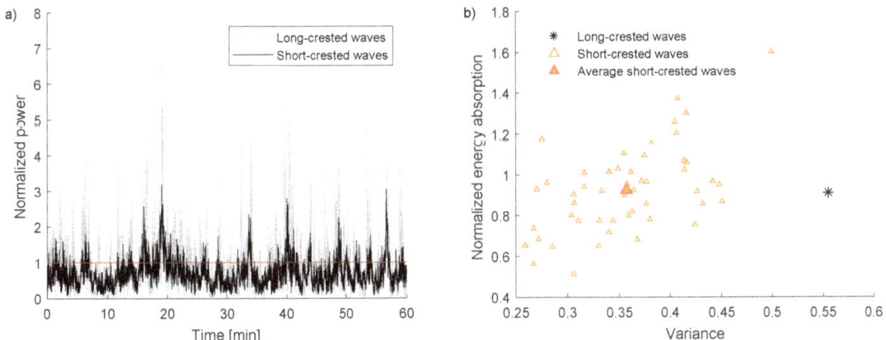

Figure 9. Performance of a wave energy park of 16 devices in long-crested, unidirectional waves, as compared to short-crested waves. The short-crested waves are composed of waves travelling in 15 directions with spreading function $s = 5$, corresponding to the directional spreading function and surface elevation shown in Figure 4b. (**a**) power for the full park in long-crested and short-crested irregular waves. The power has been divided by $N \cdot \bar{P}^i_{\text{isolated}}$ to be displayed in terms of the normalized power. For clarity, $q = 1$ is highlighted with a red line; (**b**) normalized energy absorption versus variance for the long-crested and short-crested waves, respectively. The average values for all the 50 simulations of short-crested waves is highlighted with a filled triangle.

To evaluate this, both the variance and the energy absorption have been plotted in Figure 9b. Ideal performance would end up in the upper left part of the figure, with high energy absorption and low variance. The performance in long-crested waves is shown by a black asterisk, and the performance in short-crested waves is shown by orange triangles, one for each simulation (corresponding to different random phases). The average of all the short-crested waves is shown with a larger filled triangle. As can be seen from the figure, the energy absorption in short-crested waves has a significantly lower variance, i.e., the power fluctuations are significantly lower. The spread in normalized energy absorption is large and ranges from 0.512 to 1.60, but the average of all simulations in short-crested waves has a normalized energy absorption of 0.925, which is similar to and even slightly higher than the value in long-crested waves, $q = 0.910$. To summarize, the power fluctuations are lower in short-crested waves, and the energy absorption is comparable. In short, the performance is better in short-crested waves as compared to long-crested waves.

3.3. Effect of Varying Wave Directions and Spreading Parameter

In Section 3.2, results were presented for short-crested waves with 15 wave directions and spreading parameter $s = 5$, corresponding to directional wave spectrum and corresponding surface elevation shown in Figure 4b. In this section, we evaluate the performance in short-crested waves of different number of wave directions and different spreading parameter values. In Figure 10, the extension of Figure 9b is shown for a larger class of short-crested waves. Short-crested waves with spreading parameter $s = 5$ consisting of 3, 15 and 63 wave directions (corresponding to the waves in Figure 4) and short-crested waves with spreading parameter $s = 1, s = 5$ and $s = 25$ consisting of waves with 15 wave directions (corresponding to the waves in Figure 3b, 4b and 5b) have been considered.

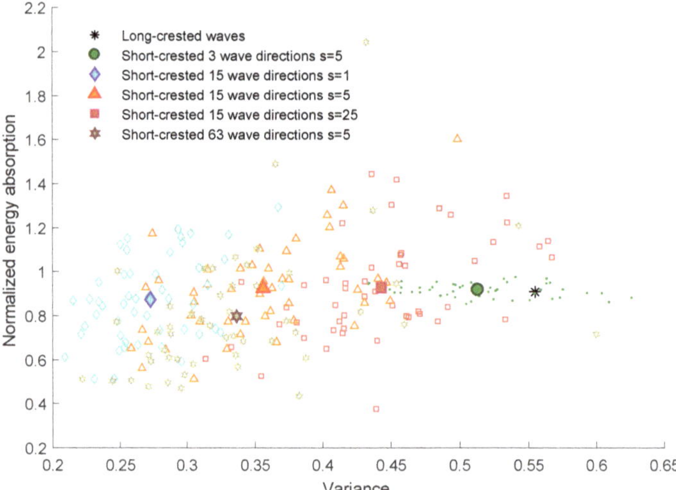

Figure 10. Performance of short-crested waves with different spreading parameters and number of wave directions, as compared to performance in long-crested waves. The variance (measure of the power fluctuations) and the normalized energy absorption of the park are shown. All 50 of the simulations with different random phases are shown as small unfilled markers, whereas the mean values are shown as larger filled markers.

From Figure 10, it is clear that all short-crested waves show a similar performance with lower power fluctuations and similar energy absorption, as compared to the performance in long-crested waves. The more the short-crested waves tend towards long-crested waves (i.e., the fewer the wave directions and the higher the spreading parameter, as discussed in Section 3.1 and shown in Figures 3–6), the more the resulting performance converges to the unidirectional case. An ideal wave energy park would have low fluctuations and high power absorption and end up in the upper left corner. The simulations show that the spread in the results is large, but when an average is taken over all 50 simulations with different random phases, the trend is clear; the parks in short-crested waves have lower fluctuations, and no significant change in power absorption as compared to long-crested waves.

4. Conclusions

Realistic ocean waves can be short-crested and consist of irregular waves travelling in several directions simultaneously. However, most studies on wave energy parks have been carried out in long-crested, unidirectional waves. The few that have considered short-crested waves have indicated that the performance of the wave energy parks might be different in short-crested as compared to long-crested waves, and that the subject needs further study.

In this paper, theory for including short-crested (multidirectional) irregular waves in a multiple scattering method has been developed and implemented in a semi-analytical MATLAB code. The directional dependence of the waves is described by a directional spreading function, and the impact of different spreading parameters and number of wave directions has been analysed. The short-crested waves converge to unidirectional, long-crested waves with increasing spreading parameter and decreasing number of wave directions.

The model has then been used to evaluate the performance of wave energy parks in short-crested waves. The performance is evaluated according to two important parameters: the energy absorption and the power fluctuations. An optimal wave energy park will have a large energy absorption but low power fluctuations, in order to ensure a good electricity quality to the electric grid.

Since all the wave direction components have different random phases, each simulation will produce different results due to the different waves produced at each location in the park. A convergence study was performed and showed that the energy absorption in short-crested waves converges to the long-crested case; an average of 50 simulations differed only 1.6% in energy absorption from the long-crested waves case. The power fluctuations, however, are clearly lower in the short-crested waves. The conclusion, therefore, is that the performance of wave energy parks in short-crested waves can be better than what has been anticipated from simulations using long-crested waves, with similar energy absorption and lower power fluctuations.

Acknowledgments: The research in this paper was supported by the Swedish Research Council (VR, grant number 2015-04657), the Swedish Energy Authority (project number 40421-1), Lundström-Åman scholarship fund, Miljöfonden, Jubelfeststipendiet and the Wallenius foundation. Most of the computations of parks consisting of 16 WECs in short-crested waves were performed on resources provided by the Swedish National Infrastructure for Computing (SNIC) at Uppsala Multidisciplinary Center for Advanced Computational Science (UPPMAX).

Author Contributions: Malin Göteman developed the method, carried out the modelling, analysed the results and wrote the paper. Cameron McNatt provided insightful suggestions to the theory of short-crested waves and multiple scattering. Marianna Giassi, Jens Engström and Jan Isberg provided input regarding the analysis and presentation of the results.

Conflicts of Interest: The authors declare no conflict of interest.

Abbreviations

The following abbreviations are used in this manuscript:

WEC   Wave energy converter
DSF   Directional spreading function

Appendix A. Defined Functions and Parameters

The vertical eigenfunctions are defined as

$$Z_0(z) = \tfrac{1}{N_0}\cosh(k(z+h)), \quad N_0^2 = \tfrac{1}{2}\left(1 + \frac{\sinh(2kh)}{2kh}\right),$$
$$Z_m(z) = \tfrac{1}{N_m}\cos(k_m(z+h)), \quad N_m^2 = \tfrac{1}{2}\left(1 + \frac{\sin(2k_m h)}{2k_m h}\right). \tag{A1}$$

Integrating the vertical coordinate z along the fluid column gives rise to the constants

$$c_{s0}^i = \frac{(-1)^s k}{k^2 + \lambda_s^{i\,2}}\frac{1}{L^i N_0}\sinh(kL^i), \quad c_{sm}^i = \frac{(-1)^s k_m}{k_m^2 - \lambda_s^{i\,2}}\frac{1}{L^i N_m}\sin(k_m L^i), \tag{A2}$$

where the distance between the float bottom and the seabed at equilibrium is $L^i = h - d^i$ and $\lambda_m^i = \pi m / L^i$. The expressions needed for Graf's addition theorems are

$$T_{0ln}^{ij} = \frac{1}{H_l(kR^j)}H_{l-n}(kR_{ij})e^{i\theta_{ij}(l-n)}, \quad T_{mln}^{ij} = \frac{I_n(k_m R^i)}{K_l(k_m R^j)}K_{l-n}(k_m R_{ij})e^{i\theta_{ij}(l-n)}(-1)^n, \tag{A3}$$

and R_{ij}, θ_{ij} are the distance and the angle between the two cylinders, respectively. The matrices involved in solving for the coefficients in the exterior domain in the multiple scattering problem are

$$D_{smn}^i = q_{mn}^i\delta_{ms} - L_{smn}^i, \quad \widetilde{D}_{smn}^i = \widetilde{q}_{mn}^i\delta_{ms} - L_{smn}^i\left(J_n(kR^i)\delta_{m0} + 1 - \delta_{m0}\right),$$
$$L_{smn}^i = \frac{|n|L^i}{R^i}c_{0s}^i c_{0m}^i + 2\sum_{r=1}^{\infty} p_{rn}^i c_{rs}^i c_{rm}^i, \quad R_s = -\frac{R^i}{2L^i}c_{0s^i} - 2\sum_{r=1}^{\infty} p_{r0}^i c_{rs}^i\frac{(-1)^r}{\pi^2 r^2},$$
$$q_{0n}^i = \frac{khH_n'(kR^i)}{H_n(kR^i)}, \quad q_{mn}^i = \frac{k_m h K_n'(k_m R^i)}{K_n(k_m R^i)}, \quad \widetilde{q}_{0n}^i = khJ_n'(kR^i), \quad \widetilde{q}_{mn}^i = \frac{k_m h I_n'(k_m R^i)}{I_n(k_m R^i)}, \tag{A4}$$
$$p_{mn}^i = \frac{\pi m I_n'(\lambda_m^i R^i)}{I_n(\lambda_m^i R^i)}.$$

When the parameters α have been solved from the diffraction matrix equation, the coefficients in the potential in the interior domain can be solved for from

$$\gamma_{sn}^i = \left[\left(-\frac{1}{6} + \frac{(R^i)^2}{4(L^i)^2}\right)\delta_{s0} - \frac{(-1)^s}{\pi^2 s^2}(1 - \delta_{s0})\right]V^i L^i \delta_{n0}$$
$$+ \left(\alpha_{0n}^i + J_n(kR)\left(A_{0n}^i + \sum_{j\neq i}\sum_{l=-\infty}^{\infty} T_{0ln}^{ij}\alpha_{0l}^j\right)\right)c_{s0}^i + \sum_{m=1}^{\infty}\left(\alpha_{mn}^i + \sum_{j\neq i}\sum_{l=-\infty}^{\infty} T_{mln}^{ij}\alpha_{ml}^j\right)c_{sm}^i, \tag{A5}$$

with $V^i = 0$ for the scattering problem and A_{0n}^i for the radiation problem.

References

1. Budal, K. Theory for absorption of wave power by a system of interacting bodies. *J. Ship Res.* **1977**, *21*, 248.
2. Falnes, J. Radiation impedance matrix and optimum power absorption for interacting oscillators in surface waves. *Appl. Ocean Res.* **1980**, *2*, 75.
3. Thomas, G.; Evans, D. Arrays of three-dimensional wave energy absorbers. *J. Fluid Mech.* **1981**, *108*, 67–88.
4. Bozzi, S.; Giassi, M.; Miquel, A.; Antonini, A.; Bizzozero, F.; Gruosso, G.; Archetti, R.; Passoni, G. Wave energy farm design in real wave climates: The Italian offshore. *Energy* **2017**, *122*, 378–389.
5. Penalba, M.; Touzón, I.; Lopez-Mendia, J.; Nava, V. A numerical study on the hydrodynamic impact of device slenderness and array size in wave energy farms in realistic wave climates. *Ocean Eng.* **2017**, *142*, 224–232.
6. Xu, D.; Stuhlmeier, R.; Stiassnie, M. Harnessing wave power in open seas II: Very large arrays of wave-energy converters for 2D sea states. *J. Ocean Eng. Mar. Energy* **2017**, *3*, 151–160.

7. Tay, Z.; Venugopal, V. Hydrodynamic interactions of oscillating wave surge converters in an array under random sea state. *Ocean Eng.* **2017**, *145*, 382–394.
8. López-Ruiz, A.; Bergillos, R.; Raffo-Caballero, J.; Ortega-Sánchez, M. Towards an optimum design of wave energy converter arrays through an integrated approach of life cycle performance and operational capacity. *Appl. Energy* **2018**, *209*, 20–32.
9. Fàbregas Flavià, F.; McNatt, C.; Rongère, F.; Babarit, A.; Clément, A. A numerical tool for the frequency domain simulation of large arrays of identical floating bodies in waves. *Ocean Eng.* **2018**, *148*, 299–311.
10. Thorburn, K.; Leijon, M. Farm size comparison with analytical model of linear generator wave energy converters. *Ocean Eng.* **2007**, *34*, 908–916.
11. Folley, M.; Whittaker, T. The effect of sub-optimal control on the spectral wave climate on the performance of wave energy converter arrays. *Appl. Ocean Res.* **2009**, *31*, 260–266.
12. Rahm, M.; Svensson, O.; Boström, C.; Waters, R.; Leijon, M. Experimental results from the operation of aggregated wave energy converters. *Renew. Power Gener. IET* **2012**, *6*, 149–160.
13. Child, B.; Laporte Weywada, P. Verification and validation of a wave farm planning tool. In Proceedings of the 10th EWTEC Conference, Aalborg, Denmark, 2–5 September 2013.
14. Göteman, M.; Engström, J.; Eriksson, M.; Isberg, J.; Leijon, M. Methods of reducing power fluctuations in wave energy parks. *J. Renew. Sustain. Energy* **2014**, *6*, 043103.
15. Giassi, M.; Göteman, M.; Thomas, S.; Engström, J.; Eriksson, M.; Isberg, J. Multi-Parameter Optimization of Hybrid Arrays of Point Absorber Wave Energy Converters. In Proceedings of the 12th European Wave and Tidal Energy Conference (EWTEC), Cork, Ireland, 27 August–1 September 2017.
16. Göteman, M. Wave energy parks with point-absorbers of different dimensions. *J. Fluids Struct.* **2017**, *74*, 142–157.
17. Borgarino, B.; Babarit, A.; Ferrant, P. Impact of wave interactions effects on energy absorption in large arrays of wave energy converters. *Ocean Eng.* **2012**, *41*, 79–88.
18. Engström, J.; Eriksson, M.; Göteman, M.; Isberg, J.; Leijon, M. Performance of a large array of point-absorbing direct-driven wave energy converters. *J. Appl. Phys.* **2013**, *114*, 204502.
19. Sjolte, J.; Tjensvoll, G.; Molinas, M. Power collection from wave energy farms. *Appl. Sci.* **2013**, *3*, 420–436.
20. Sarkar, D.; Renzi, E.; Dias, F. Wave farm modelling of oscillating wave surge converters. *Proc. R. Soc. A* **2014**, *470*, 20140118.
21. Göteman, M.; Engström, J.; Eriksson, M.; Isberg, J. Optimizing wave energy parks with over 1000 interacting point-absorbers using an approximate analytical method. *Int. J. Mar. Energy* **2015**, *10*, 113–126.
22. Babarit, A. On the park effect in arrays of oscillating wave energy converters. *Renew. Energy* **2013**, *58*, 68–78.
23. Dias, F.; Renzi, E.; Gallagher, S.; Sarkar, D.; Wei, Y.; Abadie, T.; Cummins, C.; Rafiee, A. Analytical and computational modelling for wave energy systems: the example of oscillating wave surge converters. *Acta Mech. Sin.* **2017**, *33*, 647–662.
24. Stratigaki, V.; Troch, P.; Stallard, T.; Forehand, D.; Kofoed, J.; Folley, M.; Benoit, M.; Babarit, A.; Kirkegaard, J. Wave Basin Experiments with Large Wave Energy Converter Arrays to Study Interactions between the Converters and Effects on Other Users in the Sea and the Coastal Area. *Energies* **2014**, *7*, 701–734.
25. Sun, L.; Zang, J.; Eatock Taylor, R.; Taylor, P. Effects of wave spreading on performance of a wave energy converter. In Proceedings of the 29th International Workshop on Water Waves and Floating Bodies, Osaka, Japan, 30 March–2 April 2014.
26. Ji, X.; Liu, S.; Bingham, H.; Li, J. Multi-directional random wave interaction with an array of cylinders. *Ocean Eng.* **2015**, *110*, 62–77.
27. Troch, P.; Beels, C.; De Rouck, J.; De Backer, G. Wake effects behind a farm of wave energy converters for irregular long-crested and short-crested waves. *Coast. Eng. Proc.* **2011**, *1*, 53.
28. Göteman, M.; Engström, J.; Eriksson, M.; Isberg, J. Fast Modeling of Large Wave Energy Farms Using Interaction Distance Cut-Off. *Energies* **2015**, *8*, 13741–13757.
29. Giassi, M.; Göteman, M. Layout design of wave energy parks by a genetic algorithm. *Ocean Eng.* **2018**, *154*, 252–261.
30. Tyrberg, S.; Waters, R.; Leijon, M. Wave Power Absorption as a Function of Water Level and Wave Height: Theory and Experiment. *IEEE J. Ocean. Eng.* **2010**, *35*, 558–564.
31. Hong, Y.; Eriksson, M.; Boström, C.; Waters, R. Impact of Generator Stroke Length on Energy Production for a Direct Drive Wave Energy Converter. *Energies* **2016**, *9*, 730.
32. Linton, C.; McIver, P. *Handbook of Mathematical Techniques for Wave/Structure Interactions*; CRC Press: Boca Raton, FL, USA, 2001.

33. Cartwright, D. The use of directional spectra in studying output of a wave recorder on a moving ship. In Proceedings of the Ocean Wave Spectra: Proceedings of a Conference, Easton, Maryland, 1–4 May 1963; Prentice-Hall, Inc.: Upper Saddle River, NJ, USA, 1963; pp. 203–218.
34. Mitsuyasu, H.; Tasai, F.; Suhara, T.; Mizuno, S.; Ohkusu, M.; Honda, T.; Rikiishi, K. Observations of the directional spectrum of ocean WavesUsing a cloverleaf buoy. *J. Phys. Oceanogr.* **1975**, *5*, 750–760.
35. Falnes, J. *Ocean Waves and Oscillating Systems: Linear Interactions Including Wave-Energy Extraction*; Cambridge University Press: Cambridge, UK, 2002.

Article

Hydrodynamic Performance of an Array of Wave Energy Converters Integrated with a Pontoon-Type Breakwater

De Zhi Ning [1,*], Xuan Lie Zhao [1], Li Fen Chen [1,2] and Ming Zhao [1,3]

[1] State Key Laboratory of Coastal and Offshore Engineering, Dalian University of Technology, Dalian 116024, China; zhaoxuanlie@163.com (X.L.Z.); chenlifen239@163.com (L.F.C.); M.Zhao@westernsydney.edu.au (M.Z.)

[2] Centre for Offshore Foundation Systems, The University of Western Australia, 35 Stirling Highway, Perth, WA 6009, Australia

[3] School of Computing, Engineering and Mathematics, University of Western Sydney, Penrith, NSW 2751, Australia

* Correspondence: dzning@dlut.edu.cn

Received: 10 February 2018; Accepted: 15 March 2018; Published: 18 March 2018

Abstract: The cost of wave energy converters (WECs) can be reduced significantly by integrating WECs into other marine facilities, especially in sea areas with a mild wave climate. To reduce the cost and increase the efficiency, a hybrid WEC system, comprising a linear array (medium farm) of oscillating buoy-type WECs attached to the weather side of a fixed-type floating pontoon as the base structure is proposed. The performance of the WEC array is investigated numerically using a boundary element method (BEM) based on the linear potential flow theory. The linear power take-off (PTO) damping model is used to calculate the output power of the WEC array. The performance of the WEC array and each individual WEC device is balanced by using the mean interaction factor and the individual interaction factor. To quantify the effect of the pontoon, the hydrodynamic results of the WEC arrays with and without a pontoon are compared with each other. Detailed investigations on the influence of the structural and PTO parameters are performed in a wide wave frequency range. Results show that the energy conversion efficiency of a WEC array with a properly designed pontoon is much higher than that without a pontoon. This integration scheme can achieve the efficiency improvement and construction-cost reduction of the wave energy converters.

Keywords: wave energy converter; pontoon; efficiency improvement; hybrid system; linear potential flow theory

---

1. Introduction

Although many concepts of wave energy converters (WECs) have been developed, few of them have been realized, mainly because the high construction cost prevents them from being implemented [1,2]. For wave energy converters, reducing the construction-cost and improving the energy conversion efficiency are of great significance.

The construction-cost reduction can be achieved by combining WECs with other marine facilities, such as breakwaters, offshore platforms, ships, etc. [3–6]. Recently, many studies have been aimed at investigating the performance of hybrid systems, such as WEC-wind turbine [3,7] and WEC-breakwater hybrid systems [6,8–10]. Hybrid structures are very important for promoting wave energy harvesting because they reduce the cost.

The energy conversion efficiency of WECs can be improved through various methods, such as replacing large-size buoys with many small buoys [11], latching control [12,13], using multi-stable mechanisms [14], nonlinear model predictive control methodology [15], etc. Modifications of the shape of the WEC devices or power take-off (PTO) control systems are required in the aforementioned methods. To achieve the construction cost reduction and the improvement of the energy conversion performance of WECs simultaneously, it is of practical importance to design a hybrid-structure that can improve the energy conversion performance of the WECs significantly without modifying the original WEC devices.

Utilizing the energy from both the incident waves and the reflected waves from offshore structures has a great potential to improve the performance of WECs. The superposition of the incident waves and the reflected waves at the weather side of a pontoon-type breakwater can be beneficial to the improvement of the performance of the WECs located in this region. Viviano et al. [16] experimentally investigated the reflection coefficient of a land-based oscillating water column (OWC) WEC and their results showed that an OWC structure integrated into the vertical wall of a breakwater can also serve as a wave absorber for reducing wave reflection. Howe and Nader [17] demonstrated that the efficiency of a breakwater-integrated OWC device is significantly higher than that of an isolated OWC type WEC. McIver and Evans [18], Mavrakos et al. [19], and Schay et al. [20] found that the power obtained by point absorbers in front of a reflecting seawall is much more than that obtained by the same point absorbers without a seawall. Through analytical analysis, Zhao et al. [21] proved that the capture width ratio of a two-dimensional oscillating buoy-type WEC device in front of a fixed pontoon could be increased significantly compared to that of the same WEC without a pontoon. In the above-mentioned studies, the efficiency of WEC devices was improved because of the superposition of the incident and reflected waves from a seawall or a pontoon. In addition, attaching a WEC array to the weather side of a breakwater can provide power to the offshore operation conveniently. Meanwhile, the breakwater can provide the supporting structure for the WEC array. For the convenience of discussion, a hybrid system of a WEC array attached to a pontoon is named as a WEC-pontoon system in this paper.

The need to reduce costs drives the design of WEC-pontoon systems in the form of an array, so that the components of the WECs can be shared [22–24]. To the best knowledge of the authors, the performance of an array of WECs attached to a pontoon-type structure has not been studied systematically. Thorough investigation of the performance of a WEC array installed at the weather side of a pontoon-type structure will further advance the technology of WECs. Although the hydrodynamics of an array of WECs has been investigated widely [25], little research has been conducted to study a breakwater-WEC system comprising an array of WECs (at the medium level at least). In this paper, a breakwater-WEC system with an array of heaving-mode WECs that are installed at the weather side of a pontoon structure is proposed, and its hydrodynamic performance is investigated. The heaving-mode WECs are selected in the proposed system [26].

The commonly used methods for evaluating hydrodynamic performance of WECs include frequency-domain models, time-domain models, and spectral-domain models. Detailed reviews of these methods can be found in Refs. [25,27,28]. Frequency domain methods based on the linear potential flow theory are often adopted because of their high computational efficiency. Frequency domain methods include boundary element methods (BEM) [29–32] and matching eigen-function methods [33–38]. Compared with the BEM, the matching eigen-function method is more efficient, but not suitable for problems with complex geometries. Thus, as a preliminary study, the performance of a heaving-mode WEC array installed at the weather side of a pontoon is investigated using a higher order boundary element method (HOBEM) code package WAFDUT. The linear PTO damping model is used to calculate the output power.

The rest of the paper is structured as follows: In Section 2, the integrated system and the selected numerical method is introduced. In Section 3, the validation of the proposed numerical method, results for the parametric study, and some discussions are presented. Finally, conclusions are drawn in Section 4.

2. Hybrid System and Numerical Method

2.1. Hybird System

Figure 1 shows a sketch of a WEC-pontoon system, which comprises a collinear array of circular cylindrical heaving-mode WECs with PTO systems and a fixed pontoon. The similar collinear WEC array without a pontoon has been investigated by Bellew et al. [39]. The kinetic energy of the heave motion of each WEC is converted into energy by a PTO system. The fixed pontoon is pile-supported. The PTO system is fixed between the WECs and the pontoon. All the WEC devices in the array are identical and equally spaced. The radius and draft of each WEC device are defined as a and d, respectively, and the distance between two adjacent WEC devices (WEC-WEC spacing) is s_1. The distance of the collinear array and weather side of the pontoon (WEC-pontoon spacing) is s_2. The rear pontoon is defined by the length D, the breadth B, and the draft T. A medium-size WEC array generally has 10–30 WEC devices [40]. A WEC array with 11 heaving-type oscillating buoy devices is considered in this study.

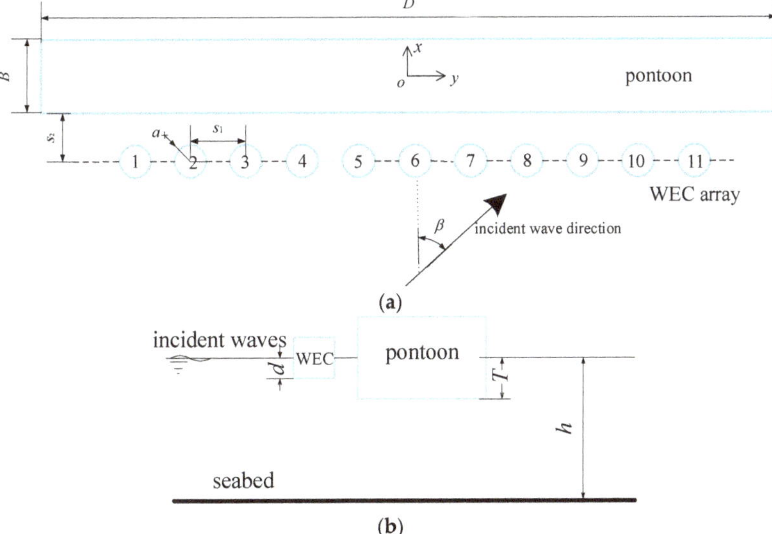

Figure 1. The sketch of the proposed WEC-pontoon system: (**a**) the top view; and (**b**) the side view (the WECs in the array are labelled as devices #1–#11, respectively, on the top view).

2.2. Numerical Method

To investigate the hydrodynamic performance of the pontoon-WEC system, the HOBEM code package WAFDUT is used to solve the diffraction and radiation problems of the multi-body system. The program WAFDUT was developed based on the linear potential flow theory and the HOBEM is used to implement the integral equation to solve the hydrodynamic problems [41]. This HOBEM model computes six-degrees-of-freedom wave-induced motions of multiple bodies with arbitrary shapes in frequency domain. More applications of the program WAFDUT can be found in [42,43].

The rear pontoon is stationary and the buoys of the WECs only move vertically. The equation of motion of the WEC devices in the frequency domain can be written as:

$$\left\{ -\omega^2 \left(\begin{bmatrix} M_1 & & \\ & \ddots & \\ & & M_N \end{bmatrix} + \begin{bmatrix} \mu_1^1 & \cdots & \mu_1^N \\ \vdots & \ddots & \vdots \\ \mu_N^1 & \cdots & \mu_N^N \end{bmatrix} \right) - i\omega \left(\begin{bmatrix} \lambda_1^1 & \cdots & \lambda_1^N \\ \vdots & \ddots & \vdots \\ \lambda_N^1 & \cdots & \lambda_N^N \end{bmatrix} + \begin{bmatrix} \lambda_{PTO,1} & & \\ & \ddots & \\ & & \lambda_{PTO,N} \end{bmatrix} \right) + \begin{bmatrix} K_1 & & \\ & \ddots & \\ & & K_N \end{bmatrix} \right\} \begin{pmatrix} A_{R,1} \\ \vdots \\ A_{R,N} \end{pmatrix} = \begin{pmatrix} F_{z,1} \\ \vdots \\ F_{z,N} \end{pmatrix} \quad (1)$$

where ω is the wave angular frequency, i the imaginary unit, M_n and K_n are the mass and stiffness of the nth body, respectively, and μ_i^j and λ_i^j are the added mass and damping coefficient of the ith device in the heave mode due to the heave motion of the jth device, respectively. The diagonal element $\lambda_{\text{PTO},n}$ in the PTO damping matrix λ_{PTO} is the PTO damping acting on the nth device. $A_{R,n}$ and $F_{z,n}$ are the heave response amplitude and wave exciting force in heave mode of the nth device ($n = 1{\sim}N$), respectively. N is the total number of the WECs in the array.

The power $P_n(\omega)$ produced by the nth device at frequency of ω is calculated by:

$$P_n(\omega) = \frac{1}{2}\omega^2 \lambda_{\text{PTO},n} |A_{R,n}|^2 \tag{2}$$

Then the total power absorbed by the WEC array at the frequency of ω is:

$$P_{\text{total}}(\omega) = \sum_{n=1}^{N} P_n(\omega) \tag{3}$$

To quantify the influence of the hydrodynamic interactions on the efficiency of the WEC array, mean interaction factor q_{mean} is introduced:

$$q_{\text{mean}}(\omega) = \frac{P_{\text{total}}(\omega)}{N \times P_{\text{isolated}}(\omega)} \tag{4}$$

where $P_{\text{isolated}}(\omega)$ denotes the maximum absorbed power of an isolated WEC device at the frequency of ω, which can be obtained by using the optimal PTO damping [38]. q_{mean} reflects the hydrodynamic interactions on the performance of an array: a value $q_{\text{mean}} > 1$ reflects that there is positive interaction in an array, whereas $q_{\text{mean}} < 1$ shows a destructive interaction.

Especially, the individual interaction factor $q_{\text{ind},n}$ is used to investigate the performance of the individual buoy (nth device) in the array, which is defined as [41]:

$$q_{\text{ind},n}(\omega) = \frac{P_n(\omega)}{P_{\text{isolated}}(\omega)} \tag{5}$$

where P_n denotes the absorbed power by the nth device in the array.

The dimensionless total extracted power of the WEC array ($P_{\text{total,d}}$) and the dimensionless extracted power of the isolated WEC ($P_{\text{isolated,d}}$) are defined as $P_{\text{total}}/P_{\text{in}}$ and $P_{\text{isolated}}/P_{\text{in}}$, respectively. P_{in} indicates the incident wave energy with a width of $2a$.

The mean interaction factor and the individual interaction factor can be used to balance the performance of the WEC array and each individual device.

3. Results and Discussion

3.1. Validations

The accuracy of the HOBEM model in predicting the hydrodynamic performance of the WEC array presented in Section 2.2 is validated by comparing the present results with the published benchmark results of the mean interaction factor q_{mean} for a 5×1 WEC array [39]. The WEC array consists of five hemispheres with a same radius a that only oscillate in the heave direction. The WEC-WEC spacing of $4a$ and a water depth of $7a$ were considered. The WEC array was under the action of beam seas. The hydrodynamic coefficients and wave forces were calculated by using the commercial software WAMIT in [44]. The mass of each device was twice of the displaced mass of the device. In the HOBEM simulations using WAFDUT, 150 elements are used for each hemisphere during the calculations (see Figure 2). The calculated mean interaction factor q_{mean} are compared with the numerical results in Bellow [39] in Figure 3 and a good agreement is achieved. This provides confidence in the HOBEM model for investigating the hydrodynamics of the proposed complex WEC-pontoon system, a new concept of WECs.

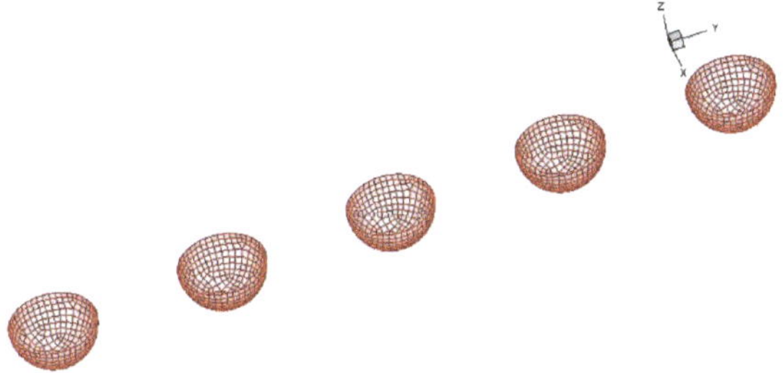

Figure 2. 3D mesh of the calculation model for the five hemispherical WEC devices in the validation test.

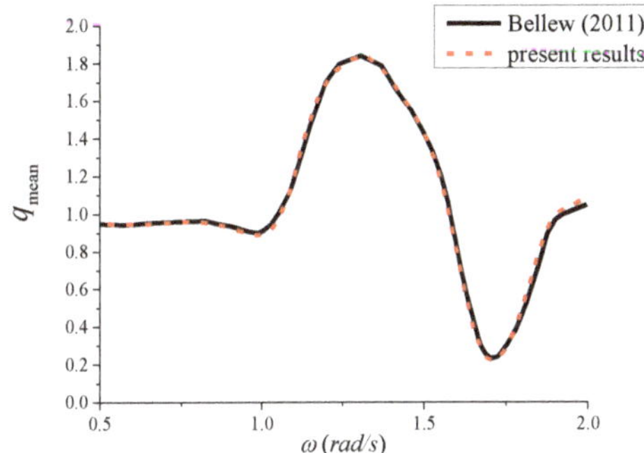

Figure 3. Comparison of the calculated mean interaction factor q_{mean} with the numerical results in [39].

3.2. Parametric Study

The shape of the considered WEC devices is circular cylinder. Three parameters remain unchanged: the water depth of $h = 10$ m, the draft $d/h = 0.2$, and the radius $a/h = 0.135$. The effects of the following parameters on the performance of the WEC array are investigated: the WEC-pontoon spacing s_2, WEC-WEC distance s_1, the wave incident angle β, the dimensions of the rear pontoon (draft T, length D, and breadth B) and PTO damping $\lambda_{PTO,n}$. Eighty-four elements for each front buoy and 700 elements for the pontoon are used throughout the calculations after the spatial convergence tests (see Figure 4). For comparisons, results of the corresponding conventional isolated WEC array without the pontoon are also included. Note that, for the China coastal areas, the frequency range of $1 < kh < 4.5$ is of interest. The frequency range for the calculations is selected as $0 < kh < 6$ in the present study.

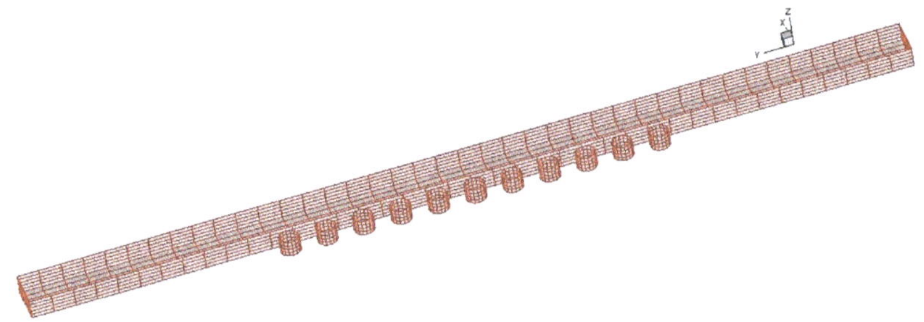

Figure 4. 3D mesh of the calculation model for the parametric study. ($T/h = 0.25$, $D/h = 12$, $B/h = 0.6$, $s_1/h = 0.5$, $s_2/h = 0.2$)

3.2.1. Effect of the WEC-Pontoon Spacing s_2

Five WEC-pontoon spacings are considered in this subsection. The parameters of draft T/h, the length D/h, and the breadth B/h of the pontoon, WEC-WEC spacing s_1/h, and WEC-pontoon spacing s_2/h are summarized in Table 1. The parameters of the pontoon used here will also be used in Sections 3.2.2 and 3.2.3, and Section 3.2.5. In Sections 3.2.1–3.2.4, the diagonal element ($\lambda_{\text{PTO},n}$) of PTO damping matrix is chosen as the optimal PTO damping corresponding to the nth device in isolation, which can be calculated as:

$$\lambda_{\text{PTO},n} = \sqrt{(K_n/\omega - \omega(M_n + \mu_n))^2 + \lambda_n^2} \tag{6}$$

where μ_n and λ_n represent the added mass and damping coefficient of the nth buoy in the isolated case, respectively [45]. For an isolated WEC in the open water, the maximum produced power can be achieved at a wave frequency ω by using the optimal PTO damping $\lambda_{\text{PTO},n}$. Note that the values of diagonal elements of the PTO damping matrix are the same. The mass term and stiffness term of the PTO system are not considered in this study.

Table 1. Parameters considered in Section 3.2.1.

T/h	D/h	B/h	s_1/h	s_2/h
0.25	12	0.6	0.5	0.2, 0.25, 0.3, 0.35, 0.4

From Figure 5 it can be seen that two peak values and one trough value can be found for the WEC-pontoon system with smaller s_2/h (=0.2, 0.25, 0.3) within the frequency range of $0 < kh < 6$, and both the first and the second peak values decrease with increasing s_2. The second peak value is greater than the first one for cases with $s_2/h = 0.2$ and the reverse trend can be found for cases with $s_2/h = 0.25$ and 0.3. However, the second peak value approaches zero for cases with greater s_2 (i.e., $s_2/h = 0.35, 0.4$). The peak and the trough values of q_{mean} shift towards the lower frequency range with increasing s_2 (except for the cases where the second peak value vanishes). To illustrate the trend of the q_{mean}, which is characterized by the two peaks and a trough value between them, the dimensionless quantity of the numerator ($P_{\text{total},d}$) and the denominator ($P_{\text{isolated},d}$) in Equation (4) are presented in Figure 6. From Figure 6, the parabolic trend with a slight oscillation can be observed for $P_{\text{total},d}$ vs. kh. Comparatively, for $P_{\text{isolated},d}$ vs. kh, the unique peak value corresponding to the slight oscillation of $P_{\text{total},d}$ vs. kh can be found. The presence of the trough value at a critical wavenumber k_c is attributed to the fact that the denominator of the q_{mean} occurs the maximum value at k_c (see Figure 6). Therefore, the location of the trough value in Figure 5 is corresponding to the natural frequency in heave mode of the isolated WEC device. And the presence of the trough value lead to the two apparent peak values on its two sides. Furthermore, it can be seen that the energy conversion performance of the pontoon-integrated array (with proper parameter design) is obviously better than that of the isolated WEC array (i.e., without the pontoon) in a wide frequency range. Specifically, $q_{\text{mean}} > 3$ can be found in some frequency range.

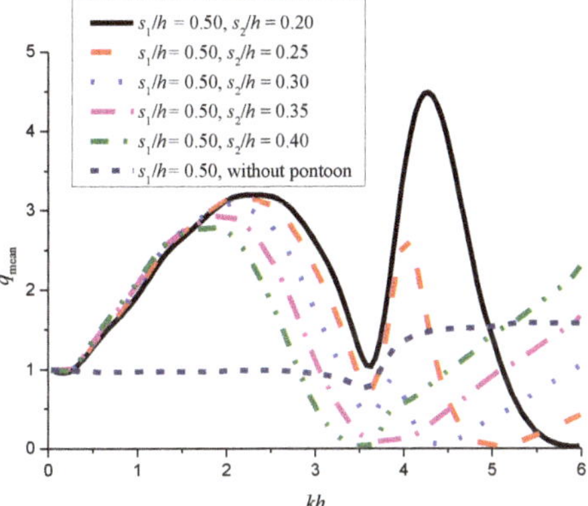

Figure 5. Variations of mean interaction factor q_{mean} with the dimensionless wave number *kh* for different WEC-pontoon spacing s_2.

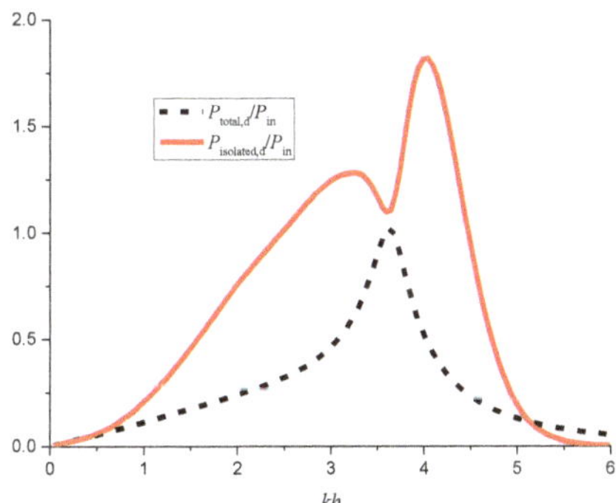

Figure 6. Variations of dimensionless total extracted power of the WEC array ($P_{\text{total,d}}$) and the dimensionless extracted power of the isolated WEC ($P_{\text{isolated,d}}$) with the dimensionless wave number *kh*.

The dimensionless exciting wave force in the heave-mode ($F_z/\rho\pi gAa^2$), the added mass ($\mu_n{}^n/\rho\pi a^2 d$) and the damping coefficient ($\lambda_n{}^n/\rho\omega\pi a^2 d$) of individual devices for both the isolated and integrated WEC array (i.e., with and without pontoon) are shown in Figure 7.

The symbols ρ and g are the water density and the gravitational acceleration, respectively. Note that only the diagonal elements of the added mass matrix and damping coefficient matrix are presented. Since the WEC array is under the action of beam waves ($\beta = 0°$), the hydrodynamic coefficients of all WEC devices are symmetrical about the central point of the array. Thus, only the results for the WEC devices #1–#6 are given. The dimensionless added mass of each device of the WEC array without the pontoon decreases with increasing *kh*. The dimensionless added masses of the WEC-pontoon system have minimum values at about *kh* = 4.75. However, the added mass of each WEC device of the array without the pontoon is smaller than that of the WEC-pontoon system. The damping coefficients of all the WEC devices in both WEC arrays with and without a pontoon increase firstly and then decrease with increasing *kh*. Additionally, the damping coefficient of the WEC array without the pontoon is less than that of the WEC array with the pontoon in the frequency range of *kh* < 5. For both the WEC array without and with the pontoon, the damping coefficients of the WEC devices at the edges of the array (i.e., device #1 and #11) are smaller than the others. The exciting wave force of each WEC device decreases with increasing kh for the array without the pontoon. However, the exciting forces of the pontoon-integrated array increase firstly and then decrease with increasing kh. The exciting force of each device in the pontoon-integrated array is greater than that in the array without the pontoon in the range of $kh < 5$ approximately. However, the reverse trend is presented in the range of $kh > 5$. For the pontoon-integrated array, the improvement of the capture efficiency is mainly attributed to the incensement of the exciting wave force acting on the WEC devices, which is ultimately due to the wave reflection from the pontoon.

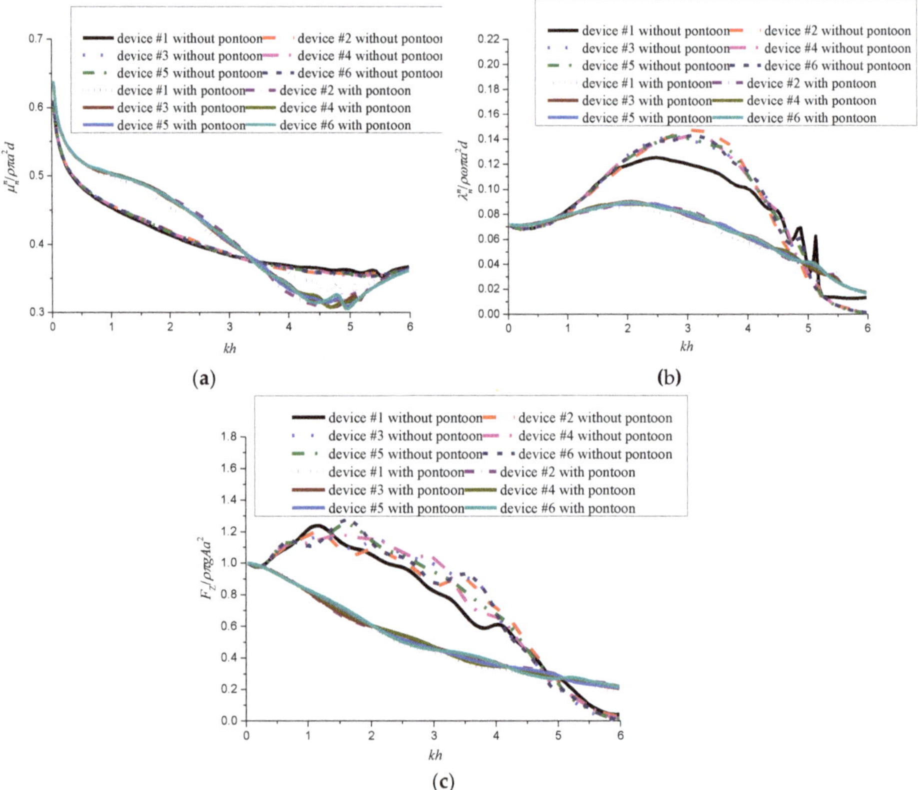

Figure 7. Variations of dimensionless heave added mass (**a**), dimensionless heave damping coefficient (**b**), and dimensionless heave wave exciting force (**c**) of individual WECs with and without the pontoon. ($s_2/h = 0.2$, $\beta = 0°$).

3.2.2. Effect of the WEC-WEC Spacing s_1

The arrays with WEC-WEC spacing of $s_1/h = 0.3$, 0.5, 0.7, and 0.9 are considered, while the WEC-pontoon distance remains $s_2/h = 0.20$. The corresponding results for the array without the pontoon with the same WEC-WEC spacing are also calculated. The results in Figure 8 show that, for the WEC-pontoon system, the variations of q_{mean} with kh for different s_1 are similar to each other and they are very different from the corresponding array without a pontoon. The q_{mean} of a WEC array without the pontoon is nearly 1 until $kh = 3$, beyond which q_{mean} slightly increases with increasing kh. The significant increase in the efficiency of the WEC array with the pontoon can be found at the two frequency ranges (high-efficiency ranges), which are defined as range-1 and range-2 for the low- and high-frequency ranges, respectively. A minimum value of q_{mean} occurs between these two ranges of kh. The maximum q_{mean} in the range-1 is lower than that in range-2. However, range-1 is much wider than range-2. When $s_1/h > 0.50$, even the minimum value of q_{mean} between the range-1 and range-2 is greater than 1, indicating that the efficiency of the WEC-array with a pontoon is higher than that without a pontoon at a wide range of kh. A WEC-array with a very small s_1/h (=0.3) is found to have a smaller q_{mean} than any other larger s_1/h considered. Between these two high-efficiency ranges, q_{mean} reaches its minimum value. In practice, the minimum value of q_{mean} should be avoided in order to gain as much energy as possible.

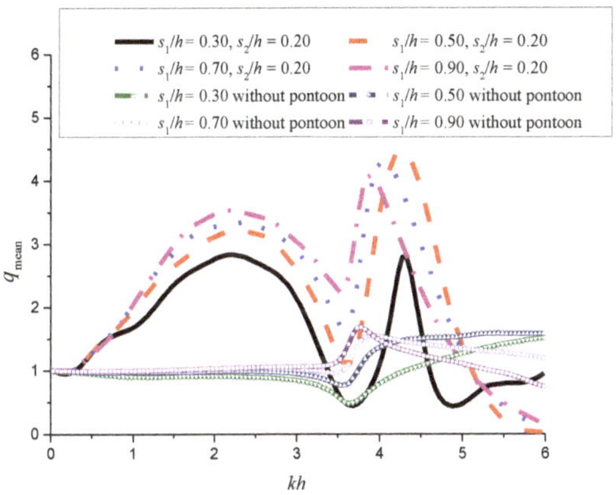

Figure 8. Variations of q_{mean} with kh for WEC arrays with different WEC-WEC spacing of $s_1/h = 0.3$, 0.5, 0.7, and 0.9. The WEC-pontoon spacing is $s_2/h = 0.2$.

3.2.3. Effect of the Wave Incident Angle β

The WEC arrays with $\beta = 0°$, $30°$, $60°$, and $90°$ are considered. The WEC-WEC spacing of the array and the WEC-pontoon distance are $s_1/h = 0.5$ and $s_2/h = 0.2$, respectively. Figure 9 shows the variation of q_{mean} with kh for all the wave incident angles considered. Furthermore, to investigate the performance of each individual WEC device, the results of individual interaction factors of all WEC devices are presented in Figure 10.

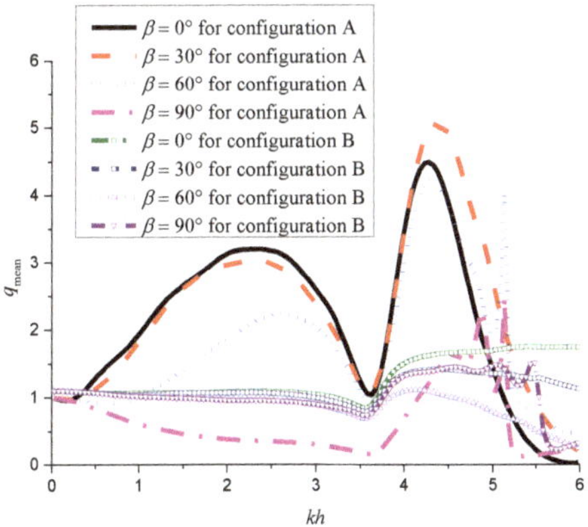

Figure 9. Variations of q_{mean} with kh for WEC array with (configuration A) and without (configuration B) pontoon with different wave incident angles.

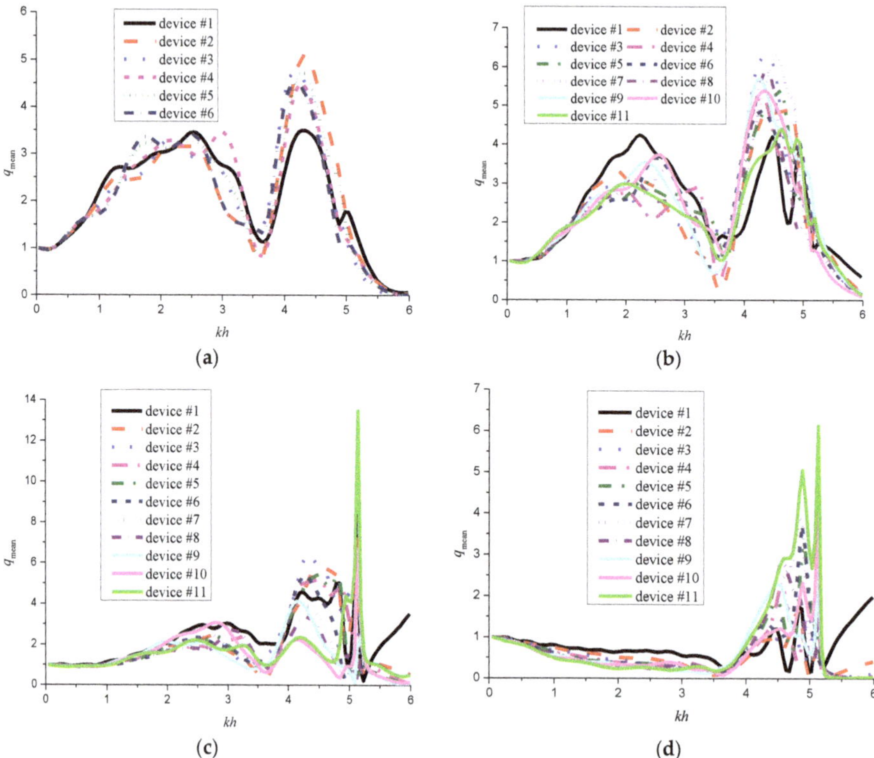

Figure 10. Individual interaction factors of all WEC devices for the WEC-pontoon system with $\beta = 0°$ (**a**), 30° (**b**), 60° (**c**), and 90° (**d**).

It can be seen from Figure 9 that the effect of the wave-incident angle on q_{mean} is quite significant. The variation trends of q_{mean} with kh for $\beta = 0°$, 30°, and 60° are similar with each other. For these three incident angles, two high-efficiency ranges are found. As $\beta = 60°$, q_{mean} in the first high-efficiency range is significantly reduced, and as $\beta = 90°$ the first high-efficiency range disappears and the q_{mean} in the second high-efficiency range significantly reduces. This is due to a weaker interaction between the WEC array and the pontoon when the incident wave is parallel with the pontoon. The effect of the wave incident angle on q_{mean} in the second high-efficiency range is the smallest for $\beta \leq 60°$. The strong wave-multibody structure interactions in shorter waves lead to the spiking observed at the high frequency range for larger wave incidence angle (i.e., $\beta = 60°$ and 90°). Similarly, the spiking phenomenon can be observed for the wave exciting forces (see Figure 11).

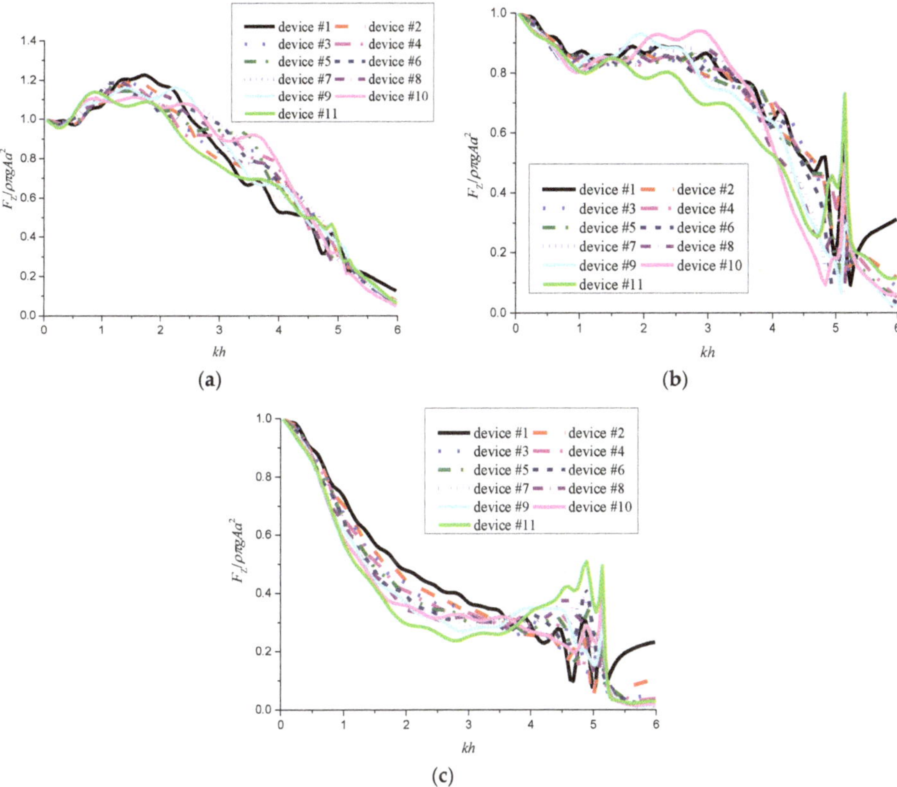

Figure 11. Dimensionless heave wave exciting forces on each WEC device for $\beta = 30°$ (**a**), $60°$ (**b**), and $90°$ (**c**).

For the WEC array without a pontoon, the effect of the β on the efficiency is weak when $kh < 3.75$. When $kh > 3.75$, the minimum q_{mean} occurs when $\beta = 60°$. For the pontoon-integrated array, as $kh > 5.5$, q_{mean} of the array becomes lower than that of a WEC array without the pontoon, which is due to the fact that the standing waves can be formed at this frequency band. The standing waves are not beneficial for the energy extraction of the oscillating buoy WEC.

In Figure 10, the variations of the q_{ind} with kh follow the trend of the mean interaction factor q_{mean}. The difference between the q_{ind} of all the WEC devices increases with increasing β. At large values of β the interaction factor varies among the WEC devices mainly because the WEC devices in the lee side of the array are significantly affected by the weather side WEC devices. For $\beta = 60°$ and $90°$, a 'spike' of q_{ind} can be found for devices close to the weather sea side at the higher frequency range in Figure 10.

This spike is corresponding to the spike in the wave exciting force at the corresponding location for $\beta = 60°$ and $90°$ shown in Figure 11b,c. In addition, it can also be seen from Figures 11a–c and 7c that the variation of wave exciting force against kh of the WEC devices for $\beta = 0°$ and $30°$ is different from that for $\beta = 60°$ and $90°$. It is mainly reflected in that the decreasing trend with increasing kh and 'spikes' in the high-frequency range can be found for cases with $\beta = 60°$ and $90°$. At the locations of the 'spikes', the wave exciting force of the WEC devices at the shadow side is greater than that of the others. By contrast, the trend that increases first, and then decreases can be found and no 'spikes' are found for cases with $\beta = 0°$ and $30°$. Strong high-frequency oscillations of the exciting forces can be observed for the multi-bodies under action of (quasi) beam seas. This may lead to the spikes in at the high frequency range while the pontoon-WEC system is under action of the oblique waves with $\beta = 60°$ and $90°$.

3.2.4. Effect of the Dimension of the Rear Pontoon

Compared with the WEC array without a pontoon, the improvement of the energy conversion performance of the WEC-pontoon system is attributed to the reflected waves caused by the rear pontoon. It is believed that the dimensions of a pontoon determine the reflection coefficient and transmission coefficient. The influence of the draft T, breadth B, and length D of the rear pontoon on system performance is investigated in this section. While both parameters s_1 and s_2 are fixed as $0.5\,h$, calculations are performed for $B/h = 0.6$, $D/h = 12$, $T/h = 0.15$, 0.20, 0.25, 0.30, 0.35, and 0.40. The corresponding results are presented in Figure 12. It can be seen that the variations of q_{mean} with kh for different drafts of the rear pontoon are similar with each other. With increasing T, the first peak value of q_{mean} increases and second peak value decreases. The kh corresponding to both peak values shift towards a lower frequency range with increasing T. For cases in the low-frequency range, the increase in the draft leads to an increase of the reflection coefficient. Thus, the increase of the efficiency can be found. For cases in the high frequency range, the increase in the draft leads to strong standing waves. This may have a negative impact on the wave energy extraction of the heaving WEC devices. i.e., reduce the efficiency of the system. However, the minimum value and its location changes little with increasing T. Although the modification of the q_{mean} occurs for different cases, the q_{mean} is still greater than 1 *as* $kh < 5$.

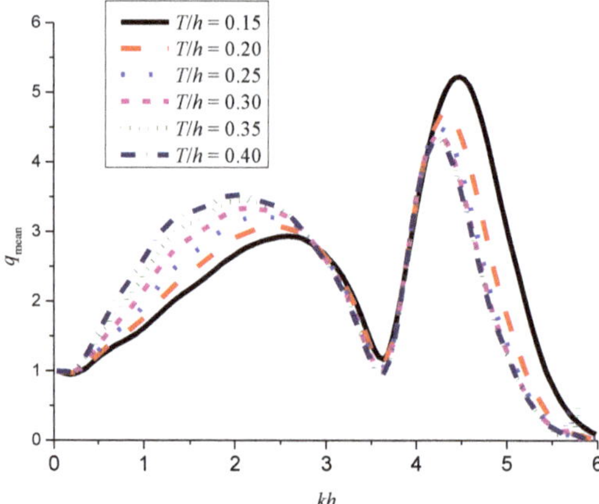

Figure 12. Variations of q_{mean} vs. kh for the WEC array with different drafts T of the rear pontoon.

The effect of pontoon breadth is investigated by carrying out calculations with $D/h = 12$, $T/h = 0.25$ and four pontoon breadths of $B/h = 0.4$, 0.6, 0.8, and 1.0. The corresponding results are given in Figure 13. The breadth is found to affect q_{mean} in the lower frequency range $0.5 < kh < 2.6$, where q_{mean} increases with increasing breadth. q_{mean} in the range of $kh > 2.6$ are nearly independent on the pontoon breadth. This is due to the fact that the change of the pontoon width only affects the hydrodynamic interactions of the pontoon-WEC system in the longer waves.

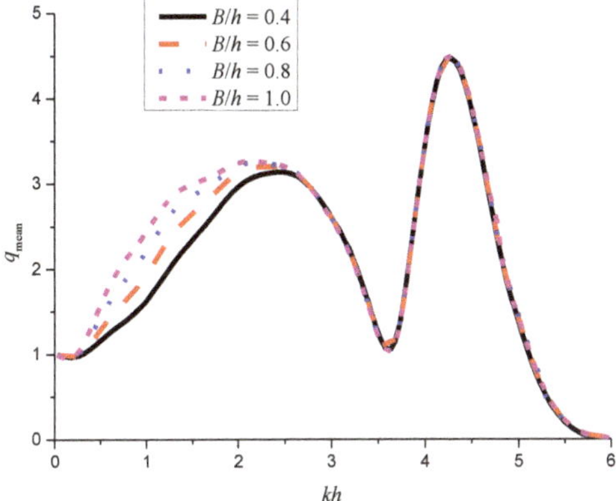

Figure 13. Variations of q_{mean} vs. *kh* for the WEC array with different breadths *B* of the rear pontoon.

Finally, the effect of length of the pontoon is investigated by the calculations with $T/h = 0.25$, $B/h = 0.6$ and four pontoon lengths of $D/h = 12, 16, 20$, and 24. From the corresponding results in Figure 14, it can be seen that the length of the pontoon has little effect on the variations of q_{mean}.

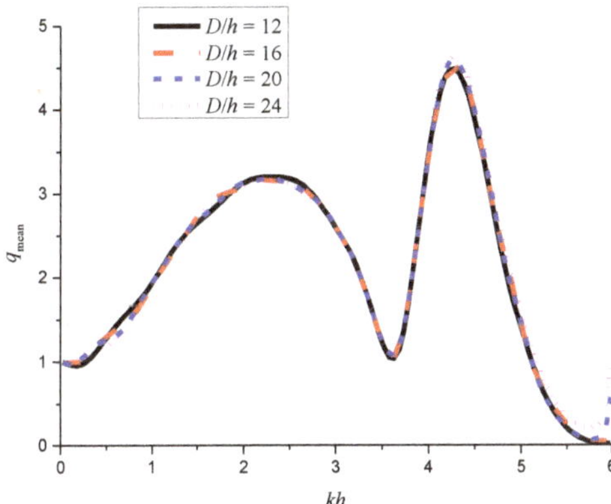

Figure 14. Variations of q_{mean} vs. *kh* for the WEC array with different lengths *D* of the rear pontoon.

3.2.5. Effect of the PTO Damping

PTO system is an essential part of WECs. In the aforementioned sections, the diagonal elements $(\lambda_{\text{PTO},n})$ of the PTO damping matrix are chosen as the optimal PTO damping of an isolated WEC device. The PTO damping can be calculated as:

$$\lambda_{\text{PTO},n} = C \cdot \sqrt{(K_n/\omega - \omega(M_n + \mu_n))^2 + \lambda_n{}^2} \tag{7}$$

where C is a constant. To investigate the effects of the coefficient C, calculations for various $C = 0.5, 1, 1.5, 2$, and 2.5 are performed. The WEC-WEC spacing and the WEC-pontoon spacing are $s_1/h = 0.5$ and $s_2/h = 0.2$, and the wave incidence angle $\beta = 0°$ is considered. It is clear from Figure 15 that the PTO damping affects the output power significantly and its influences are different for different frequency range. It is mainly reflected in that, within the test cases, the q_{mean} with $C = 1$ is optimal at a lower frequency range (i.e., $kh < 3$, approximately). The valley value increases with increasing PTO damping. However, the system with light PTO damping performs better at the frequency range of $4.0 < kh < 5.0$.

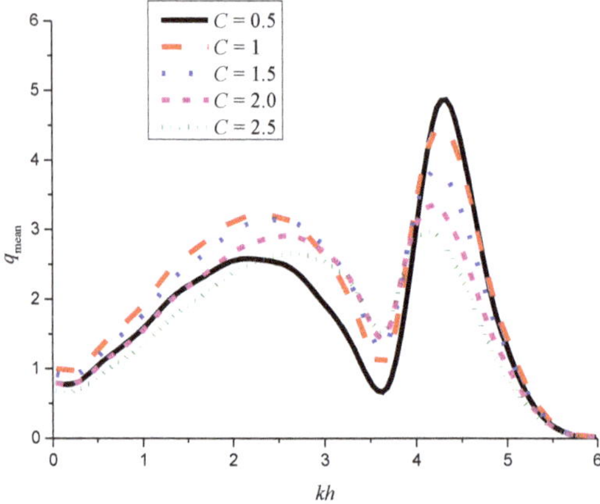

Figure 15. Variations of q_{mean} vs. kh for the WEC array with various PTO damping.

For a WEC array without a pontoon with the same value of diagonal elements of the PTO damping matrix, the maximum output power can be achieved by choosing the optimal diagonal elements of the PTO damping matrix (i.e., $C = 1$) [45]. However, this may be not correct for the WEC-pontoon system due to the interference between the WECs and the pontoon. In this paper, although the effect of the PTO damping is investigated, the maximum output power cannot be obtained. Other optimization strategies, such as using the damping value determined by iterative approach [45], could improve the energy conversion performance of the proposed WEC array. Although the optimal PTO damping matrix is not obtained at the whole tested frequency range, the present results give a guide for choosing PTO damping at different frequency range.

3.2.6. Discussions

One aim of the present work is to compare the energy conversion performance of the pontoon-integrated WEC array and the conventional WEC array (i.e., the array without the pontoon). The coefficient q_{mean} is used to indicate the energy conversion performance of the WEC array. From Sections 3.2.1 and 3.2.3, it can be found that the q_{mean} of the pontoon-integrated WEC array is greater than that of the conventional array with different parameters (WEC-WEC spacing, WEC-poontoon spacing, wave incidence angle) at a wide frequency range. It is interesting that a trough value of q_{mean} (between two peak values) is observed, and the location of the trough value is corresponding to the location of the peak value of the $P_{isolated}(\omega)$ (i.e., the optimal power produced by an isolated WEC device). Despite this, the trough value of the q_{mean} for the pontoon integrated array is greater than that of the corresponding conventional WEC array with same parameters. The q_{mean} of the pontoon-integrated WEC array is less than that of the conventional array at the high frequency range. This is due to the fact that the standing waves form in front of the pontoon for certain frequencies, which may lead to the significant reduction of the heave response of the bodies in front of the pontoon. Anyway, the present integration scheme with proper design may provide a way to improve the energy conversion performance of an array of OB-WECs.

Interestingly, a spike of q_{mean} can be found at the high frequency range while the pontoon-WEC system under action of oblique waves (with $\beta = 60°$ and $90°$). This is due to the fact that a distinct spike of q_{ind} can be found for the WECs in the shadow side. Additionally, at the frequencies corresponding to the spikes, the q_{ind} of WECs in the shadow side is greater than that of the WECs in the front side.

However, the reverse trend can be observed in the lower frequency range. For cases with $\beta = 90°$, the q_{mean} of the pontoon-integrated WEC array is less than that of the conventional WEC array, which remind us that the proposed system should be configured in beam seas or slight oblique seas.

4. Conclusions

The hydrodynamic performance of a hybrid system consisting of an oscillating buoy WEC array and a fixed rear pontoon have been investigated theoretically using linear potential flow theory. The HOBEM program (i.e., WAFDUT) is used to solve the diffraction/radiation problem of the multi-body system. The linear PTO damping is used to calculate the absorbed power. In the validation of the WAFDUT, the calculated mean interaction factor of a linear heaving WEC array agree well with previous published data.

The numerical results for a range of configurations are presented to investigate the influence of the different structural and PTO parameters on the performance of the integrated system. By comparing the energy conversion performance of the pontoon integrated array with that of the conventional array under a variety of parameters, the following conclusions can be found:

(1) The efficiency a properly designed WEC-pontoon system is much higher to that of the WEC array without pontoons over a wide frequency range.

(2) The q_{mean} of the pontoon integrated array is smaller than the convention array in the high frequency range because of the standing waves in front of the pontoon at certain frequencies.

(3) With regard to the dimension of the pontoon, the draft of the pontoon is an important factor that affects the performance of the integrated system.

(4) The performance of the integrated system is sensitive to the PTO damping at a wide frequency range.

(5) Standing waves formed in the front of the pontoon are not beneficial to the energy extraction of WECs. This phenomenon shall be avoided while designing such a system.

The proposed system improves the energy extraction over a wide frequency range and reduces the construction cost through a cost-sharing strategy.

Acknowledgments: The authors would like to acknowledge the financial support the National Natural Science Foundation of China (grant Nos. 51679036, 51628901, and 51761135011) and the Royal Academy of Engineering under the UK-China Industry Academia Partnership Programme (grant no. UK-CIAPP\73).

Author Contributions: Dezhi Ning, Xuanlie Zhao, Lifen Chen and Ming Zhao draft the work, make substantial contributions to the design of the work, make final approval of the version to be published and make agreement to be accountable for all aspects of the work in ensuring that questions related to the accuracy or integrity of any part of the work are appropriately investigated and resolved.

Conflicts of Interest: The authors declare no conflict of interest.

Nomenclature

Abbreviation	Appellation
WEC	wave energy converter
PTO	power take-off
OWC	oscillating water column
CWR	capture width ratio
BEM	boundary element methods
HOBEM	higher order boundary element method
WAFDUT	BEM solver used in this paper
WAMIT	Commercial BEM solver

References

1. Ferro, B.D. Wave and tidal energy: Its emergence and the challenges it faces. *Refocus* **2006**, *7*, 46–48. [CrossRef]
2. Pecher, A.; Kofoed, J.P. *Handbook of Ocean Wave Energy*; Springer International Publishing: Cham, Switzerland, 2017.
3. Astariz, S.; Iglesias, G. Enhancing wave energy competitiveness through co-located wind and wave energy farms. A review on the shadow effect. *Energies* **2015**, *8*, 7344–7366. [CrossRef]
4. Mustapa, M.A.; Yaakob, O.B.; Ahmed, Y.M.; Rheem, C.-K.; Koh, K.K.; Adnan, F.A. Wave energy device and breakwater integration: A review. *Renew. Sustain. Energy Rev.* **2017**, *77*, 43–58. [CrossRef]
5. Vicinanza, D.; Margheritini, L.; Kofoed, J.P.; Buccino, M. The SSG Wave Energy Converter: Performance, Status and Recent Developments. *Energies* **2012**, *5*, 193–226. [CrossRef]
6. Zhao, X.L.; Ning, D.Z.; Zhang, C.W.; Kang, H.G. Hydrodynamic Investigation of an Oscillating Buoy Wave Energy Converter Integrated into a Pile-Restrained Floating Breakwater. *Energies* **2017**, *10*, 712. [CrossRef]
7. Pérez-Collazo, C.; Greaves, D.; Iglesias, G. A review of combined wave and offshore wind energy. *Renew. Sustain. Energy Rev.* **2015**, *42*, 141–153.
8. He, F.; Huang, Z.; Law, A.W. An experimental study of a floating breakwater with asymmetric pneumatic chambers for wave energy extraction. *Appl. Energy* **2013**, *106*, 222–231. [CrossRef]
9. Iuppa, C.; Contestabile, P.; Cavallaro, L.; Foti, E.; Vicinanza, D. Hydraulic performance of an innovative breakwater for overtopping wave energy conversion. *Sustainability* **2016**, *8*, 1226. [CrossRef]
10. Naty, S.; Viviano, A.; Foti, E. Wave Energy Exploitation System Integrated in the Coastal Structure of a Mediterranean Port. *Sustainability* **2016**, *8*, 1342. [CrossRef]
11. Garnaud, X.; Mei, C.C. Comparison of wave power extraction by a compact array of small buoys and by a large buoy. *IET Renew. Power Gener.* **2010**, *4*, 519–530. [CrossRef]
12. Babarit, A.; Duclos, G.; Clément, A.H. Comparison of latching control strategies for a heaving wave energy device in random sea. *Appl. Ocean Res.* **2004**, *26*, 227–238. [CrossRef]
13. Babarit, A.; Clément, A.H. Optimal latching control of a wave energy device in regular and irregular waves. *Appl. Ocean Res.* **2006**, *28*, 77–91. [CrossRef]
14. Younesian, D.; Alam, M. Multi-stable mechanisms for high-efficiency and broad band ocean wave energy harvesting. *Appl. Energy* **2017**, *197*, 292–302. [CrossRef]
15. Son, D.; Yeung, R.W. Optimizing ocean-wave energy extraction of a dual coaxial-cylinder WEC using nonlinear model predictive control. *Appl. Energy* **2017**, *187*, 746–757. [CrossRef]
16. Viviano, A.; Naty, S.; Foti, E.; Bruce, T.; Allsop, W.; Vicinanza, D. Large-scale experiments on the behaviour of a generalised Oscillating Water Column under random waves. *Renew. Energy* **2016**, *99*, 875–887. [CrossRef]
17. Howe, D.; Nader, J. OWC WEC integrated within a breakwater versus isolated: Experimental and numerical theoretical study. *Int. J. Mar. Energy* **2017**, *20*, 165–182. [CrossRef]
18. McIver, P.; Evans, D.V. An approximate theory for the performance of a number of wave-energy devices set into a reflecting wall. *Appl. Ocean Res.* **1988**, *10*, 58–65. [CrossRef]
19. Mavrakos, S.A.; Katsaounis, G.M.; Nielsen, K.; Lemonis, G. Numerical performance investigation of an array of heaving wave power converters in front of a vertical breakwater. In Proceedings of the International Offshore and Polar Engineering Conference, Toulon, France, 23–28 May 2004; pp. 238–245.
20. Schay, J.; Bhattacharjee, J.; Soares, C.G. Numerical Modelling of a Heaving Point Absorber in Front of a Vertical Wall. In Proceedings of the 32nd International Conference on Ocean, Offshore and Arctic Engineering, Nantes, France, 9–14 June 2013.
21. Zhao, X.L.; Ning, D.Z.; Zhang, C.W.; Liu, Y.Y.; Kang, H.G. Analytical study on an oscillating buoy wave energy converter integrated into a fixed box-type breakwater. *Math. Probl. Eng.* **2017**, *2017*, 3960401. [CrossRef]
22. Collins, K.M.; Howey, B.; Hann, M.; Iglesias, G.; Greaves, D.; Vicente, P.; Harnois, V. Breakthroughs in WEC arrays: shared moorings and cablings in the WETFEET project. In Proceedings of the 6th International Conference on Ocean Energy, Edinburgh, UK, 23–25 February 2016.
23. Vicente, P.C.; António, F.D.O.; Gato, L.M.C.; Justino, P.A.P. Dynamics of arrays of floating point-absorber wave energy converters with inter-body and bottom slack-mooring connections. *Appl. Ocean Res.* **2009**, *31*, 267–281. [CrossRef]

24. Mercadé Ruiz, P.; Ferri, F.; Kofoed, P.J. Experimental Validation of a Wave Energy Converter Array Hydrodynamics Tool. *Sustainability* **2017**, *9*, 115. [CrossRef]
25. Folley, M. *Numerical Modelling of Wave Energy Converters: State-of-the-Art Techniques for Single Devices and Arrays*; Academic Press: New York, NY, USA, 2016.
26. Falnes, J. A review of wave-energy extraction. *Mar. Struct.* **2007**, *4*, 185–207. [CrossRef]
27. Folley, M.; Babarit, A.; Child, B.; Forehand, D.; O'Boyle, L.; Silverthorne, K.; Spinneken, J.; Stratigaki, V.; Troch, P. A review of numerical modelling of wave energy converter arrays. In Proceedings of the 31st International Conference on Offshore Mechanics and Arctic Engineering, Rio de Janeiro, Brazil, 1–6 June 2012.
28. Mavrakos, S.A.; McIver, P. Comparison of methods for computing hydrodynamic characteristics of arrays of wave power devices. *Appl. Ocean Res.* **1997**, *19*, 283–291. [CrossRef]
29. Chen, W.X.; Gao, F.; Meng, X.D.; Fu, J.X. Design of the wave energy converter array to achieve constructive effects. *Ocean Eng.* **2016**, *124*, 13–20. [CrossRef]
30. Taghipour, R.; Arswendy, A.; Devergez, M.; Moan, T. Structural analysis of a multi-body wave energy converter in the frequency domain by interfacing WAMIT and ABAQUS. In Proceedings of the 27th International Conference on Offshore Mechanics and Arctic Engineering, Estoril, Portugal, 15–20 June 2008.
31. Fàbregas, F.; Babarit, A.; Clément, A.H. On the numerical modeling and optimization of a bottom-referenced heave-buoy array of wave energy converters. *Int. J. Mar. Energy* **2017**, *19*, 1–15. [CrossRef]
32. Wolgamot, H.A.; Taylor, P.H.; Taylor, R.E. The interaction factor and directionality in wave energy arrays. *Ocean Eng.* **2012**, *47*, 65–73. [CrossRef]
33. Göteman, M. Wave energy parks with point-absorbers of different dimensions. *J. Fluids Struct.* **2017**, *74*, 142–157. [CrossRef]
34. Göteman, M.; Engström, J.; Eriksson, M.; Isberg, J. Optimizing wave energy parks with over 1000 interacting point-absorbers using an approximate analytical method. *Int. J. Mar. Energy* **2015**, *10*, 113–126. [CrossRef]
35. Child, B.F.M.; Venugopal, V. Optimal configurations of wave energy device arrays. *Ocean Eng.* **2010**, *37*, 1402–1417. [CrossRef]
36. Simon, M.J. Multiple scattering in arrays of axisymmetric wave-energy devices. Part 1. A matrix method using a plane-wave approximation. *J. Fluid Mech.* **1982**, *120*, 1–25. [CrossRef]
37. Siddorn, P.; Taylor, R.E. Diffraction and independent radiation by an array of floating cylinders. *Ocean Eng.* **2008**, *35*, 1289–1303. [CrossRef]
38. Konispoliatis, D.N.; Mavrakos, S.A. Hydrodynamic analysis of an array of interacting free-floating oscillating water column (OWC's) devices. *Ocean Eng.* **2016**, *111*, 179–197. [CrossRef]
39. Bellew, S. Investigation of the Response of Groups of Wave Energy Devices. Ph.D. Thesis, University of Manchester, Manchester, UK, 2011.
40. Penalba, M.; Touzón, I.; Lopez-Mendia, J.; Nava, V. A numerical study on the hydrodynamic impact of device slenderness and array size in wave energy farms in realistic wave climates. *Ocean Eng.* **2017**, *142*, 224–232. [CrossRef]
41. Teng, B.; Taylor, R.E. New higher-order boundary element methods for wave diffraction/radiation. *Appl. Ocean Res.* **1995**, *17*, 71–77. [CrossRef]
42. Teng, B.; Gou, Y.; Wang, G.; Cao, G. Motion response of hinged multiple floating bodies on local seabed. In Proceedings of the 24th International Offshore and Polar Engineering Conference, Busan, Korea, 15–20 June 2014.
43. Li, B.B.; Ou, J.P. Concept design of a new deep draft platform. *J. Mar. Sci. Appl.* **2010**, *9*, 241–249. [CrossRef]
44. WAMIT User Manual, WAMIT, Inc. 2008.
45. Bellew, S.; Stallard, T.; Stansby, P.K. Optimisation of a Heterogeneous Array of Heaving Bodies. In Proceedings of the 8th European Wave and Tidal Energy Conference, Uppsala, Sweden, 7–10 September 2009.

Article

State-Space Approximation of Convolution Term in Time Domain Analysis of a Raft-Type Wave Energy Converter

Changhai Liu, Qingjun Yang * and Gang Bao

Department of Fluid Control and Automation, Harbin Institute of Technology, Harbin 150001, China; liuchanghaihit@hotmail.com (C.L.); baog@hit.edu.cn (G.B.)
* Correspondence: 13B908002@hit.edu.cn; Tel.: +86-451-8641-8318

Received: 22 November 2017; Accepted: 3 January 2018; Published: 10 January 2018

Abstract: Two methods, frequency domain analysis and time domain analysis, are widely applied to modeling wave energy converters (WECs). Frequency domain analysis can evaluate the performance of WECs quickly and efficiently, while it refers to a linear model. When it comes to investigations on nonlinear characteristics of the power take-off (PTO) unit of WECs or control for improving the WECs' performance, time domain analysis based on a state-space approximation for the convolution term is more desirable. In this paper, a state-space approximation of the convolution term in a time domain analysis of a raft-type WEC consisting of two rafts and a PTO unit is presented. The state-space model is identified through regression in the frequency domain. Verification of such a type of time domain analysis is conducted by comparison of its simulation results with those calculated by using a frequency domain analysis, and there is a good agreement. Finally, the effects of PTO parameters, wave frequency, surge and heave motions of the joint, and quadratic damping PTO on the power capture ability of the raft-type WEC are investigated.

Keywords: raft-type wave energy converter (WEC); power take-off (PTO); frequency domain; time domain; power capture ability; capture width ratio

---

1. Introduction

Wave energy with its renewable and non-polluting properties is wildly explored in many countries [1,2]. In order to extract and utilize wave energy, many scholars have proposed varieties of energy conversion devices. The most notable device is the point-absorbing wave energy converter (WEC), which has been studied since the late 1970s [3]. Based on the mechanism of the point-absorbing WEC, other types of WECs appeared, such as oscillating-water column WEC, pendulum WEC, and raft-type WEC [4].

The raft-type WEC is a semi-submerged offshore structure, composed of a series of rafts, which are articulated end to end by the joints containing hydraulic power take-off (PTO) units [5]. The device is deployed in a direction parallel to the propagation direction of the incident waves. Under the action of the incident waves, each raft outputs surge, heave, and pitch motions, while only the relative pitch motion between the rafts around the joint forces the hydraulic cylinders to pump high-pressure (HP) oil. Then, the HP oil drives the hydraulic motor shaft to rotate, and this rotation forces the generator to produce electricity [1].

Research on the raft-type WEC originated in 1979, when Haren and Mei [6–8] adopted a two dimensional fluid-structure coupling model to study the contouring rafts by modeling the PTO unit as a linear PTO system. Their contribution included the power absorption efficiency of one raft hinged at a wall in long wave conditions, multiple rafts with variable lengths in shallow water wave conditions, and a three-raft system in both shallow and deep-water wave conditions. By assuming that the raft was a "limp-beam" model, Farley [9] explored the energy conversion ability of a flexible resonant raft.

In later years, the McCabe Wave Pump, another type of raft-type device, composed of three rectangular barges hinged together, where the relative pitch motion between the fore, aft, and the middle barges is used to generate electricity or produce potable water, was studied by Kraemer [10] and Nolan et al. [11]. Kraemer's finding showed that the pitch motion of the hinged barges could be maximized by altering the ratio of the length of barges and the wavelength. Nolan et al. adopted the bond graph method to model a more complicated PTO unit including a non-linear damper for the McCabe Wave Pump. In order to solve the dynamics of multi-body WECs, Paparella et al. [12] applied pseudo-spectral methods to the McCabe Wave Pump using both differential and algebraic equations (DAEs) formulation and ordinary differential equations (ODEs) formulation. Their results showed that pseudo-spectral methods are computationally more stable and require less computational effort for short time steps.

The Pelamis WEC manufactured by Pelamis Wave Power Company is another successful application of the raft-type device. The company has conducted many valuable researches on the Pelamis device including testing on a laboratory scale and under real sea conditions [13–15]. The Pelamis device can use both the relative pitch and relative yaw motions around its joints to generate electricity. The current model, the Pelamis P2 composed of five cylindrical modules, with a total length of 180 m, has a maximum capture width as large as 150% of its 'displacement width' [15].

With the concept of Pelamis or Pelamis-like, Thiam and Pierce [16–18] investigated the radiation loss in a Pelamis-like WEC by assuming an infinitely long flexible circular cylinder floating on the surface of an infinitely deep ocean. Their researches showed that the influence of radiation losses is significant if the overall length of the device is much larger than the wavelength of the incident wave.

In recent years, Zheng et al. [19,20] have presented a six-degree-of-freedom (six-DOF) frequency domain model for a raft-type wave energy conversion device of an elliptical cross section. In their research, the effects of the raft radius of gyration and axis ratio of the raft on the wave energy conversion ability were explored, and they conducted a theoretical study on the maximum power that the raft-type device could extract by using a linear PTO model. To investigate the effect of latching control, a six-DOF time domain model based on Cummins' equation for the raft-type device was also presented by Zheng et al. [20].

By using the Lagrange multiplier technique, Sun et al. [21] provided a frequency domain linear diffraction model to investigate the responses of interconnected floating bodies. Then, Sun et al. [22,23] applied this frequency domain linear diffraction model to a raft-type-like WEC M4 in regular waves and irregular multi-directional waves to predict relative pitch rotation and power capture. Their results showed a good agreement of relative rotation and power capture with experiments. More recently, Stansby et al. [24] presented a time domain linear diffraction model based on Cummins' method for the raft-type-like WEC M4 to investigate large capacity multi-float configurations.

So far, most of the previous researches on the raft-type WEC have been conducted by using a frequency domain analysis with the assumption of a linear model, or using a time domain analysis based on Cummins' equation with a convolution term. When the research includes control to improve the WEC's performance, a time domain analysis with the convolution term approximated by a state-space model is more desirable, since such a type of time domain analysis is well suited for controller design and simulation [25,26]. Recent literature [27] has presented such a type of time domain analysis for a raft-type WEC consisting of two rafts and a PTO unit. However, the details of how to reformulate the convolution term into a state-space model and the realization of such a state-space model were not reported. This paper focuses on providing the details regarding the state-space modeling procedure for approximating the convolution term in the time domain analysis of the raft-type WEC. Besides, based on such a type of time domain analysis, the effects of PTO parameters, wave frequency, surge and heave motions of the joint, and quadratic damping PTO on the power capture ability of the raft-type WEC are explored.

2. Description of the Device

Figure 1 presents a raft-type WEC composed of two floating rafts and a hydraulic PTO unit. The two rafts with a cylindrical shape are articulated by a joint with a gap d_0 between them, as shown in Figure 1b. The length, diameter, and density are denoted as L, D, and ρ_0, respectively. The motion characteristic of the device is formulated in a Cartesian coordinate (x, y and z) system with its origin O coincident with the center of the joint, where the z-axis is in the vertical direction, while the x- and y-axes are taken along the length and the diameter direction of the rafts in still water, respectively, as shown in Figure 1a. For the fore raft, the displacements are labeled as x_1-surge, z_1-heave, and θ_1-pitch. For the aft raft, the displacements are labeled as x_2-surge, z_2-heave, and θ_2-pitch. The surge and heave displacements of the joint are denoted as x_0 and z_0, as shown in Figure 1c.

The forces acting on each raft are shown in Figure 1c. The wave force on mode j of the rafts is denoted as $f_{\mathrm{w},j}(t)$ (here, subscript j = 1, 3, and 5 indicate the surge, heave, and pitch modes of the fore raft, respectively; j = 1′, 3′, and 5′ indicate the surge, heave, and pitch modes of the aft raft, respectively). As the wave passes along the length of the rafts, the rafts output a relative pitch motion around the joint, which drives the hydraulic PTO unit to convert wave energy.

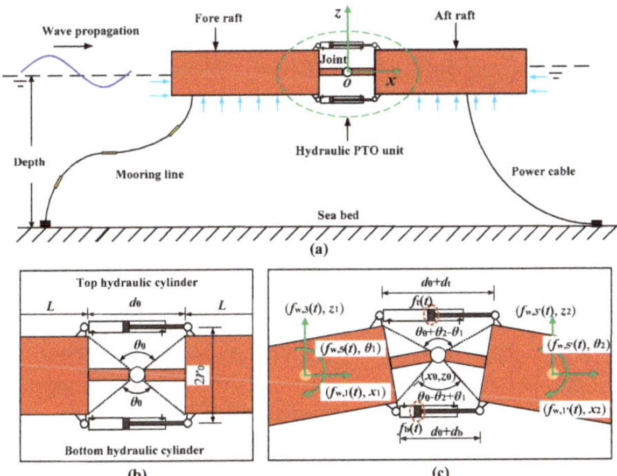

Figure 1. Schematic of a raft-type WEC consisting of two rafts and a PTO unit: (**a**) Front view (**b**) Initial position and (**c**) Position after motion.

3. Frequency Domain Analysis

By assuming the hydraulic cylinder as a linear damper plus a linear stiffness, Liu et al. [27] adopted a generalized displacement vector $\mathbf{x}(t) = [x_0 \ z_0 \ \theta_1 \ \theta_2]^{\mathrm{T}}$ to describe the motion characteristic of the device and given a four-DOF frequency domain model for waves in the positive x-direction based on the Lagrange's equations, which can be written as:

$$\left[-\omega^2(\mathbf{M} + \mathbf{A}_{\mathrm{add}}(\omega)) - \mathrm{i}\omega(\mathbf{B}(\omega) + \mathbf{C}_{\mathrm{pto}}) + (\mathbf{K}_{\mathrm{pto}} + \mathbf{K})\right]\mathbf{X}(\omega) = \mathbf{F}_{\mathrm{e}}(\omega), \tag{1}$$

where ω is the wave frequency; "i" is the imaginary unit; $\mathbf{M}$ is the generalized mass matrix; $\mathbf{A}_{\mathrm{add}}(\omega)$ is the generalized added mass matrix; $\mathbf{B}(\omega)$ is the generalized radiation damping matrix; $\mathbf{K}$ is the generalized hydrostatic restoring stiffness matrix; $\mathbf{C}_{\mathrm{pto}}$ and $\mathbf{K}_{\mathrm{pto}}$ are the damping matrix and stiffness matrix of the PTO unit, respectively; $\mathbf{X}(\omega)$ is the complex amplitude of the generalized displacement vector $\mathbf{x}(t)$; and $\mathbf{F}_{\mathrm{e}}(\omega)$ is the complex amplitude of the generalized wave excitation force vector. Their expressions are given in Appendix A.

3.1. Frequency Response Function

The frequency response function (FRF) of the wave amplitude-to-complex amplitude of the generalized displacement vector, $\mathbf{G}(\omega)$, can be described by [27]:

$$\mathbf{G}(\omega) = \frac{\mathbf{X}(\omega)}{a_{\mathrm{w}}} = \frac{\boldsymbol{\Gamma}_{\mathrm{e}}(\omega)}{-\omega^2(\mathbf{M} + \mathbf{A}_{\mathrm{add}}(\omega)) - \mathrm{i}\omega(\mathbf{B}(\omega) + \mathbf{C}_{\mathrm{pto}}) + (\mathbf{K} + \mathbf{K}_{\mathrm{pto}})}, \tag{2}$$

where a_{w} is the wave amplitude and $\boldsymbol{\Gamma}_{\mathrm{e}}(\omega)$ is the complex generalized excitation force coefficient vector (force vector per unit incident wave amplitude). The expression of $\boldsymbol{\Gamma}_{\mathrm{e}}(\omega)$ is given in Appendix A.

The FRF of the wave amplitude-to-complex amplitude of relative pitch velocity, $G_{\mathrm{v\_rp}}$, is:

$$G_{\mathrm{v\_rp}}(\omega) = (-\mathrm{i}\omega) \cdot \boldsymbol{\zeta} \cdot \mathbf{G}(\omega), \tag{3}$$

where $\boldsymbol{\zeta} = [0\ 0\ 1\ -1]$.

3.2. Power Capture Ability

The average power captured by the PTO unit from the incident wave is:

$$P_{\mathrm{ave\_cap}}(\omega) = \frac{1}{2} 2r_0{}^2 c_{\mathrm{pto}} |a_{\mathrm{w}} \cdot G_{\mathrm{v\_rp}}(\omega)|^2. \tag{4}$$

The time-average flux of energy transported by a regular wave per unit wave crest length in a finite water depth is [28]:

$$P_{\mathrm{w}} = \frac{\rho g \omega a_{\mathrm{w}}^2}{4k} \left(1 + \frac{2kh}{\sinh(2kh)}\right), \tag{5}$$

where ρ is the density of sea water; g is the gravity acceleration; h is the water depth; and k is the wave number.

The capture width ratio is a significant parameter to evaluate the power capture ability of a WEC. In the study of the raft-type-like WEC M4, Sun et al. [22,23] normalized the capture width by wavelength to obtain the capture width ratio. Here, we also adopt the same definition, since this definition of the capture width ratio enables a comparison with theoretical maxima, e.g., $3/2\pi$ in heave and pitch and/or surge for a point absorber [28] and $4/3\pi$ for a slender two-raft WEC [29]. Thus, the capture width ratio is:

$$\eta_{\mathrm{cap}} = \frac{P_{\mathrm{ave\_cap}}}{P_{\mathrm{w}} \cdot \lambda}, \tag{6}$$

where λ is the wavelength at frequency ω.

4. Time Domain Analysis

When the model refers to a linear model, the frequency domain analysis shown in the above section is convenient and efficient to evaluate the performance of the device in a prescribed sea state. However, sometimes, due to the relatively low capture width ratio of a WEC without control, it is imperative to do more thorough researches on its performance, for instance, incorporating the nonlinear characteristics of the hydraulic PTO unit or the control; at this time, the frequency domain analysis is limited, and therefore, a time domain analysis is more desirable. Taking the inverse Fourier transform of Equation (1), the time-domain model of the device is shown as below:

$$(\mathbf{M} + \mathbf{A}_{\mathrm{add}}(\infty))\ddot{\mathbf{x}}(t) + \int_0^t \mathbf{h}(t - \tau)\dot{\mathbf{x}}(\tau)\mathrm{d}\tau + \mathbf{K}\mathbf{x}(t) = \mathbf{f}_{\mathrm{e}}(t) + \mathbf{f}_{\mathrm{pto}}(t), \tag{7}$$

where $\mathbf{A}_{\mathrm{add}}(\infty)$ is the limiting value of the generalized added mass matrix $\mathbf{A}_{\mathrm{add}}(\omega)$ for $\omega = \infty$; $\mathbf{f}_{\mathrm{e}}(t)$ is the generalized wave excitation force vector; $\mathbf{f}_{\mathrm{pto}}(t)$ is the generalized force vector applied by the PTO unit; and $\mathbf{h}(t)$ is the retardation function matrix. The time and frequency domain representations of the retardation function matrix are [28]:

$$\mathbf{h}(t) = \frac{2}{\pi} \int_0^{\infty} \mathbf{B}(\omega) \cos(\omega t) d\omega, \tag{8}$$

$$\mathbf{H}(-i\omega) = \mathbf{B}(\omega) + (-i\omega) \cdot [\mathbf{A}_{\text{add}}(\omega) - \mathbf{A}_{\text{add}}(\infty)]. \tag{9}$$

In regular waves, the wave excitation force acting on the mode j of the rafts is [28]:

$$f_{e,j}(t) = a_{\text{w}} \cdot |\Gamma_j(\omega)| \cdot \cos(-\omega t + \angle\Gamma_j(\omega)). \tag{10}$$

In irregular waves, the wave excitation force acting on the mode j of the rafts can be expressed as [28]:

$$f_{e,j}(t) = \sum_{n=1}^{N} a_{\text{w}}(\omega_n) \cdot |\Gamma_j(\omega_n)| \cdot \cos(-\omega_n t + \angle\Gamma_j(\omega_n) + \varepsilon_n), \tag{11}$$

where ω_n, ε_n, and $a_{\text{w}}(\omega_n)$ are the wave frequency, random phase angle, and wave amplitude of the n-th wave component, respectively.

The time domain generalized excitation force vector can be obtained by [27]:

$$\mathbf{f}_e(t) = \begin{bmatrix} f_{e,1}(t) + f_{e,1\prime}(t) \\ f_{e,3}(t) + f_{e,3\prime}(t) \\ f_{e,5}(t) + f_{e,3}(t)l \\ f_{e,5\prime}(t) - f_{e,3\prime}(t)l \end{bmatrix}, \tag{12}$$

where l is the distance between the mass center of the raft and the joint.

The force vector applied by the PTO unit can be written as:

$$\mathbf{f}_{\text{pto}}(t) = -\mathbf{C}_{\text{pto}}\dot{\mathbf{x}}(t) - \mathbf{K}_{\text{pto}}\mathbf{x}(t). \tag{13}$$

4.1. State-Space Model of Convolution Term

The time domain motion equation presented above contains a convolution term, which denotes the fluid memory effect. Performing the simulation of such a type of time domain model can be time-consuming and may require significant amounts of computer memory, as demonstrated in Taghipour et al. [25,30]. What is worse is that it is not suited for controller analysis and design. For these reasons, different methods of approximating the convolution term have been proposed in many literatures [25,30,31]. One approach is to use a linear-time-invariant parametric model in a state-space form to substitute the convolution term by frequency domain identification [25,30,31]. The convolution term in Equation (7) for calculating the fluid-memory effect can be described by $\mu(t)$:

$$\mu(t) = \int_0^t \mathbf{h}(t - \tau)\dot{\mathbf{x}}(\tau)d\tau. \tag{14}$$

For the convolution term $\mu(t)$, it also has a form as:

$$\mu(t) = \begin{bmatrix} \mu_1(t) = \sum_{q=1}^{4} \mu_{1,q}(t) \\ \mu_2(t) = \sum_{q=1}^{4} \mu_{2,q}(t) \\ \mu_3(t) = \sum_{q=1}^{4} \mu_{3,q}(t) \\ \mu_4(t) = \sum_{q=1}^{4} \mu_{4,q}(t) \end{bmatrix}. \tag{15}$$

For each element $\mu_{p,q}(t)$ in the convolution term $\mu(t)$ can be replaced by a state-space model [29,32].

$$\mu_{p,q}(t) = \int_0^t h_{p,q}(t-\tau)\dot{x}^{(q)}(\tau)d\tau \Rightarrow \begin{cases} \dot{\mathbf{x}}_s^{(p,q)} = \mathbf{A}_s^{(p,q)}\mathbf{x}_s^{(p,q)} + \mathbf{B}_s^{(p,q)}\dot{x}^{(q)}(t) \\ \mu_{p,q}(t) = \mathbf{C}_s^{(p,q)}\mathbf{x}_s^{(p,q)} \end{cases}, \tag{16}$$

where p and q vary from 1 to 4, and $x^{(q)}(t)$ is the q-th element of the generalized displacement vector $\mathbf{x}(t)$; the sizes of matrices $\mathbf{A}_s^{(p,q)}$, $\mathbf{B}_s^{(p,q)}$, and $\mathbf{C}_s^{(p,q)}$ are $(n_{p,q} \times n_{p,q})$, $(n_{p,q} \times 1)$, and $(1 \times n_{p,q})$, respectively; and $n_{p,q}$ is the number of states corresponding to the state vector $\mathbf{x}_s^{(p,q)}$.

By using Equation (16) to replace all the elements in the convolution term $\mu(t)$, sixteen state-space models are presented. Then, for reducing the model complexity and computational load, four state-space models can be assembled by these sixteen state-space models to replace each convolution term $\mu_p(t)$, and finally, a total state-space model can be assembled by the four state-space models. Figure 2 shows the general idea of assembling the state-space model. Thus, the convolution term $\mu_p(t)$ can be rewritten as:

$$\mu_p(t) = \int_0^t \mathbf{h}_p(t-\tau)\dot{\mathbf{x}}(\tau)d\tau \Rightarrow \begin{cases} \dot{\mathbf{x}}_s^{(p)} = \mathbf{A}_s^{(p)}\mathbf{x}_s^{(p)} + \mathbf{B}_s^{(p)}\dot{\mathbf{x}}(t) \\ \mu_p(t) = \mathbf{C}_s^{(p)}\mathbf{x}_s^{(p)} \end{cases}, \tag{17}$$

where matrices $\mathbf{A}_s^{(p)}$, $\mathbf{B}_s^{(p)}$, and $\mathbf{C}_s^{(p)}$ are assembled by matrices $\mathbf{A}_s^{(p,q)}$, $\mathbf{B}_s^{(p,q)}$, and $\mathbf{C}_s^{(p,q)}$, respectively; the state vector $\mathbf{x}_s^{(p)}$ is assembled by state vectors $\mathbf{x}_s^{(p,q)}$; the sizes of matrices $\mathbf{A}_s^{(p)}$, $\mathbf{B}_s^{(p)}$, $\mathbf{C}_s^{(p)}$, and $\mathbf{x}_s^{(p)}$ are $(\sum_{q=1}^4 n_{p,q} \times \sum_{q=1}^4 n_{p,q})$, $(\sum_{q=1}^4 n_{p,q} \times 4)$, $(1 \times \sum_{q=1}^4 n_{p,q})$, and $(\sum_{q=1}^4 n_{p,q} \times 1)$, respectively. The expressions of $\mathbf{A}_s^{(p)}$, $\mathbf{B}_s^{(p)}$, $\mathbf{C}_s^{(p)}$, and $\mathbf{x}_s^{(p)}$ are shown in Appendix A.

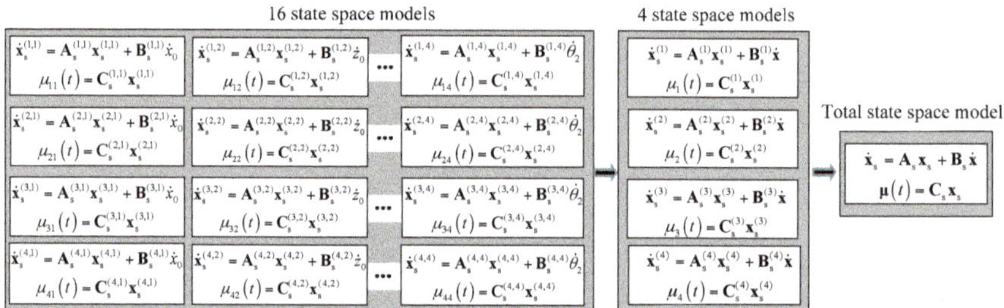

Figure 2. General idea of assembling the state-space model.

Therefore, the total state-space model of Equation (14) can be written as:

$$\mu(t) = \int_0^t \mathbf{h}(t-\tau)\dot{\mathbf{x}}(\tau)d\tau \Rightarrow \begin{cases} \dot{\mathbf{x}}_s = \mathbf{A}_s\mathbf{x}_s + \mathbf{B}_s\dot{\mathbf{x}}(t) \\ \mu(t) = \mathbf{C}_s\mathbf{x}_s \end{cases}, \tag{18}$$

where matrices $\mathbf{A}_s$, $\mathbf{B}_s$, and $\mathbf{C}_s$ are assembled by matrices $\mathbf{A}_s^{(p)}$, $\mathbf{B}_s^{(p)}$, and $\mathbf{C}_s^{(p)}$, respectively; the state vector $\mathbf{x}_s$ is assembled by state vectors $\mathbf{x}_s^{(p)}$; and the sizes of matrices $\mathbf{A}_s$, $\mathbf{B}_s$, $\mathbf{C}_s$, and $\mathbf{x}_s$ are $(\sum_{p=1}^4 \sum_{q=1}^4 n_{p,q} \times \sum_{p=1}^4 \sum_{q=1}^4 n_{p,q})$, $(\sum_{p=1}^4 \sum_{q=1}^4 n_{p,q} \times 4)$, $(4 \times \sum_{p=1}^4 \sum_{q=1}^4 n_{p,q})$, and $(\sum_{p=1}^4 \sum_{q=1}^4 n_{p,q} \times 1)$, respectively. The expressions of $\mathbf{A}_s$, $\mathbf{B}_s$, $\mathbf{C}_s$, and $\mathbf{x}_s$ are shown in Appendix A.

The next task is to obtain matrices $\mathbf{A}_s^{(p,q)}$, $\mathbf{B}_s^{(p,q)}$, and $\mathbf{C}_s^{(p,q)}$. It can be seen from Equation (16) that the Laplace transform of the retardation function $h_{p,q}(t)$ is the transfer function of the system with scalar input $\dot{x}^{(q)}(t)$ and scalar output $\mu_{p,q}(t)$. Thus, the relationship between the state-space model and the transfer function is:

$$H_{pq}(s) = \mathbf{C}_s^{(p,q)}\left(s \cdot \mathbf{I} - \mathbf{A}_s^{(p,q)}\right)^{-1}\mathbf{B}_s^{(p,q)}. \tag{19}$$

The transfer function can also be described by:

$$H_{pq}(s) = \frac{N_{pq}(s)}{Q_{pq}(s)} = \frac{\kappa_\gamma s^\gamma + \kappa_{\gamma-1} s^{\gamma-1} + \cdots + \kappa_1 s + \kappa_0}{s^\nu + \lambda_{\nu-1} s^{\nu-1} + \cdots + \lambda_1 s + \lambda_0}, \tag{20}$$

where $s = -\mathrm{i}\omega$.

Therefore, the transfer function matrix of the total system with vector input $\dot{x}(t)$ and vector output $\mu(t)$ (see Equation (18)) can be expressed by:

$$\mathbf{H}(s) = \begin{bmatrix} H_{11}(s) & H_{12}(s) & H_{13}(s) & H_{14}(s) \\ H_{21}(s) & H_{22}(s) & H_{23}(s) & H_{24}(s) \\ H_{31}(s) & H_{32}(s) & H_{33}(s) & H_{34}(s) \\ H_{41}(s) & H_{42}(s) & H_{43}(s) & H_{44}(s) \end{bmatrix}. \tag{21}$$

Equations (9), (19), and (20) show a general idea of how to obtain the constant matrices $\mathbf{A}_s^{(p,q)}$, $\mathbf{B}_s^{(p,q)}$, and $\mathbf{C}_s^{(p,q)}$ by identification using the generalized added mass matrix $\mathbf{A}_{\mathrm{add}}(\omega)$ and radiation damping matrix $\mathbf{B}(\omega)$. Moreover, for the two rafts with the same geometric parameter, it should be mentioned that some elements of the symmetrical transfer function matrix $\mathbf{H}(s)$ would satisfy the following relationships: $H_{12}(s) = H_{21}(s) = 0$, $H_{13}(s) = H_{14}(s)$, $H_{23}(s) = -H_{24}(s)$, and $H_{33}(s) = H_{44}(s)$. Therefore, the identification only need to be carried out for $H_{11}(s)$, $H_{13}(s)$, $H_{22}(s)$, $H_{23}(s)$, $H_{33}(s)$, and $H_{34}(s)$, i.e., to determine the transfer function matrix $\mathbf{H}(s)$ of the total system, the identification procedure only needs to be carried out six times.

4.2. Transfer Function Estimation Using Regression in Frequency Domain

For a single input single output transfer function $H_{pq}(s)$, a rational parametric model $\hat{H}_{pq}(s)$ can be estimated by using the frequency domain regression from non-parametric data of the FRF $H_{pq}(-\mathrm{i}\omega)$. Assuming the estimated $\hat{H}_{pq}(s)$ has a form of:

$$\hat{H}_{pq}(s) = \frac{N_{pq}(s, \boldsymbol{\varphi})}{Q_{pq}(s, \boldsymbol{\varphi})} = \frac{\kappa_\gamma s^\gamma + \kappa_{\gamma-1} s^{\gamma-1} + \cdots + \kappa_1 s + \kappa_0}{s^\nu + \lambda_{\nu-1} s^{\nu-1} + \cdots + \lambda_1 s + \lambda_0}, \tag{22}$$

where $\boldsymbol{\varphi}$ is a vector containing the estimated parameters, and defined as:

$$\boldsymbol{\varphi} = [\kappa_\gamma, \ldots, \kappa_0, \lambda_{\nu-1}, \ldots, \lambda_0]^{\mathrm{T}}. \tag{23}$$

It can be seen from Equation (22) that the problem of frequency domain identification is to find the order ν and the relative degree $\nu - \gamma$ of $\hat{H}_{pq}$, and to determine the parameter vector $\boldsymbol{\varphi}$ that gives the best non-linear least square (NL-LS) fitting to the frequency response.

$$\boldsymbol{\varphi}^* = \operatorname*{argmin}_{\boldsymbol{\varphi}} \sum_n w_n \left| H_{pq}(-\mathrm{i}\omega_n) - \frac{N_{pq}(-\mathrm{i}\omega_n, \boldsymbol{\varphi})}{Q_{pq}(-\mathrm{i}\omega_n, \boldsymbol{\varphi})} \right|^2, \tag{24}$$

where w_n is the weighting coefficient and "arg min" represents the minimizing argument.

Once the order ν and the relative degree $\nu - \gamma$ are chosen, the NL-LS fitting problem can be solved by an iterative method proposed by Sanathanan and Koerner [33]. The iterative method runs by using the polynomial corresponding to the previous iteration as the weight, and thus, the NL-LS fitting problem defined in Equation (24) is simplified as:

$$\boldsymbol{\varphi}^* = \operatorname*{argmin}_{\boldsymbol{\varphi}} \sum_n \alpha_{n,g} \left| Q_{pq}(-\mathrm{i}\omega_n, \boldsymbol{\varphi}) H_{pq}(-\mathrm{i}\omega_n) - N_{pq}(-\mathrm{i}\omega_n, \boldsymbol{\varphi}) \right|^2, \tag{25}$$

where

$$\alpha_{n,g} = \frac{1}{\left|Q_{pq}\left(-\mathrm{i}\omega_n, \, \boldsymbol{\varphi}_{g-1}\right)\right|^2}.$$

The linear problem described in Equation (25) can be solved by Matlab function *invfresqs* with the option of using a vector of weighting coefficients. Usually, the iterative procedure is implemented starting with $\alpha_{n,g} = 1$, and after a few iterations $Q_{pq}\left(-\mathrm{i}\omega, \, \boldsymbol{\varphi}_g\right) \approx Q_{pq}\left(-\mathrm{i}\omega, \, \boldsymbol{\varphi}_{g-1}\right)$. Then, the optimal parameter of vector $\boldsymbol{\varphi}$ can be determined as $\boldsymbol{\varphi}^* = \boldsymbol{\varphi}_g$. Thus, the parametric model $\hat{H}_{pq}(s)$ can be obtained. Then, the constant matrices $\mathbf{A_s}^{(p,q)}$, $\mathbf{B_s}^{(p,q)}$, and $\mathbf{C_s}^{(p,q)}$ of the state-space model can be determined by using the Matlab function *tf2ss*. Thereafter, the matrices $\mathbf{A_s}$, $\mathbf{B_s}$, and $\mathbf{C_s}$ can be assembled by the formulations shown in Appendix A. Finally, the identified state-space model can be applied to replace the convolution term in the time domain model.

4.3. Power Capture Ability

The instantaneous power captured by the PTO unit is:

$$P_{\text{ins\_cap}}(t) = -\mathbf{f}_{\text{pto}}(t) \cdot \dot{\mathbf{x}}(t). \tag{26}$$

The average power captured in the time domain is:

$$P_{\text{ave\_cap}} = \frac{1}{T} \int_{t_0}^{t_0+T} P_{\text{ins\_cap}}(t)\mathrm{d}t, \tag{27}$$

where t_0 is a moment when the device has come into a steady state of motion and T is the wave period. The capture width ratio in the time domain is:

$$\eta_{\text{cap}} = \frac{P_{\text{ave\_cap}}}{P_{\text{w}} \cdot \lambda}. \tag{28}$$

5. Numerical Results and Discussion

In this section, the identification of the state-space model is firstly carried out. Then, verification of the time domain analysis is performed by comparing its simulation results with those calculated by using the frequency domain analysis. Thereafter, the effects of PTO parameters, wave frequency, surge and heave motions of the joint, and quadratic damping PTO on the performance of the device are investigated and discussed.

5.1. Identification of State-Space Model

Before performing the time domain analysis, identification of the state-space model needs to be carried out. The identification is carried out in the frequency domain by using an iterative method. The general assumptions made in the hydrodynamic analysis of the device are: (1) the rafts are considered as rigid bodies; (2) the fluid is incompressible inviscid; and (3) the flow is irrotational. A truncated frequency ranging from 0.01 rad/s to 9 rad/s is adopted to calculate the hydrodynamic parameters (i.e., the added mass $a_{uj}(\omega)$, the radiation damping $b_{uj}(\omega)$, and the complex excitation force coefficient $\Gamma_j(\omega)$) in ANSYS AQWA [34], commercial hydrodynamic software based on three dimensional radiation/diffraction theory.

Figure 3 shows the identification results of the retardation functions $H_{11}(-\mathrm{i}\omega)$, $H_{13}(-\mathrm{i}\omega)$, $H_{22}(-\mathrm{i}\omega)$, $H_{23}(-\mathrm{i}\omega)$, $H_{33}(-\mathrm{i}\omega)$, and $H_{34}(-\mathrm{i}\omega)$ for the device with structure parameters $L = 10$ m, $D = 1$ m, $d_0 = 1$ m, and $\rho_0 = 512.5$ kg/m^3. The sizes of constant matrices $\mathbf{A_s}$, $\mathbf{B_s}$, and $\mathbf{C_s}$ are 87×87, 87×4, and 4×87, respectively, which are not shown here for their large sizes. The original data of retardation functions $H_{11}(-\mathrm{i}\omega)$, $H_{13}(-\mathrm{i}\omega)$, $H_{22}(-\mathrm{i}\omega)$, $H_{23}(-\mathrm{i}\omega)$, $H_{33}(-\mathrm{i}\omega)$, and $H_{34}(-\mathrm{i}\omega)$ calculated by Equation (9) are also presented in Figure 3. It can be seen that there is a good agreement between the identification results and those obtained by Equation (9).

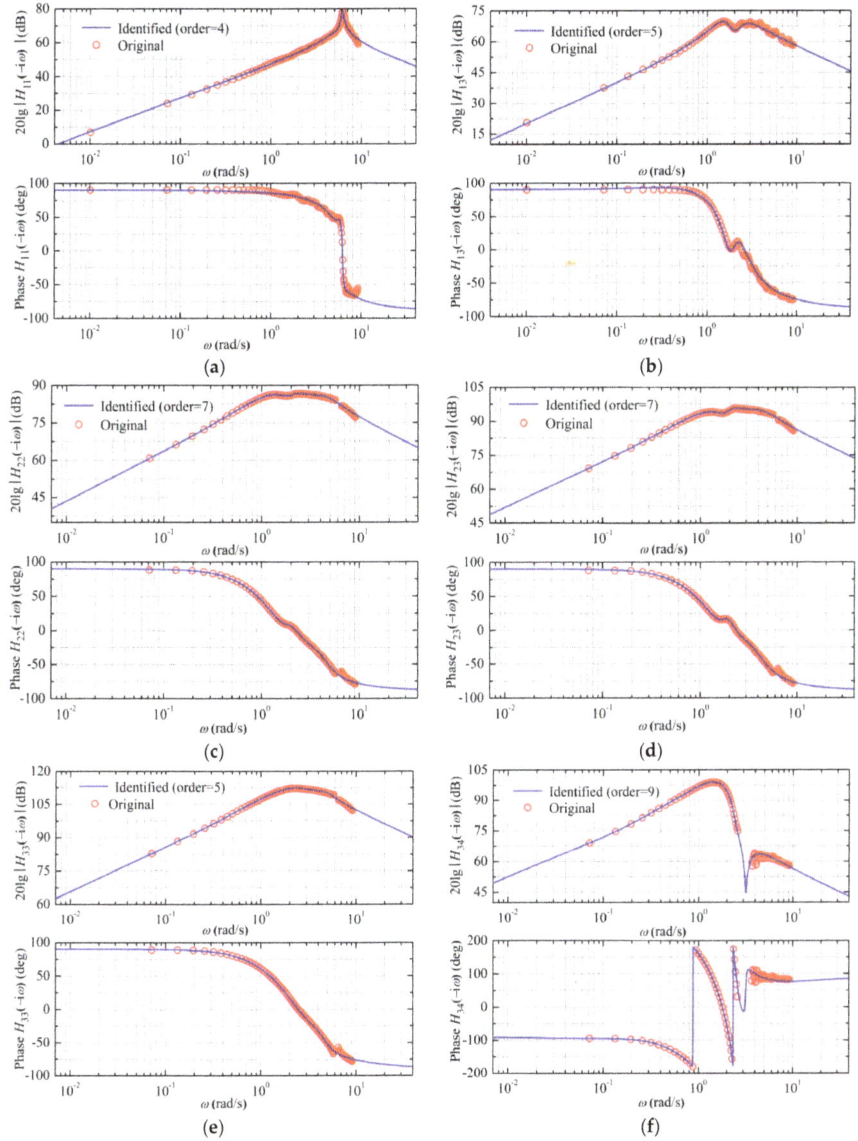

Figure 3. Identification results of retardation functions: (**a**) $H_{11}\,(-\mathrm{i}\omega)$; (**b**) $H_{13}\,(-\mathrm{i}\omega)$; (**c**) $H_{22}\,(-\mathrm{i}\omega)$; (**d**) $H_{23}\,(-\mathrm{i}\omega)$; (**e**) $H_{33}\,(-\mathrm{i}\omega)$; (**f**) $H_{34}\,(-\mathrm{i}\omega)$.

5.2. Validation of Time Domain Analysis

The time domain model (see Equation (7)) can be solved by the fourth-order Runge-Kutta method. Thus, the FRF of the wave amplitude-to-complex amplitude of relative pitch velocity and the capture width ratio can be obtained. In the frequency domain analysis, they can be calculated quickly by Equations (3) and (6), respectively. The stiffness of the PTO unit is assumed to be zero in this work, unless otherwise specified. Four types of structure parameters are considered, as shown in Table 1.

Table 1. Four types of parameters.

Parameter	Type-One	Type-Two	Type-Three	Type-Four
Raft length L (m)	10	20	20	30
Raft diameter D (m)	1	1	2	2
Raft space d_0 (m)	1	1	1	1
Raft density ρ_0 (kg/m^3)	512.5	512.5	512.5	512.5
Raft volume V/m^3	7.8540	15.7080	62.8319	94.2478
Damping coefficient c_{pto} (kN/m/s)	500	500	500	500
Mounting position r_0 (m)	0.5	0.5	1	1
Optimal ratio kL	3.1163	3.6184	3.4987	3.6715
Optimal ratio of raft length to wavelength	0.4960	0.5759	0.5568	0.5843
Resonant frequency of relative pitch velocity ω_{rp} (rad/s)	1.7485	1.3322	1.3100	1.0957
Resonant frequency of heave velocity ω_{h} (rad/s)	1.4078	1.2354	1.2100	1.0179
Resonant frequency of surge velocity ω_{s} (rad/s)	1.0144	0.7370	0.7322	0.6021
Amplitude of FRF of wave amplitude to relative pitch velocity at ω_{rp} (rad/s/m)	0.4281	0.3181	0.3054	0.1799
Amplitude of FRF of wave amplitude to heave velocity at ω_{h} (m/s/m)	1.2623	1.5099	1.4735	1.2589
Amplitude of FRF of wave amplitude to surge velocity at ω_{s} (m/s/m)	0.7881	0.5624	0.5655	0.4674

Figure 4 shows the comparison of the performance of the devices with these four types of structure parameters between the frequency domain calculations and the time domain simulations. It can be seen from Figure 4 that there is a good agreement between the results obtained by using the time domain analysis and those calculated by using the frequency domain analysis.

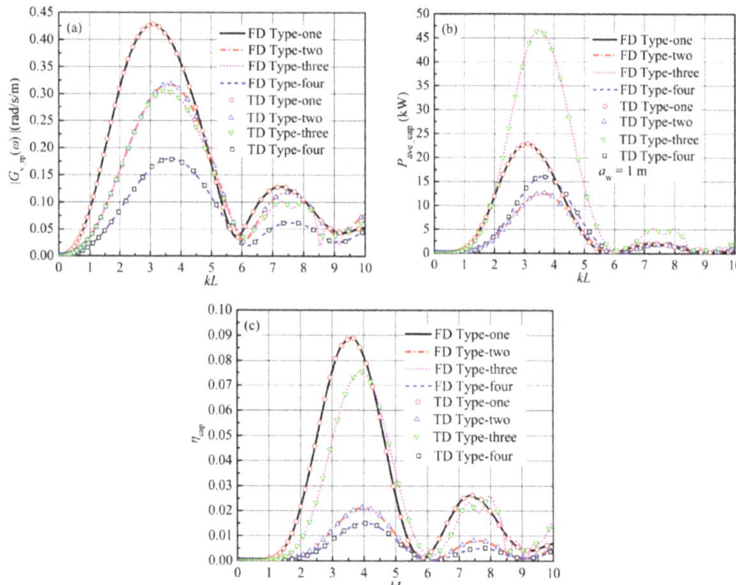

Figure 4. Performance of the device in regular waves (FD = Frequency domain TD = Time domain): (**a**) Amplitude of $G_{\mathrm{v\_rp}}$; (**b**) $P_{\mathrm{ave\_cap}}$; and (**c**) η_{cap}.

5.3. Numerical Results in Regular Waves

5.3.1. Influence of Wave Frequency

One can also see from Figure 4b that, for the devices with these four types of structure parameters, the optimal ratio *kL* corresponding to the peak captured power ranges from 3.1163 to 3.6715, which is shown explicitly in Table 1. The optimal ratio of raft length to wavelength varies from 0.4960 to 0.5843, whereas the corresponding resonant frequency of relative pitch velocity varies from 1.7485 rad/s to 1.0957 rad/s. Figure 5 shows the variation of average captured power and capture width ratio with wave frequency.

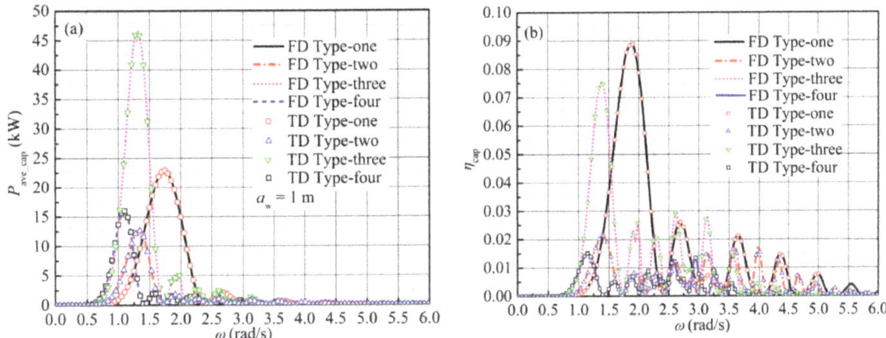

Figure 5. Influence of wave frequency in regular waves (FD = Frequency domain TD = Time domain): (a) $P_{\text{ave\_cap}}$; and (b) η_{cap}.

It can be seen from Figure 5 that the captured power and the capture width ratio are relatively large when the wave frequency gets close to the corresponding resonant frequency of the device, whereas they are rather small at the wave frequency far away from the resonant frequency. Therefore, in order to maximize the captured power and the capture width ratio, the resonant frequency of the designed device should be close to the considered wave frequency. The resonant frequency depends on the raft mass, the rotary inertia about the mass centre, the added mass, the radiation damping, the hydrostatic restoring stiffness, and the damping coefficient and stiffness of the PTO unit, among which the added mass, the hydrostatic restoring stiffness, the raft mass, and rotary inertia mainly depend on the raft size when the material of the raft is chosen. If there is no control included in the PTO unit, the arrival of resonance is usually at the cost of a relatively large raft size, which is shown explicitly in Table 1. The resonant frequency of the device with a type-one parameter is 1.7485 rad/s, whereas the resonant frequency of that with a type-four parameter is 1.0957 rad/s; however, the raft volume of the device with a type-four parameter is twelve times as large as that with a type-one parameter.

5.3.2. Influence of Mounting Position r_0

As shown in Equations (4) and (6), the mounting position r_0 of hydraulic cylinder is a key parameter influencing the capture width ratio of the device. It is assumed that the mounting position r_0 is not confined to the diameter of the rafts. The power capture ability of the device is examined by using the time domain analysis over a wide range of mounting positions r_0. Figure 6a shows the variation of the capture width ratio of the device with a type-four parameter with the mounting position normalized by the diameter of the raft at four different wave periods. It can be seen that the capture width ratio η_{cap} increases with an increasing normalized mounting position r_0/D, and then decreases after reaching a peak value. This is not surprising. As shown in Equation (A.6), there is a quadratic function relationship between the rotational damper $2r_0{}^2 c_{\text{pto}}$ and the mounting poison

r_0. For any specified wave period T and any specified damping coefficient c_{pto}, a too large mounting position r_0 (i.e., too large rotational damping $2r_0^2 c_{pto}$) would lead to a small relative pitch motion, and consequently, the device outputs a little power; whereas a too small mounting position r_0 (i.e., too small rotational damping $2r_0^2 c_{pto}$) would induce a small PTO force, which also results in a little captured power. Therefore, for any specified wave period and any specified damping coefficient c_{pto}, there exists an optimal normalized mounting position r_0^*/D, which corresponds to a peak capture width ratio η_{cap}^*.

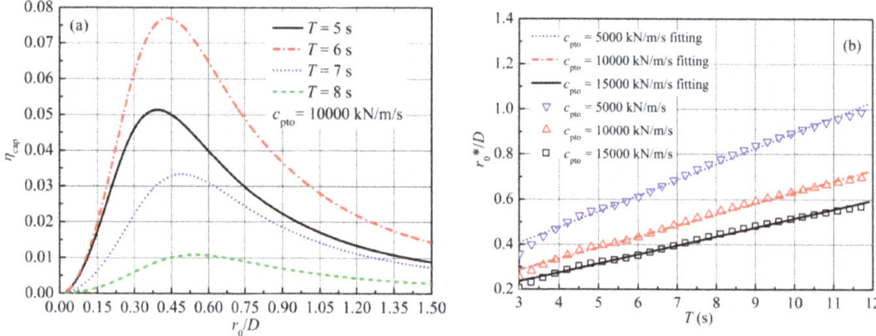

Figure 6. Influence of mounting position in regular waves: (**a**) Variation of η_{cap} with r_0/D; and (**b**) Variation of r_0^*/D with T.

The variation of the optimal normalized mounting position r_0^*/D with wave period T is illustrated in Figure 6b. It is found that the optimal normalized mounting position r_0^*/D presents an approximately linear relationship with wave period, and a smaller damping coefficient c_{pto} gives a larger gradient of the approximate linear relationship.

5.3.3. Influence of Damping Coefficient c_{pto} and Stiffness k_{pto}

In the above sections, the stiffness k_{pto} is assumed to be zero, while in fact, the stiffness plays a significant role in power extraction. In order to see how the stiffness k_{pto} and the damping coefficient c_{pto} affect the power capture ability of the device, the capture width ratio is examined by using the time domain analysis over a wide range of damping coefficients c_{pto} and stiffnesses k_{pto}. Figure 7 shows the influence of the damping coefficient c_{pto} and stiffness k_{pto} on the capture width ratio of the device with a type-four parameter.

It can be seen from Figure 7a,b that, for a specified wave period T, there exists an optimal stiffness k_{pto}^*, which only depends on the wave period T. For a specified wave period T and a specified stiffness k_{pto}, there exists an optimal damping coefficient c_{pto}^*. The optimal damping coefficient c_{pto}^* decreases with increasing stiffness k_{pto}, and then increases after reaching a minimum value, and is symmetric to the optimal stiffness k_{pto}^*, as shown in Figure 7b. When the optimal damping coefficient c_{pto}^* is obtained in the condition of optimal stiffness k_{pto}^*, an optimal combination of the optimal damping coefficient and the optimal stiffness is achieved. This optimal combination could improve the power extraction ability significantly. As is shown in Figure 7a,b, the capture width ratio obtains a value of 0.0769 at $k_{pto} = 0$ kN/m and the corresponding optimal damping coefficient, whereas it can reach 0.2868 at the optimal combination, i.e., an increase of more than 270%. Therefore, if the stiffness k_{pto} of the PTO unit can be adjusted to be an optimal stiffness k_{pto}^* in the varying wave states, which makes the resonant frequency close to the considered wave frequency, the capture width ratio of the device can be improved dramatically. However, as it is revealed in [35], the nonlinear viscous damping would play an important role in the dynamic response of a WEC, especially in the vicinity of the resonant frequency. Hence, the present numerical model based on inviscid flow theory may overestimate the captured power, and the capture width ratio may thus be overpredicted.

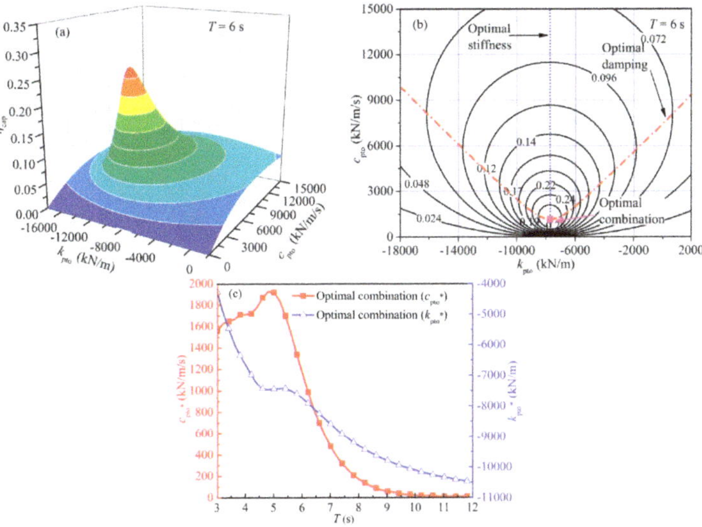

Figure 7. Influence of damping coefficient and stiffness in regular waves: (**a**) 3D η_{cap} in $T = 6$ s; (**b**) η_{cap} contour in $T = 6$ s; and (**c**) Optimal damping coefficient and optimal stiffness in optimal combination.

The variation of the optimal damping coefficient $c_{pto}{}^*$ and the optimal stiffness $k_{pto}{}^*$ in the optimal combination with wave period is presented in Figure 7c. It can be learned from Figure 7c that the optimal stiffness $k_{pto}{}^*$ in the optimal combination is normally negative, which means that the PTO unit may output power to the rafts in some period of its cycle. To be scientific, it is a kind of reactive control [28], but it is difficult to implement practically without a complicated PTO unit. It can also be seen from Figure 7c that the optimal damping coefficient $c_{pto}{}^*$ in the optimal combination increases with the increase of wave period, then decreases after reaching a peak value, whereas, generally, the optimal stiffness $k_{pto}{}^*$ in the optimal combination decreases monotonously.

5.3.4. Influence of Surge and Heave Motions of Joint

To investigate the effect of surge and heave motions of the joint on the performance of the device, we remove the surge and heave motions of the joint, and take the motion system as a two-DOF model and a three-DOF model, and then compare their results with those of the full four-DOF model. Figure 8a presents the FRF of the wave amplitude-to-complex amplitude of the relative pitch velocity G_{v_rp} for the device with a type-one parameter. Figure 8b shows the corresponding capture width ratio η_{cap}, obtained by the model with and without a consideration of surge and heave motions of the joint, whereas the variation of capture width ratio η_{cap} with the damping coefficient c_{pto} is presented in Figure 8c.

It can be seen from Figure 8 that the results obtained by the full-four DOF model are almost overlaid by those obtained by the three-DOF model only removing the surge motion of the joint. However, the results obtained by the three-DOF model only removing the heave motion of the joint and those obtained by the two-DOF model removing both the surge and heave motions of the joint deviate greatly from those obtained by the full-four DOF model. This indicates that the surge motion of the joint has little effect on the performance of the device, whereas the heave motion of the joint makes a huge difference. This is, perhaps, not surprising based on physical intuition. For a deeper understanding, we can see from Table 1 that the resonant frequency of heave velocity is very close to that of relative pitch velocity, while the resonant frequency of surge velocity is far away from that of relative pitch velocity. In addition, the amplitude of FRF of the wave amplitude to heave velocity at the resonant frequency of heave velocity is largely higher than that of the wave amplitude to surge velocity at the resonant frequency of surge velocity. The information in Table 1 and Figure 8 reveals that the relative pitch mode shows stronger coupling with the heave mode than the surge mode. Therefore, the surge motion of the joint could be neglected. Thus, the motion equation of the device could be reduced to a three-DOF model only with the consideration of the heave motion of the joint and two pitch motions of the two rafts. Besides, by using the three-DOF model removing the surge motion of the joint to obtain the total state-space model for substituting the convolution term, the identification procedure only need to be carried out four times, which is more concise than the full four-DOF model.

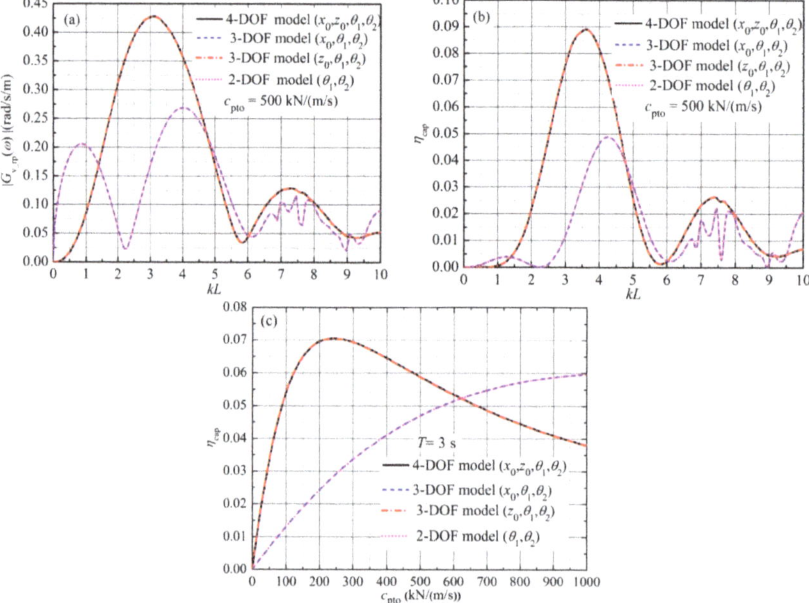

Figure 8. Influence of surge and heave motions of joint in regular waves: (**a**) Variation of Amplitude of G_{v_rp} with kL; (**b**) Variation of η_{cap} with kL; and (**c**) Variation of η_{cap} with c_{pto}.

5.3.5. Influence of Quadratic Damping PTO

In this section, the force applied by the PTO system is assumed to be quadratic damping. It can be written as:

$$\mathbf{f}_{\text{pto}}(t) = \begin{bmatrix} f_{x0}(t) \\ f_{z0}(t) \\ f_{\theta_1}(t) \\ f_{\theta_2}(t) \end{bmatrix} = \begin{bmatrix} 0 \\ 0 \\ -\text{sign}\left(\dot{\theta}_1 - \dot{\theta}_2\right) \cdot \beta \cdot r_0^2 \cdot \left(\dot{\theta}_1 - \dot{\theta}_2\right)^2 \\ \text{sign}\left(\dot{\theta}_1 - \dot{\theta}_2\right) \cdot \beta \cdot r_0^2 \cdot \left(\dot{\theta}_1 - \dot{\theta}_2\right)^2 \end{bmatrix}, \tag{29}$$

where $f_{x0}(t)$, $f_{z0}(t)$, $f_{\theta1}(t)$, and $f_{\theta2}(t)$ are the PTO forces acting on the generalized modes x_0, z_0, θ_1, and θ_2, respectively; and β is the quadratic damping coefficient.

We apply the time domain analysis to explore the effect of quadratic damping on the performance of the device with a type-one parameter. Figure 9a shows how relative pitch velocity and PTO force vary with time, whereas the variation of capture width ratio with quadratic damping coefficient β is presented in Figure 9b.

It can be seen from Figure 9a that the relative pitch velocity presents nonlinear characteristics due to the nonlinear PTO unit. As shown in Figure 9b, the capture width ratio η_{cap} increases with an increasing quadratic damping coefficient β, and then decreases after reaching a peak value. For the wave states with the same period, a larger wave amplitude gives a higher capture width ratio at small quadratic damping coefficients β, while the peak capture width ratio η_{cap}^* has the same value at different wave amplitudes.

To recognize the difference in behaviour exhibited between linear damping and quadratic damping, we compare the peak capture width ratios obtained by using linear damping and quadratic damping. Figure 10 presents the comparison of peak capture width ratios η_{cap}^* obtained by using linear damping and quadratic damping. It can be seen that the peak capture width ratio η_{cap}^* obtained by using quadratic damping is slightly larger than that obtained by using linear damping, especially in the vicinity of wave period $T = 3.3$ s. For 2 s < T < 2.6 s or 2.6 s < T < 5 s, the difference in peak capture width ratios obtained by using linear damping and quadratic damping increases with increasing wave period, and then decreases after reaching a local maximum value. The maximum peak capture width ratio η_{cap}^* obtained by using quadratic damping is 0.1086, whereas that obtained by using linear damping is 0.1031, and they are all obtained at wave period $T = 3.3$ s.

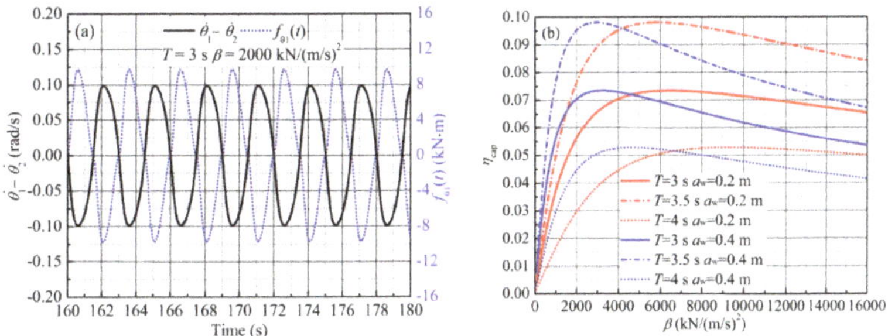

Figure 9. Performance of the device with quadratic damping PTO in regular waves: (**a**) Variation of $\dot{\theta}_1 - \dot{\theta}_2$ and $f_{\theta 1}(t)$ with time; and (**b**) Variation of η_{cap} with β.

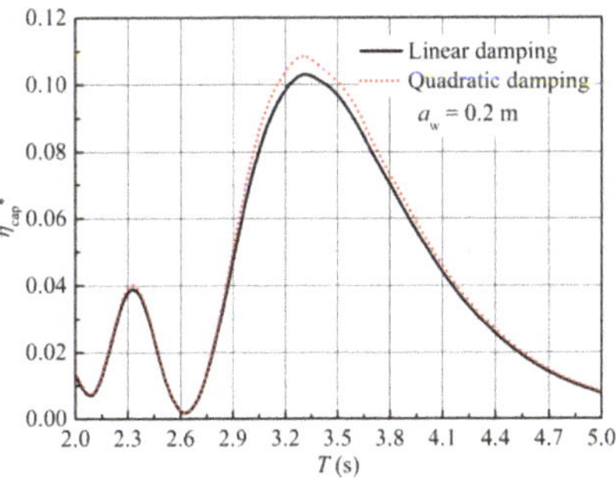

Figure 10. Variation of η_{cap}^* with wave period for the device with linear damping or quadratic damping.

5.4. Numerical Results in Irregular Waves

We also investigate the performance of the device in irregular waves by using the time domain analysis. For the irregular wave computations, the Pierson–Moskowitz energy density spectrum as a function of the angular frequency is used to provide a good representation of the sea states. The Pierson–Moskowitz spectrum has the form [36]:

$$S(\omega) = 5\pi^4 \frac{H_s^2}{T_p^4} \frac{1}{\omega^5} \exp\left[-\frac{20\pi^4}{T_p^4} \frac{1}{\omega^4}\right],$$
(30)

where H_s is the significant wave height and T_p is the peak wave period.

5.4.1. Influence of Mounting Position R_0

In regular waves, it is found that there exists an optimal normalized mounting position r_0^*/D, and the optimal normalized mounting position r_0^*/D presents an approximately linear relationship with wave period. In this section, we intend to investigate how the capture width ratio varies with mounting position r_0 in irregular waves. A wide range of mounting positions r_0 are examined. Figure 11a shows the variation of capture width ratio η_{cap} with normalized mounting position r_0/D for the device with a type-four parameter, whereas the variation of the optimal normalized mounting position r_0^*/D with peak wave period is presented in Figure 11b.

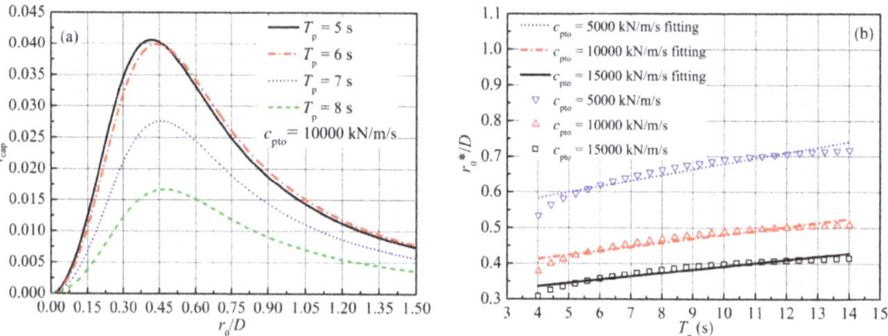

Figure 11. Influence of mounting position in irregular waves: (**a**) Variation of η_{cap} with r_0/D; and (**b**) Variation of r_0^*/D with T_p.

One can see that, similar to the results in regular waves, for a specified peak wave period T_p and a specified damping coefficient c_{pto}, there exists an optimal normalized mounting position r_0^*/D. However, unlike the results in regular waves, the relationship between the optimal normalized mounting position r_0^*/D and the peak wave period T_p presents obvious nonlinear characteristics in irregular waves. This is perhaps due to the random nature of the irregular waves.

5.4.2. Influence of Damping Coefficient c_{pto} and Stiffness k_{pto}

As it has been demonstrated that in regular waves the damping coefficient c_{pto} and the stiffness k_{pto} play a significant role in the capture width ratio, we wonder whether the damping coefficient c_{pto} and the stiffness k_{pto} play a similar role in the capture width ratio in irregular waves. Figure 12a,b show the influence of the damping coefficient c_{pto} and stiffness k_{pto} on the capture width ratio of the device with a type-four parameter in irregular waves. Unlike the results in regular waves, it can be seen from Figure 12b that the optimal stiffness k_{pto}^* not only depends on the peak wave period but also increases slightly with an increasing damping coefficient c_{pto}, and the optimal damping coefficient c_{pto}^* is not symmetric to the optimal stiffness k_{pto}^*. The variation of the optimal damping coefficient c_{pto}^* and the optimal stiffness k_{pto}^* in the optimal combination with peak wave period is presented in Figure 12c. Similar to the results in regular waves, the optimal stiffness k_{pto}^* in the optimal combination is negative and decreases with increasing peak wave period, whereas the optimal damping coefficient c_{pto}^* in the optimal combination increases with increasing peak wave period, and then decreases after reaching a peak value.

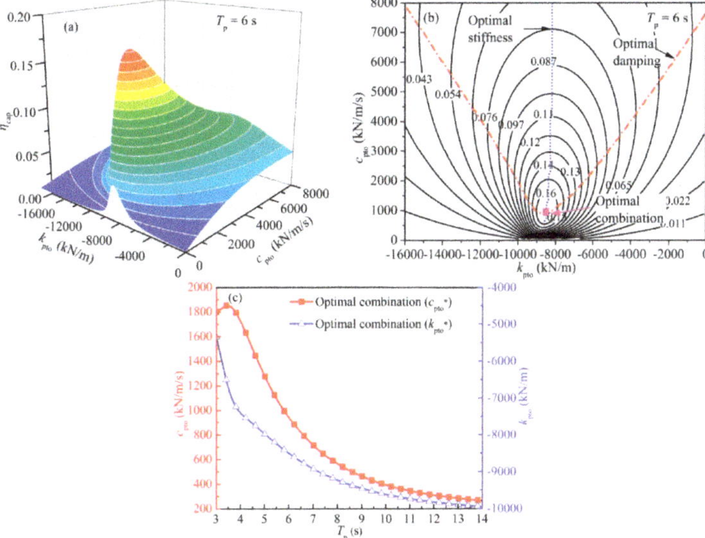

Figure 12. Influence of damping coefficient and stiffness in irregular waves: (**a**) 3D η_{cap} in $T_p = 6$ s; (**b**) η_{cap} contour in $T_p = 6$ s; and (**c**) Optimal damping coefficient and optimal stiffness in optimal combination.

5.4.3. Influence of Surge and Heave Motions of Joint

To see whether surge and heave motions of the joint affect the power capture ability of the device in irregular waves, the capture width ratio of the device with a type-one parameter is examined by using the model with and without a consideration of surge and heave motions of the joint. Figure 13a shows the variation of capture width ratio η_{cap} with peak wave period, whereas the variation of capture width ratio η_{cap} with damping coefficient c_{pto} is presented in Figure 13b.

Similar to the results in regular waves, it can be seen from Figure 13 that the surge motion of the joint has no effect on the capture width ratio, whereas the heave motion of the joint makes much difference to the capture width ratio.

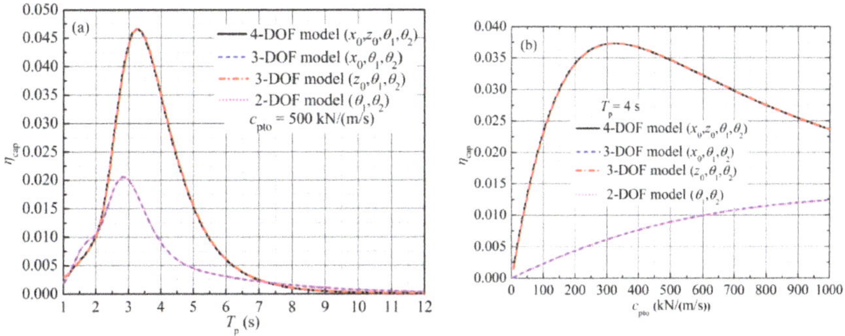

Figure 13. Influence of surge and heave motions of joint in irregular waves: (**a**) Variation of η_{cap} with T_p; and (**b**) Variation of η_{cap} with c_{pto}.

5.4.4. Influence of Quadratic Damping PTO

As it was demonstrated in the previous section that the peak capture width ratio $\eta_{\text{cap}}{}^*$ obtained by using quadratic damping is slightly larger than that obtained by using linear damping in regular waves, it is necessary to investigate how quadratic damping β influences the capture width ratio η_{cap} in irregular waves. Figure 14a shows the variation of capture width ratio η_{cap} with quadratic damping coefficient β, whereas the variation of peak capture width ratio $\eta_{\text{cap}}{}^*$ with peak wave period is presented in Figure 14b. As shown in Figure 14a, there exists an optimal quadratic damping coefficient β^*, which corresponds to a peak capture width ratio $\eta_{\text{cap}}{}^*$. However, unlike the results in regular waves, it can be seen from Figure 14b that the peak capture width ratio $\eta_{\text{cap}}{}^*$ obtained by using quadratic damping is almost the same as that obtained by using linear damping.

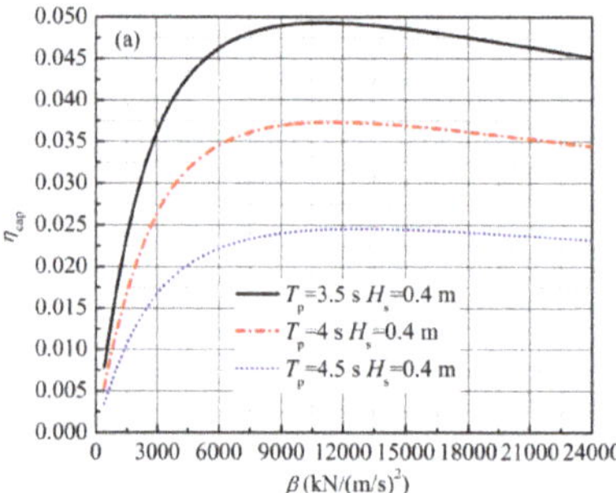
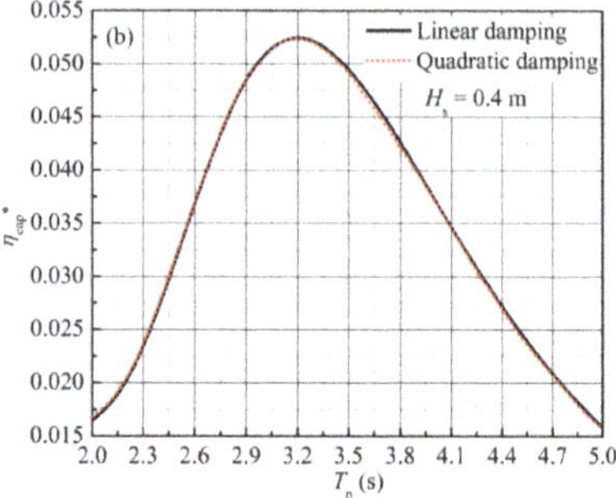

Figure 14. Influence of quadratic damping PTO in irregular waves: (**a**) Variation of η_{cap} with β; and (**b**) Variation of $\eta_{\text{cap}}{}^*$ with T_{p}.

6. Conclusions

In this paper, a state-space approximation of the convolution term in a time domain analysis of a raft-type WEC consisting of two rafts and a PTO unit is presented. The state-space model is identified through regression in the frequency domain. Verification of such a type of time domain analysis is conducted by comparison of its simulation results with those calculated by using frequency domain analysis, and there is a good agreement. The effects of PTO parameters, wave frequency, surge and heave motions of the joint, and quadratic damping PTO on the power capture ability of such a raft-type WEC are investigated. From the above investigations, some conclusions can be drawn as follows:

(1) State-space approximation of the convolution term through regression in the frequency domain has sufficient accuracy. The time domain analysis with the convolution term approximated by a state-space model could be used to investigate the performance of the device.

(2) To obtain a relatively large capture width ratio, the resonant frequency of the designed device should be as close to the considered wave frequency as possible. When there is no control included in the PTO unit, the arrival of resonance is usually at the cost of a relatively large raft size.

(3) For a certain wave period (or peak wave period), there exists an optimal mounting position $r_0{}^*$, corresponding to a peak capture width ratio $\eta_{\text{cap}}{}^*$. In regular waves, the relationship between the optimal normalized mounting position $r_0{}^*/D$ and the wave period is approximately linear, and a smaller damping coefficient c_{pto} gives a larger gradient of the approximately linear relationship; however, this relationship presents nonlinear characteristics in irregular waves.

(4) In regular waves, the optimal stiffness $k_{\text{pto}}{}^*$ only depends on wave period, and the optimal damping coefficient $c_{\text{pto}}{}^*$ relies on wave period and stiffness k_{pto}, and is symmetric to the optimal stiffness $k_{\text{pto}}{}^*$. However, in irregular waves, the optimal stiffness $k_{\text{pto}}{}^*$ depends on not only the wave period, but also the damping coefficient c_{pto}, and the optimal damping coefficient $c_{\text{pto}}{}^*$ is

not symmetric to the optimal stiffness $k_{pto}{}^{*}$. The optimal damping coefficient $c_{pto}{}^{*}$ in the optimal combination increases with increasing wave period (or peak wave period), and then decreases after reaching a peak value, whereas the optimal stiffness $k_{pto}{}^{*}$ in the optimal combination is usually negative and decreases monotonously.

(5) The surge motion of the joint could be neglected. The motion equation of the device can be reduced to a three-DOF model only with the consideration of heave motion of the joint and two pitch motions of the two rafts.

(6) In regular waves, the peak capture width ratio $\eta_{cap}{}^{*}$ obtained by using quadratic damping is slightly larger than that obtained by using linear damping; however, this advantage vanishes in irregular waves.

The present model based on potential theory is relatively convenient to predict the performance of the raft-type WEC. However, without any consideration of energy dissipation due to the fluid viscosity, vortex shedding, and turbulence, it may over-assess the captured power and consequently leads to an over-estimation of the capture width ratio. This problem will be studied in the near future.

Acknowledgments: The authors gratefully acknowledge the support of National Natural Science Foundation of China (51075081).

Author Contributions: Changhai Liu derived the model, performed the simulation, and wrote the paper; Qingjun Yang was responsible for supervising this research and was involved in exchanging ideas and reviewing the article draft; Gang Bao contributed analysis methods.

Conflicts of Interest: The authors declare no conflict of interest.

Appendix A

Expressions of terms used in Equation (1) [27]:

$$\mathbf{M} = \begin{bmatrix} 2m & 0 & 0 & 0 \\ 0 & 2m & ml & -ml \\ 0 & ml & I_c + ml^2 & 0 \\ 0 & -ml & 0 & I_c + ml^2 \end{bmatrix} \tag{A1}$$

$$\mathbf{A}_{add}(\omega) = \begin{bmatrix} a_{11} + a_{11\prime} + a_{1\prime1} + a_{1\prime1\prime} & a_{1\prime3} + a_{13\prime} & a_{15} + a_{1\prime5} + a_{1\prime3}l & a_{1\prime5\prime} + a_{15\prime} - a_{13\prime}l \\ a_{31\prime} + a_{3\prime1} & a_{33} + a_{33\prime} + a_{3\prime3} + a_{3\prime3\prime} & (a_{33} + a_{3\prime3})l + a_{3\prime5} & -(a_{3\prime3\prime} + a_{33\prime})l + a_{35\prime} \\ a_{51} + a_{5\prime1} + a_{3\prime1}l & (a_{33} + a_{33\prime})l + a_{53\prime} & a_{55} + a_{33}l^2 & a_{55\prime} - (a_{53\prime} - a_{35\prime})l - a_{33\prime}l^2 \\ (a_{5\prime1\prime} + a_{5\prime1}) - a_{3\prime1}l & -(a_{3\prime3\prime} + a_{3\prime3})l + a_{5\prime3} & a_{5\prime5} - (a_{3\prime5} - a_{5\prime3})l - a_{3\prime3}l^2 & a_{5\prime5\prime} + a_{3\prime3\prime}l^2 \end{bmatrix} \tag{A2}$$

$$\mathbf{B}(\omega) = \begin{bmatrix} b_{11} + b_{11\prime} + b_{1\prime1} + b_{1\prime1\prime} & b_{1\prime3} + b_{13\prime} & b_{15} + b_{1\prime5} + b_{1\prime3}l & b_{1\prime5\prime} + b_{15\prime} - b_{13\prime}l \\ b_{31\prime} + b_{3\prime1} & b_{33} + b_{33\prime} + b_{3\prime3} + b_{3\prime3\prime} & (b_{33} + b_{3\prime3})l + b_{3\prime5} & -(b_{3\prime3\prime} + b_{33\prime})l + b_{35\prime} \\ b_{51} + b_{5\prime1} + b_{3\prime1}l & (b_{33} + b_{33\prime})l + b_{53\prime} & b_{55} + b_{33}l^2 & b_{55\prime} - (b_{53\prime} - b_{35\prime})l - b_{33\prime}l^2 \\ (b_{5\prime1\prime} + b_{5\prime1}) - b_{3\prime1}l & -(b_{3\prime3\prime} + b_{3\prime3})l + b_{5\prime3} & b_{5\prime5} - (b_{3\prime5} - b_{5\prime3})l - b_{3\prime3}l^2 & b_{5\prime5\prime} + b_{3\prime3\prime}l^2 \end{bmatrix} \tag{A3}$$

$$\mathbf{K} = \begin{bmatrix} 0 & 0 & 0 & 0 \\ 0 & k_{33} + k_{3\prime3\prime} & k_{33}l & -k_{3\prime3\prime}l \\ 0 & k_{33}l & k_{55} + k_{33}l^2 & 0 \\ 0 & -k_{3\prime3\prime}l & 0 & k_{5\prime5\prime} + k_{3\prime3\prime}l^2 \end{bmatrix} \tag{A4}$$

$$\mathbf{F}_e(\omega) = \mathbf{\Gamma}_e(\omega) \cdot a_w = \begin{bmatrix} \Gamma_1(\omega) + \Gamma_{1\prime}(\omega) \\ \Gamma_3(\omega) + \Gamma_{3\prime}(\omega) \\ \Gamma_5(\omega) + l \cdot \Gamma_3(\omega) \\ \Gamma_{5\prime}(\omega) - l \cdot \Gamma_{3\prime}(\omega) \end{bmatrix} \cdot a_w \tag{A5}$$

$$\mathbf{C}_{pto} = \begin{bmatrix} 0 & 0 & 0 & 0 \\ 0 & 0 & 0 & 0 \\ 0 & 0 & 2r_0{}^2 c_{pto} & -2r_0{}^2 c_{pto} \\ 0 & 0 & -2r_0{}^2 c_{pto} & 2r_0{}^2 c_{pto} \end{bmatrix}, \mathbf{K}_{pto} = \begin{bmatrix} 0 & 0 & 0 & 0 \\ 0 & 0 & 0 & 0 \\ 0 & 0 & 2r_0{}^2 k_{pto} & -2r_0{}^2 k_{pto} \\ 0 & 0 & -2r_0{}^2 k_{pto} & 2r_0{}^2 k_{pto} \end{bmatrix} \tag{A6}$$

Here, m is the mass of each raft; I_c is the rotary inertia about the centre of mass; a_{uj} and b_{uj} are the frequency-dependent added mass and radiation damping at mode u due to the motion of mode j ($u = 1, 3, 5, 1', 3', 5'; j = 1, 3, 5, 1', 3', 5'$), respectively; k_{33} and $k_{3'3'}$ are the hydrostatic restoring stiffnesses of the fore and aft rafts caused by the their heave motions, respectively; k_{55} and $k_{5'5'}$ are the hydrostatic restoring stiffnesses of the fore and aft rafts caused by the their pitch motions, respectively; $\Gamma_j(\omega)$ is the complex excitation force coefficient at mode j (force per unit incident wave amplitude); the modulus and argument of $\Gamma_j(\omega)$ are $|\Gamma_j(\omega)|$ and $\angle\Gamma_j(\omega)$, respectively; r_0 is half the distance of the mounting position between the top and bottom hydraulic cylinders, as shown in Figure 1b; and c_{pto} and k_{pto} are the equivalent damping coefficient and stiffness of the hydraulic cylinder, respectively.

Formulations for resembling the total state-space model:

$$\mathbf{A}_s^{(p)} = \begin{bmatrix} \mathbf{A}_s^{(p,1)} & 0 & 0 & 0 \\ 0 & \mathbf{A}_s^{(p,2)} & 0 & 0 \\ 0 & 0 & \mathbf{A}_s^{(p,3)} & 0 \\ 0 & 0 & 0 & \mathbf{A}_s^{(p,4)} \end{bmatrix} \tag{A7}$$

$$\mathbf{B}_s^{(p)} = \begin{bmatrix} \mathbf{B}_s^{(p,1)} & 0 & 0 & 0 \\ 0 & \mathbf{B}_s^{(p,2)} & 0 & 0 \\ 0 & 0 & \mathbf{B}_s^{(p,3)} & 0 \\ 0 & 0 & 0 & \mathbf{B}_s^{(p,4)} \end{bmatrix} \tag{A8}$$

$$\mathbf{C}_s^{(p)} = \begin{bmatrix} \mathbf{C}_s^{(p,1)} & \mathbf{C}_s^{(p,2)} & \mathbf{C}_s^{(p,3)} & \mathbf{C}_s^{(p,4)} \end{bmatrix} \tag{A9}$$

$$\mathbf{x}_s^{(p)} = \begin{bmatrix} \mathbf{x}_s^{(p,1)} \\ \mathbf{x}_s^{(p,2)} \\ \mathbf{x}_s^{(p,3)} \\ \mathbf{x}_s^{(p,4)} \end{bmatrix} \tag{A10}$$

$$\mathbf{A}_s = \begin{bmatrix} \mathbf{A}_s^{(1)} & 0 & 0 & 0 \\ 0 & \mathbf{A}_s^{(2)} & 0 & 0 \\ 0 & 0 & \mathbf{A}_s^{(3)} & 0 \\ 0 & 0 & 0 & \mathbf{A}_s^{(4)} \end{bmatrix} \tag{A11}$$

$$\mathbf{C}_s = \begin{bmatrix} \mathbf{C}_s^{(1)} & 0 & 0 & 0 \\ 0 & \mathbf{C}_s^{(2)} & 0 & 0 \\ 0 & 0 & \mathbf{C}_s^{(3)} & 0 \\ 0 & 0 & 0 & \mathbf{C}_s^{(4)} \end{bmatrix} \tag{A12}$$

$$\mathbf{B}_s = \begin{bmatrix} \mathbf{B}_s^{(1)} \\ \mathbf{B}_s^{(2)} \\ \mathbf{B}_s^{(3)} \\ \mathbf{B}_s^{(4)} \end{bmatrix}, \mathbf{x}_s = \begin{bmatrix} \mathbf{x}_s^{(1)} \\ \mathbf{x}_s^{(2)} \\ \mathbf{x}_s^{(3)} \\ \mathbf{x}_s^{(4)} \end{bmatrix} \tag{A13}$$

References

1. McCormick, M.E. *Ocean Wave Energy Conversion*; Dover Publications: Mineola, NY, USA, 2007; Volume 45, ISBN 9780486462455.
2. Cruz, J. *Ocean Wave Energy-Current Status and Future Prespectives*; Springer Science & Business Media: Berlin, Germany, 2008; ISBN 978-3-540-74894-6.

3. Drew, B.; Plummer, A.R.; Sahinkaya, M.N. A review of wave energy converter technology. *Proc. Inst. Mech. Eng. Part A J. Power Energy* **2009**, *223*, 887–902. [CrossRef]

4. López, I.; Andreu, J.; Ceballos, S.; de Alegría, I.M.; Kortabarria, I. Review of wave energy technologies and the necessary power-equipment. *Renew. Sustain. Energy Rev.* **2013**, *27*, 413–434. [CrossRef]

5. Haren, P. Optimal Design of Hagen-Cockerell Raft. Master's Thesis, Massachusetts Institute of Technology, Cambridge, MA, USA, 1979.

6. Haren, P.; Mei, C.C. Head-sea diffraction by a slender raft with application to wave-power absorption. *J. Fluid Mech.* **1981**, *104*, 505–526. [CrossRef]

7. Haren, P.; Mei, C.C. An array of Hagen-Cockerell wave power absorbers in head seas. *Appl. Ocean Res.* **1982**, *4*, 51–56. [CrossRef]

8. Haren, P.; Mei, C.C. Wave power extraction by a train of rafts: Hydro-dynamic theory and optimum design. *Appl. Ocean Res.* **1979**, *1*, 147–157. [CrossRef]

9. Farley, F.J.M. Wave energy conversion by flexible resonant rafts. *Appl. Ocean Res.* **1982**, *4*, 57–63. [CrossRef]

10. Kraemer, D.R.B. The Motions of Hinged-Barge Systems in Regular Seas. Ph.D. Thesis, Johns Hopkins University, Baltimore, MD, USA, 2001.

11. Nolan, G.; Catháin, M.; Murtagh, J.; Ringwood, J. Modelling and Simulation of the Power Take-Off System for a Hinge-Barge Wave-Energy Converter. In Proceedings of the Fifth European Wave Energy Conference, University College Cork, Ireland, 17–20 September 2003; pp. 1–8.

12. Paparella, F.; Bacelli, G.; Paulmeno, A.; Mouring, S.E.; Ringwood, J.V. Multibody Modelling of Wave Energy Converters Using Pseudo-Spectral Methods with Application to a Three-Body Hinge-Barge Device. *IEEE Trans. Sustain. Energy* **2016**, *7*, 966–974. [CrossRef]

13. Retzler, C. Measurements of the slow drift dynamics of a model Pelamis wave energy converter. *Renew. Energy* **2006**, *31*, 257–269. [CrossRef]

14. Yemm, R.; Pizer, D.; Retzler, C.; Henderson, R. Pelamis: Experience from concept to connection. *Philos. Trans. R. Soc. A Math. Phys. Eng. Sci.* **2012**, *370*, 365–380. [CrossRef] [PubMed]

15. Henderson, R. Design, simulation, and testing of a novel hydraulic power take-off system for the Pelamis wave energy converter. *Renew. Energy* **2006**, *31*, 271–283. [CrossRef]

16. Thiam, A.G.; Pierce, A.D. Segmented and continuous beam models for the capture of ocean wave energy by Pelamis-like devices. *Proc. Meet. Acoust.* **2012**, *11*, 65001. [CrossRef]

17. Thiam, A.G.; Pierce, A.D. Radiation loss in Pelamis-like ocean wave energy conversion devices. *Proc. Meet. Acoust.* **2013**, *12*, 65003. [CrossRef]

18. Thiam, A.G.; Pierce, A.D. Damping mechanism concepts in ocean wave energy conversion: A simplified model of the Pelamis converter. *J. Acoust. Soc. Am.* **2010**, *127*, 65002. [CrossRef]

19. Zheng, S.; Zhang, Y.; Sheng, W. Maximum theoretical power absorption of connected floating bodies under motion constraints. *Appl. Ocean Res.* **2016**, *58*, 95–103. [CrossRef]

20. Zheng, S.-M.; Zhang, Y.-H.; Zhang, Y.-L.; Sheng, W.-A. Numerical study on the dynamics of a two-raft wave energy conversion device. *J. Fluids Struct.* **2015**, *58*, 271–290. [CrossRef]

21. Sun, L.; Taylor, R.E.; Choo, Y.S. Responses of interconnected floating bodies. *IES J. Part A Civ. Struct. Eng.* **2011**, *4*, 143–156. [CrossRef]

22. Sun, L.; Stansby, P.; Zang, J.; Moreno, E.C.; Taylor, P.H. Linear diffraction analysis for optimisation of the three-float multi-mode wave energy converter M4 in regular waves including small arrays. *J. Ocean Eng. Mar. Energy* **2016**, *2*, 429–438. [CrossRef]

23. Sun, L.; Zang, J.; Stansby, P.; Moreno, E.C.; Taylor, P.H.; Taylor, R.E. Linear diffraction analysis of the three-float multi-mode wave energy converter M4 for power capture and structural analysis in irregular waves with experimental validation. *J. Ocean Eng. Mar. Energy* **2017**, *3*, 51–68. [CrossRef]

24. Stansby, P.; Moreno, E.C.; Stallard, T. Large capacity multi-float configurations for the wave energy converter M4 using a time-domain linear diffraction model. *Appl. Ocean Res.* **2017**, *68*, 53–64. [CrossRef]

25. Taghipour, R.; Perez, T.; Moan, T. Hybrid frequency-time domain models for dynamic response analysis of marine structures. *Ocean Eng.* **2008**, *35*, 685–705. [CrossRef]

26. Kristiansen, E.; Hjulstad, Å.; Egeland, O. State-space representation of radiation forces in time-domain vessel models. *Model. Identif. Control* **2006**, *27*, 23–41. [CrossRef]

27. Liu, C.; Yang, Q.; Bao, G. Performance investigation of a two-raft-type wave energy converter with hydraulic power take-off unit. *Appl. Ocean Res.* **2017**, *62*, 139–155. [CrossRef]

28. Falnes, J. *Ocean Waves and Oscillating Systems: Linear Interactions Including Wave-Energy Extraction*; Cambridge University Press: Cambridge, UK, 2002; Volume 56, ISBN 0521017491.

29. Newman, J.N. Absorption of wave energy by elongated bodies. *Appl. Ocean Res.* **1979**, *1*, 189–196. [CrossRef]

30. Taghipour, R.; Perez, T.; Moan, T. Time-Domain Hydroelastic Analysis of a Flexible Marine Structure Using State-Space Models. *J. Offshore Mech. Arct. Eng.* **2008**, *131*, 11603. [CrossRef]

31. Perez, T.; Fossen, T.I. Practical aspects of frequency-domain identification of dynamic models of marine structures from hydrodynamic data. *Ocean Eng.* **2011**, *38*, 426–435. [CrossRef]

32. Duarte, T.; Sarmento, A.; Alves, M.; Jonkman, J. State-Space Realization of the Wave-Radiation Force within FAST. In Proceedings of the ASME 32nd International Conference on Ocean, Offshore and Arctic Engineering, Nantes, France, 9–14 June 2013; pp. 1–12. [CrossRef]

33. Sanathanan, C.; Koerner, J. Transfer function synthesis as a ratio of two complex polynomials. *IEEE Trans. Automat. Contr.* **1963**, *8*, 56–58. [CrossRef]

34. ANSYS AQWA. Available online: http://www.ansys.com/fr-fr/products/structures/ansys-aqwa (accessed on June 2015).

35. Wang, D.; Qiu, S.; Ye, J. An experimental study on a trapezoidal pendulum wave energy converter in regular waves. *China Ocean Eng.* **2015**, *29*, 623–632. [CrossRef]

36. Faltinsen, O.M. *Sea Loads on Ships and Offshore Structures*; Cambridge University Press: Cambridge, UK, 1990; Volume 1, ISBN 0-521-45870-6.

 *energies*

Article

Peak Forces on Wave Energy Linear Generators in Tsunami and Extreme Waves

Linnea Sjökvist [1,2,*] and Malin Göteman [1]

[1] Department of Engineering Sciences, Uppsala University, SE-752 21 Uppsala, Sweden; malin.goteman@angstrom.uu.se

[2] Center for Natural Disaster Science (CNDS), Villavägen 16, SE-752 36 Uppsala, Sweden

* Correspondence: linnea.sjokvist@angstrom.uu.se; Tel.: +46-733-497-459

Received: 30 June 2017; Accepted: 24 August 2017; Published: 2 September 2017

Abstract: The focus of this paper is the survivability of wave energy converters (WECs) in extreme waves and tsunamis, using realistic WEC parameters. The impact of a generator damping factor has been studied, and the peak forces plotted as a function of wave height. The paper shows that an increased damping decreases the force in the endstop hit, which is in agreement with earlier studies. However, when analyzing this in more detail, we can show that friction damping and velocity dependent generator damping affect the performance of the device differently, and that friction can have a latching effect on devices in tsunami waves, leading to higher peak forces. In addition, we study the impact of different line lengths, and find that longer line lengths reduce the endstop forces in extreme regular waves, but on the contrary increase the forces in tsunami waves due to the different fluid velocity fields.

Keywords: wave energy; OpenFOAM; peak forces; extreme waves; tsunami; linear damping; friction

---

1. Introduction

In order for wave energy to be a viable energy option, the survivability of wave energy converters (WECs) in harsh offshore environments must be guaranteed. The peak forces that will occur in storms and other extreme wave events, such as tsunami waves, must be studied in the design process of a wave power plant. Wave tank testing can be carried out to analyze the behavior in extreme waves, but it is expensive and time-consuming. Full scale offshore testing is even more expensive, and although it is necessary for evaluation of a WECs performance before commercialization, it does not offer the controlled environment and the possibility to repeat the same wave conditions needed to evaluate the impact of specific parameters. Computational fluid dynamics (CFD) modeling offers a complement to physical experiments, and allows the user to study specific parameters individually in predefined conditions. The turbulent flow around a WEC in high sea states can be approximated with Reynolds Average Navier-Stokes (RANS) equations together with a turbulence model [1], and the surface can be tracked with the volume of fluid (VOF) method. Several CFD models of different wave power concepts have been developed and verified in recent years [2,3]. In [4], wave load of breaking waves on an overtopping device was studied in a physical wave flume, and in [5], the results were reproduced with good agreement using a RANS-VOF model. In [6,7] a point-absorbing WEC with linear-elastic mooring, moving in six degrees of freedom, was modeled showing good agreement with wave tank experiments, and in reference [8], another point-absorber device was modeled both fixed and freely floating. In [9,10], RANS-VOF models are verified with physical wave tank data and the peak forces of WECs with limited stroke length were studied. In [9], a model was verified with physical experiments where frictional PTO-damping, assumed constant, was used. The model was then used to study the effect of wave height, H, on peak forces in irregular waves, finding a nonlinear, e^H, relation for a moderate sea state and one frictional PTO-damping. In [10], the influence of the frictional damping on the peak forces for moderate and extreme waves was studied using two RANS-VOF models, one linear model and physical wave tank experiments. The four cases of no damping, low, medium and high frictional damping were studied, and two buoy geometries were compared. The magnitude of the peak forces was seen to decrease with increased friction. No trends were fitted to the results, but studying the figures, linear trends could be suggested. It should be noted that the response of a device in an extreme wave depends on the load history as well as the surface elevation [11], and if the peak forces are plotted as function of wave height, a scattered result is expected.

In this paper, a RANS-VOF model, previously verified with physical wave tank experiment [10], is used to study the dynamical behavior and the forces on the WEC developed at Uppsala University [12]. Ultimately, the PTO-damping should be optimized both to minimize the peak forces in the connection line and maximize the energy absorption, across all sea states at the location of the device. In this paper, the response to periodic waves with constant wave period has been studied. It should be noted that the response is expected to depend on the wave period, and the results in this paper should be considered as a first step to understand the dynamical behavior of this device. Both frictional damping, assumed constant, $F_{fric} = constant$, and linear generator damping, $F_{PTO} = \gamma \dot{r}$, are studied. The WEC concept consists of a linear generator at the seabed that is directly driven by a point-absorbing buoy floating at the surface. A mooring line connects the buoy with the translator that is moving vertically inside the linear generator at the seabed. The translator has a limited stroke length, and endstop springs are installed to dampen the motion in both the upper and lower endstop. A peak force will occur in the connection line when the upper endstop spring is hit, and if the wave is high enough to fully compress the enstop spring, a second peak force is expected, corresponding to the elasticity of the connection line.

Much work, both theoretical [13–16] and experimental [17–19], has been accomplished to maximize the power absorption and increase the force in the connection line for this type of WEC. An optimal WEC with high power output and high survivability will have high mean forces and low peak forces. The energy absorption can be increased by optimizing the generator damping to the sea state, as well as the buoy geometry and size. For the generators at the Lysekil testsite on the Swedish west coast, WEC prototypes have been operated offshore with generator damping approximated around 10 to 40 kNs/m [20]. Linear theory has been used to optimize generator damping coefficient to the seastate [21–23], and it was seen that a higher generator damping, around 50–100 kNs/m dependent on buoy size, could increase the energy absorption.

Further, in the physical wave tank experiments presented in [24], and in [10], it was shown that the peak forces can be decreased by an increased generator damping. In reference [25], RANS-VOF simulations were used to numerically study the peak forces of the endstop hits. The peak forces were seen to level out even for increased wave amplitude when the buoy was overtopped by the waves, but only if the generator damping was high enough to keep the translator from hitting the upper endstop spring. Those results suggest that an increased generator damping would lead to both a higher mean power and lower peak forces for prototypes at the Lysekil site. However, it should be noticed that in these mentioned works, both in the physical and numerical modeling of the generator damping, the translator-stator overlap, the part of the stator winding covered by the translator, was assumed to be full during the full stroke length. In reference [26,27], the line force was measured onshore for the total stroke length of a linear generator, including the decreased line force during the partial translator-stator overlap. This decreased line force would allow the buoy and the translator to accelerate just before hitting the endstop, which would increase the peak force of the endstop hit. To account for this important physical property, this is included in the RANS-VOF model used in this paper, and the influence of a decreasing translator-stator overlap on the endstop forces of the WEC is studied. From the mentioned studies, it is clear that the generator damping, the friction, the partial translator-stator overlap and the length of the connection line all influence the experienced line forces, which affect the survivability of the device. In this paper, all these parameters are included and their effect studied in detail. Further; the forces and dynamics are studied when the WEC is impacted by high regular waves and compared with the impact of a tsunami wave. Both frictional damping, assumed to have constant magnitude, $F_{fric} = constant$, and linear generator damping, $F_{PTO} = \gamma \dot{r}$, are studied in this paper. The influence of line length on peak forces is modelled and compared for regular waves and tsunami waves.

2. Method

2.1. Governing Equations

The open-source software OpenFOAM v2.4.0 was used for all the simulations in this paper. The Reynolds Averaged Navier-Stokes equations (RANS) approach was used with an RNG $\kappa\epsilon$ turbulence model. The two-phase Navier-Stokes solver interDyMFoam was used, where the equations for the air and the water is written assuming a single fluid mixture:

$$\nabla \cdot \bar{u} = 0 \tag{1}$$

$$\frac{\partial}{\partial t}(\rho\bar{u}) + \nabla \cdot (\rho(\bar{u} - \bar{u}_g)\bar{u}) = -\nabla p + \nabla \cdot \bar{S} + \rho \vec{f}_b \tag{2}$$

where $\bar{u}$ is the fluid velocity and $\bar{u}_g$ is the grid velocity, ρ is the mixture density and p is the pressure. $\bar{S} = 2\mu\bar{D}$ is the viscous stress tensor where μ is the mixture viscosity and $\bar{D}$ is the strain tensor. f_b is the force from a rigid body. The volume of fluid (VOF) method uses a scalar field α to track the two fluids (water and air, respectively) at the surface boundary. A volume of only water is represented by $\alpha = 1$, only air by $\alpha = 0$, and values in between represents mixtures. The parameter α is solved from a transport equation.

$$\frac{\partial \alpha}{\partial t} + \nabla \cdot (\alpha(\bar{u} - \bar{u}_g)) = 0. \tag{3}$$

The fluid properties in the mixture is expressed using α.

$$\Phi = \alpha\Phi_{water} + (1 - \alpha)\Phi_{air} \tag{4}$$

where Φ is a fluid property such as μ or ρ.

2.2. Numerical Implementation

2.2.1. Numerical Wavetank

The simulation domain forming the wavetank was 300 m long, 75 m high and had a width of 60 m. The water depth was 26 m to resemble the depth at the Lysekil test site. A mesh resolution study was performed and is presented in Figure 1. The force in the connection line is presented as deviation from the result of the finest mesh, and it is seen to converge with increased mesh resolution. The same mesh was used in [28]. According to this, the domain was discretized using 1,988,000 hexahedral mesh element; the mesh is seen in Figure 2. Each background mesh element had a side length of 2.5 m, which was refined to 0.625 m in a 25 m high box around the water surface. In the vicinity of the buoy, each mesh element was further refined to 0.078 m.

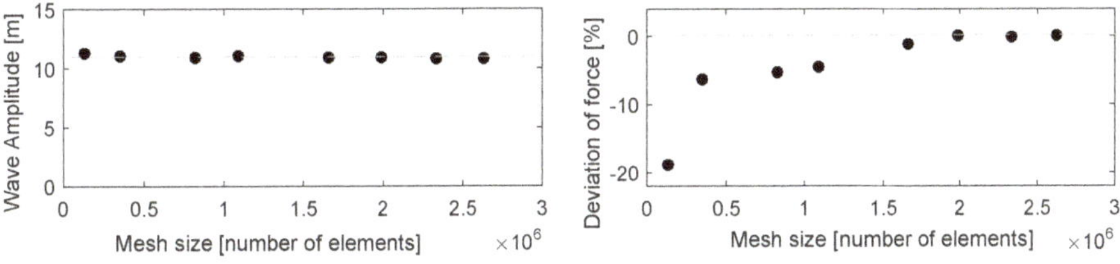

Figure 1. Convergence of the force in the connection line for increasing mesh resolution.

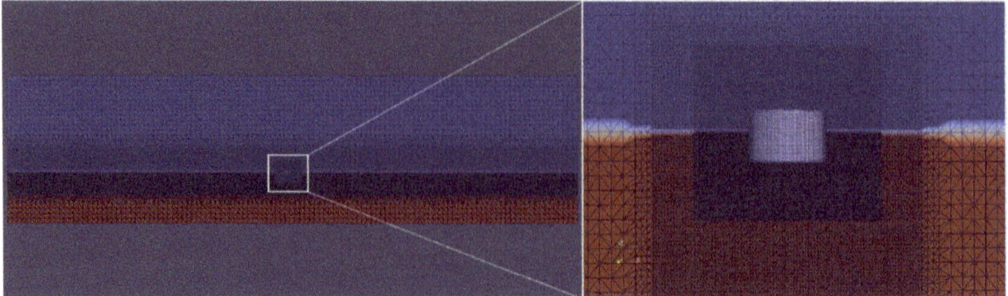

Figure 2. The mesh is refined in a box surrounding the surface, and further refined in the vicinity of the buoy. The color represents the scalar field α. Blue represents $\alpha = 0$ (air), while red represents $\alpha = 1$ (water).

2.2.2. Incident Regular Wave

The regular waves in the model were generated and absorbed using the library waves2foam, using relaxation zones to eliminate wave reflections from boundaries and internally [29]. Relaxation zones of 100 m were placed at the inlet and outlet boundaries. The 5th order Stokes wave generator of the waves2foam library was used to implement the regular waves. To study the WEC behavior in regular waves corresponding to highly energetic sea states at the Lysekil testsite, a constant wave period of 8.5 s and wave heights in the range 3 m to 7 m were chosen. These regular waves carry the same energy as irregular waves with significant wave heights of 4.2 to 9.9 m. This corresponds to the most extreme sea states measured at the Lysekil offshore site [30].

2.2.3. Incident Tsunami Wave

While a deep water wave mainly affects the surface and the fluid velocity decreases with depth, a tsunami wave affects the speed and pressure of the whole water column [31]. The dam-break approach is commonly used to simulate tsunami waves [32,33], and was chosen for the tsunami simulations in this paper. The simulation domain was elongated with 300 m, and an 8 m high water volume was placed on top of the initially still water surface, as seen in Figure 3. The water volume had the same width as the numerical wave tank, and a length of 230 m. The WEC was placed 120 m downstream the initial water volume. The water volume was released as the simulation started, and formed a propagating wave with an amplitude of 6.3 m. The initial conditions were chosen for the tsunami wave to be comparable with a storm induced extreme wave at the Lysekil research site. A relaxation zone of 20 m was placed at the outlet boundary. Since this is too short to eliminate all reflections, the simulations were terminated before reflections occurred.

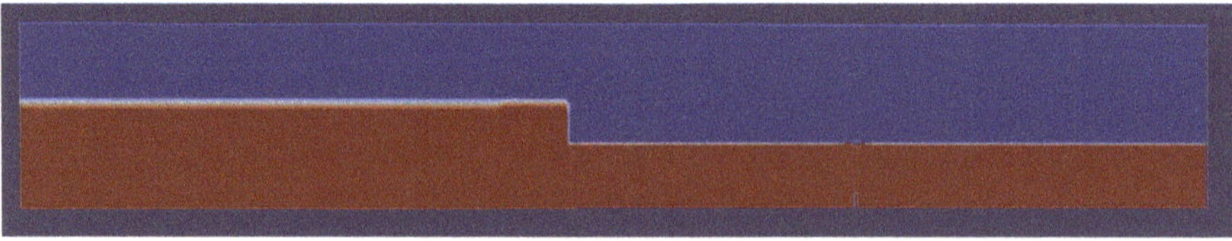

Figure 3. For the dam-break approach, the simulation domain is elongated and a water volume is placed on top of the initially still water surface.

2.3. The WEC Model

The WEC model used in this paper has previously been validated with physical wave tank data, with good agreement, in reference [10,25]. The buoy of the WEC was modeled as a floating rigid body in the wavetank, moving in six degrees of freedom, moored to a fixed position at the seabed. The unit vector $\hat{r}$ points from the buoy position to the anchoring position. The linear generator was modeled as a restraining force in the mooring line, see Figure 4a. The OpenFOAM solver sixDoFRigidBodyMotion solves for the force $\bar{F}_p$ and the torque $\bar{M}_p$ acting on the buoy by integrating the fluid pressure over the surface area. The restraints from the generator added on the buoy are called F_{line} in this paper, and are all directed along the vector $\hat{r}$, see Figure 4. The total force acting on the body, F_{tot}, is then F_p together with F_{line} and the mass of the buoy, and the equation of motion of the WEC modeled in OpenFOAM is:

$$(m_b + m_t)\ddot{\bar{r}}(t) = \int p\hat{n}dS - F_{line}\hat{r} - m_b\bar{g}, \tag{5}$$

where m_b and m_t are the mass of the buoy and of the translator and $\bar{r}$ is the position vector of the buoy. The restraining forces are the gravity force on the translator $F_{m_t g} = m_t g$, the electromagnetic damping in the generator simplified as $F_{PTO} = \gamma\dot{r}$, and frictional loss, F_{fric}. The generator damping F_{PTO} and friction forces F_{fric} are directed in the opposite direction of the translator movement. As seen in Figure 4b, the translator-stator overlap is not full during the whole stroke length, and the force factor F_{PTO} must be multiplied with A_{frac}^c, where the factor A_{frac} is the fraction of the stator that is overlapped with translator, and c is a factor 2 [27]. When the translator hits the upper endstop spring, a restraining endstop spring force is added; $F_{spring} = \kappa_s(r - r_{rest} - l_{free}^{up})$, where r_{rest} is the length of r at still water and the translator stands in the middle position. l_{free}^{up} is the free stroke length, seen in Figure 4b. If the endstop spring gets fully compressed, a second endstop force is added, corresponding to the elasticity of the connection line; $F_{endstop} = \kappa_{line}(r - r_{rest} - l_{total}^{up})$, where l_{total}^{up} is the total stroke length, seen in Figure 4b. The Heaviside functions δ_{free} and δ_{total} define when the endstop spring is hit and when it is fully compressed, and depend on the length of the the vector $\bar{r}$ as:

$$\delta_{free}/\delta_{total} = \begin{cases} 1, & r > r_{rest} + l_{free}^{up}/l_{total}^{up} \\ 0, & \text{otherwise.} \end{cases} \tag{6}$$

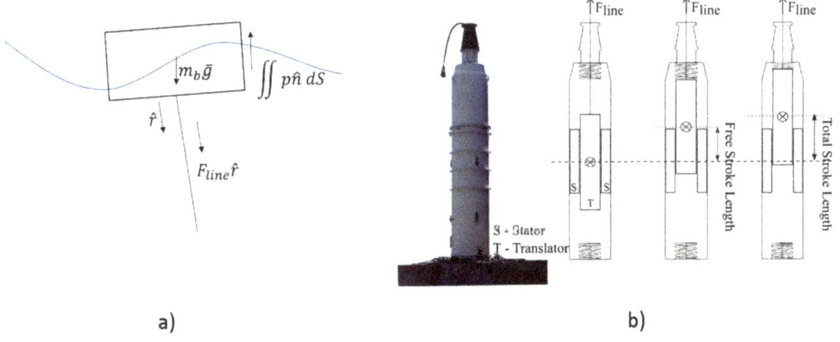

a) b)

Figure 4. (**a**) The WEC is modeled as a floating buoy, restrained by a force in the connection line. The force is directed along a vector pointing at the fixed anchoring position; (**b**) The generator has a limited stroke length. It is also seen that the translator-stator overlap is not full during the whole stroke length.

When the translator stands on the bottom of the WEC, the line is slack and F_{line} is set to zero using the Heaviside function δ_{down}. The total restraining force is described by:

$$\bar{F}_{line} = \delta_{down}(-m_t g \pm F_{PTO}A_{frac}^c - \kappa r \pm F_{fric} - F_{spring}\delta_{free} - F_{endstop}\delta_{total})\hat{r}. \tag{7}$$

If F_{line} becomes a positive number, the line would slack and F_{line} is set to zero. The WEC parameters used in the model are found in Table 1.

Table 1. WEC parameters.

Parameter	Abbrevation	Value
Buoy radius	R_{buoy}	1.7 m
Buoy height	h_{buoy}	2.1 m
Buoy mass	m_{buoy}	5700 kg
Translator mass	$m_{translator}$	6500 kg
Translator height	$h_{translator}$	3 m
Stator height	h_{stator}	2 m
Free stroke length up/down	$l_{freestroke}^{up/down}$	1 m
Total stroke length up/down	$l_{totstroke}^{up/down}$	1.25 m
Endstop spring constant	$\kappa_{endstop-spring}$	250 kN/m
Spring constant corresponding to line elasticity	κ_{line}	2600 kN/m

3. Results

3.1. Peak Forces as a Function of Wave Height

To study the impact of endstop hits and overtopping waves on peak forces, incident periodic waves with a constant wave period of 8.5 s were simulated, and the peak forces in the connection line were plotted as functions of corresponding wave height. Five incident waves of increasing heights were used; the surface elevations at the position of the WEC are seen in Figure 5, and the wave steepness (kA), of the waves are found in Table 2. This was repeated for eight levels of generator damping, ranging from $\gamma = 0$ kNs/m, to $\gamma = 140$ kNs/m with a constant increase of 20 kNs/m. Using potential linear flow theory and assuming an infinite stroke length, the optimum linear damping at this wave period was found to be 103 kNs/m. The friction was constant in all cases, $F_{fric} = 5$ kN, chosen as the measured friction of the WEC in reference [26]. In Figure 6, the resulting peak forces are plotted as a function of wave height. If all peak forces are considered, both linear and quadratic fits would be possible because of the scattering in the result. However, if only the peaks when the endstop spring is fully compressed are considered (corresponding to black markers in Figure 6), the resulting peak forces seem to increase linearly with wave height. However, the result is still scattered.

For damping factors between $\gamma = 0$ kNs/m and $\gamma = 100$ kNs/m, the peak forces decrease with increased generator damping γ, but further increased damping does not decrease the peak forces. This can be understood when the endstop hits are studied separately. If the endstop is hit (a black marker in the figure), a higher wave height results in a higher peak force. Increasing the generator damping slows down the translator, and the peak force decreases. However, when the generator damping is further increased, for γ higher than 80 kNs/m in this parametric study, the generator is strong enough to keep the translator from hitting the upper endstop, as indicated by the red markers in the figure. When the endstop is not hit, the highest force in the connection line depends on the strength of the generator instead of the force of the endstop hit, and since a higher γ means a stronger generator, $F_{PTO} = \gamma \dot{r}$, the line force increase with increased γ. The reason behind this phenomenon is further explained by Figure 7, where the total force in the connection line is decomposed and the magnitude of $F_{PTO} A_{frac}^2$, F_{fric}, F_{spring} and $F_{endstop}$ are shown for three levels of generator damping, as the WEC is impacted by a wave with H = 6 m. The total line force is a sum of these forces and the gravitational force of the translator weight. The position and speed of the translator are also shown. It is seen that for $\gamma = 60$ kNs/m and $\gamma = 80$ kNs/m, the peak force in the total connection line force mainly corresponds to a sum of the spring force F_{spring} and the endstop force $F_{endstop}$. The translator speed decreases with increased γ, which explains why $F_{endstop}$ decreases with increased γ; the impuls energy when the translator hits the upper endstop depends directly on the speed and the mass of the translator. For the higher damping of $\gamma = 100$ kNs/m, the generator is strong enough to slow down the translator enough to keep it from fully compress the upper endstop spring, as can be seen from the translator position in Figure 7. Since the endstop is not hit, $F_{endstop} = 0$ and the highest force in the connection line does instead correspond to F_{PTO}. Further increased generator damping γ cannot decrease the $F_{endstop}$, and the total force in the connection line will increase again. However, it should be noted that the increased line force due to increased F_{PTO} is desired as it can be converted to electricity, in contrast to the damaging force of the endstop hit.

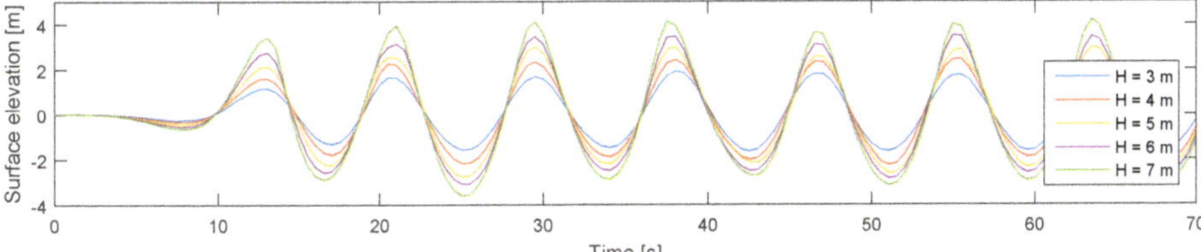

Figure 5. Surface elevation of the modeled incident waves. The wave period was 8.5 s, and five different wave heights were simulated.

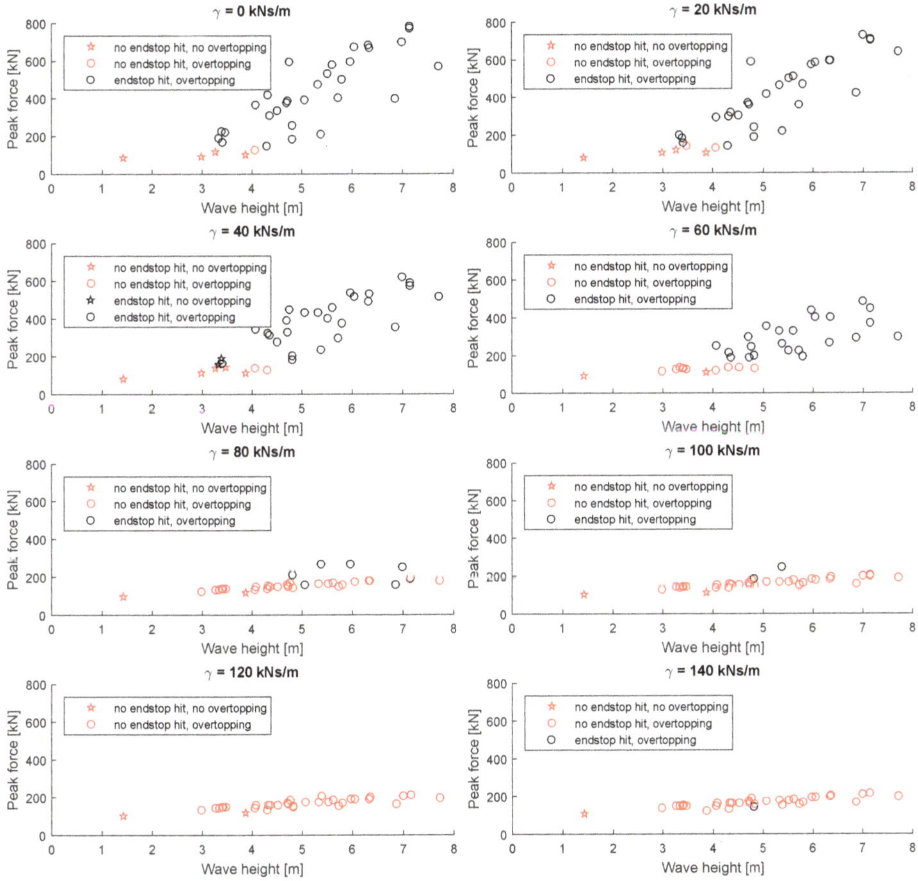

Figure 6. The peak forces plotted as a function of wave height. The peak forces decrease with increasing generator damping for damping factors up to $\gamma = 100$ kNs/m. For higher γ, the generator is strong enough to keep the translator from hitting the upper endstop and the peak forces are not further reduced.

Table 2. Wave steepness.

Wave Height	3 m	4 m	5 m	6 m	7 m
Wave Steepness (kA)	0.08	0.11	0.14	0.17	0.19

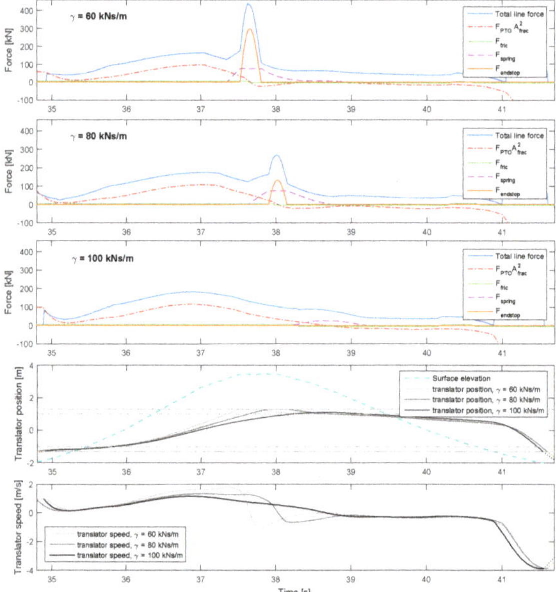

Figure 7. The total force in the connection line is a sum of of $F_{PTO}A^2_{frac}$, F_{fric}, F_{spring}, $F_{endstop}$ and the gravitational force. The line force is compared for three levels of generator damping γ. The corresponding translator position and speed are also seen. The surface elevation of the incident wave is seen together with the translator position. The horizontal dashed and solid lines mark the free and the total stroke length, respectively.

3.2. The Impact of Friction on the Survivability in Extreme Waves

Although it was seen in the physical experiments in [24] that peak forces were reduced by a high friction, it is possible that a low friction could have a latching effect, which would influence the force in the connection line. To study the impact of friction on survivability in extreme waves, both a high regular wave and a tsunami wave were modeled. Five levels of friction were studied for each wave. A friction of $F_{fric} = 5$ kN was the experimental friction measured in [26] and is considered low, and it was also the friction used in Section 3.1. The theoretical case of no friction was included for comparison. Higher friction than 35 kN is not considered relevant, since it would significantly increase the wear inside the generator. The generator damping factor was set to $\gamma = 40$ kNs/m for all simulated cases in this section, a value that has been used at the Lysekil test site [20].

In Figure 8, the line force is seen when the WEC was impacted by a regular wave with a height of 6.5 m and a wave period of 8.5 s. For comparison, the highest single wave statistically appearing on average once per 100 years at the Lysekil test site is 6.2 m [30]. The simulation was run for 70 s, which resulted in 7 wave peaks. The transient first two peaks in the front of the wave train are followed by periodic waves. No friction resulted in the highest peak force for all wave peaks. For the periodic wave peaks, peak three to seven, the highest friction resulted in the lowest peak forces, and decreasing the friction increased the peak force. However, for the two first wave peaks, the lowest peak force was achieved for the second highest friction, not the highest friction.

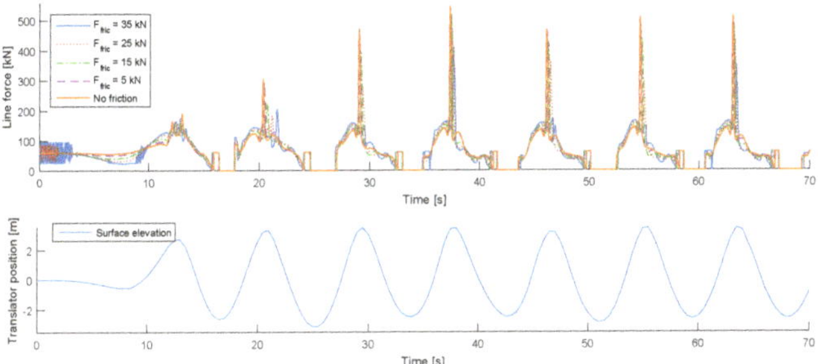

Figure 8. The force in the connection line when the WEC is impacted by a high regular wave. Five levels of friction are compared.

In Figure 9, a zoom of the line force of the first two peaks is seen together with the surface elevation of the wave and the translator position. The horizontal solid line represents the total stroke length of the WEC, when the translator stands on the bottom and the line slacks. The dashed horizontal line represents the lower endstop spring. It should be noted that the surface elevation is measured right above the anchoring position of the WEC. The buoy moves in six degrees of freedom and it can have large surge excursions. Studying the translator position, it can be seen that the friction is delaying the translator motion, and increases the phase shift between the buoy and the wave. The translator is more delayed by the highest friction in the second wave peak than in the first, but for peak three to seven, the delay was once again lower. The force of a wave on a buoy varies with this phase shift, and this is assumed to contribute to why the highest friction did not result in the lowest peak force for the first two wave peaks.

The line force and the translator position during the tsunami event is seen in Figure 10. This tsunami event, modeled with the dam break approach, has two wave peaks; one leading peak of 6.3 m, and a second peak of 4.7 m. Multiple wave peaks are a result of that the dam-break approach has been used with the wet bed condition. In Figure 11, the water velocity in the vicinity of the buoy is seen as the WEC was impacted by the first and the second tsunami peaks, and during the peak forces. At the current depth of 26 m, the wave front of the tsunami did not evolve to be very steep, but when the turbulent bore propagated over the WEC, the buoy was already submerged. The peak force in the connection line occurred before the wave peak. This is in contrast with the periodic wave, where the peak force occurred at the same time or after the wave peak. In the first wave peak of the tsunami, the highest peak force occurred for the highest level of friction, and in contrast with the results of the periodic wave, the endstop force decreased with decreased friction and was lowest for 5 kN of friction. This can be seen as a consequence of the fact that the translator motion is delayed by the friction, which gives a latching effect and increases the phase shift between the buoy and the wave, which increases the endstop force. The endstop force of the second, lower, wave peak is actually higher than of the leading wave peak. This can be understood from the fact that for a tsunami wave, the water velocity is high, and does not only depend on the current wave height. Comparing the water velocity surrounding the buoy at the times of the peak forces, t = 8.5 s and t = 14.5 s in Figure 11, it is seen that the water velocity is comparable, which explains why the peak forces of the wave peaks are comparable despite the different wave heights; when the buoy is submerged, the magnitude of the line force depends more on the surrounding water velocity than the wave height. In the second wave peak as well as for the first, a friction force of 5 kN resulted in the lowest peak force, and the second lowest peak force occurred when no friction was applied. Higher friction resulted in higher endstop force.

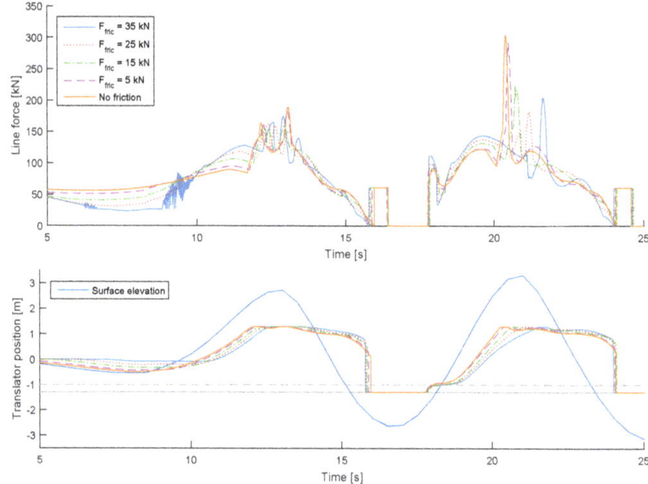

Figure 9. This is a zoom of Figure 8. The WEC is impacted by a high regular wave. The solid horizontal line marks the bottom of the generator, and the dashed horizontal line marks the lower endstop spring.

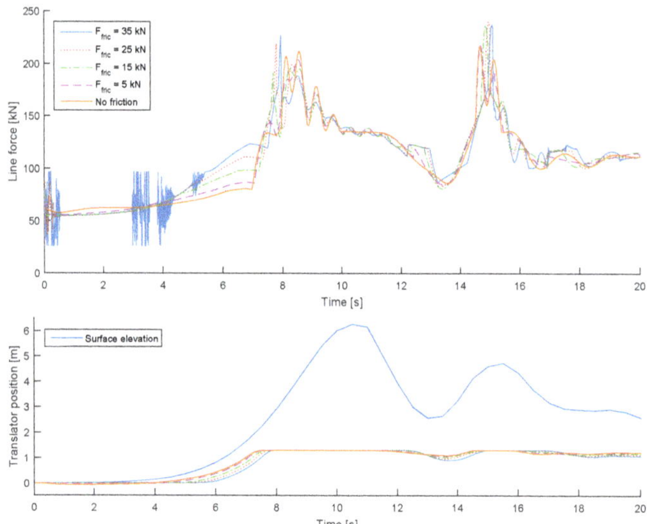

Figure 10. The force in the connection line and the translator position when the WEC is impacted by a tsunami event. Five levels of friction are compared.

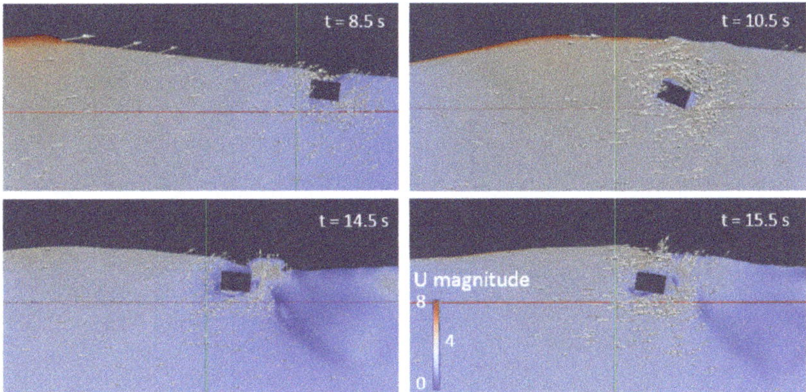

Figure 11. The WEC impacted by the tsunami wave. The figure shows the four main events: The first peak force occurred at t = 8.5 s, one second before the first wave peak propagated over the WEC at t = 10.5 s. The second peak force occurred at t = 14.5 s, one second before the second wave peak at t = 15.5 s. The origin marks the initial position of the buoy.

Even though an increased friction decreased the peak forces when the WEC was impacted by periodic waves, as expected, an increased friction did not result in an increased survivability for the tsunami wave event. This deviant behavior was also seen in the transient at the front of the regular wave train.

3.3. Endstop Forces Decrease with Increased Damping for Periodic Waves

As seen from Section 3.1, the damping of the generator has a large effect on the behavior of the device. To study this in more detail, we looked at the line forces in the time domain, and compared speed dependent generator damping with frictional damping. When all the periodic waves are studied, peak three to seven for all regular waves, it is seen that the endstop forces decrease with increased damping, both speed dependent generator dampings, $F_{PTO} = \gamma \dot{r}$, as in Section 3.1, and for constant frictional dampings, $F_{fric} = constant$, as in Section 3.2. In Figure 12, the line force is shown in the time domain for one wave peak from the study in Section 3.1, comparing the impact of different levels of γ. The incident wave height was 5.7 m, and is the fifth wave peak of the second highest incident wave in Figure 5. It is confirmed that if the endstop was hit, seen as a sharp peak, then an increased damping factor γ resulted in a decreased peak force. However, once the damping was high enough to prevent the translator from hitting the endstop, further increased damping did not reduce the line force further, but instead increased it. For this specific wave peak, endstop hits were prevented for $\gamma = 80 \, \text{kNs/m}$ or higher.

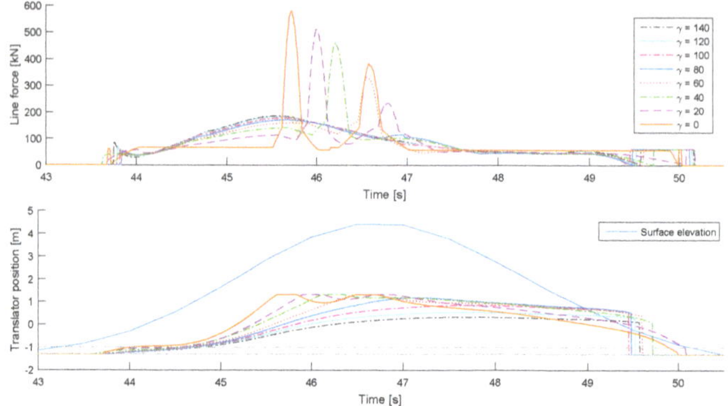

Figure 12. The line force and the translator position when the WEC is impacted by a high periodic wave, comparing the influence of generator damping factor γ. Increasing the generator damping decreases the force of the endstop hits. For $\gamma = 80$ kNs/m or higher, endstop hits were prevented.

The same behavior, decreased peak forces with increased damping, was seen for the constant, frictional, damping. In Figure 13, the line force and translator position of the third wave peak in Section 3.2 is shown, comparing the different levels of frictional damping. It is seen that for this periodic wave, the friction decreased the speed of the translator which reduced the force when the translator hits the endstop. The same behavior was seen for all the periodic waves in Section 3.2. That increasing the damping decreases the peak forces was also confirmed by the physical wave tank experiment performed in reference [24], were the influence of high frictional damping on the wave loads on a WEC in extreme waves was studied in a 1:20 scaled experiment. In Figure 14, reprinted with permission from reference [24], the line force is seen when a WEC was impacted by a regular wave with a height of 5.7 m and a period time of 10.7 s. A focused extreme wave was embedded and hit the WEC at t = 134 s. All values of Figure 14 are presented in 1:1 scale. In the left plot, a low frictional damping of 18 kN was applied. In the plot in the middle, a medium frictional damping of 59 kN was applied, and in the figure to the right, a high damping of 83 kN was applied. Increased damping decreased the force of the endstop hit, both for the high regular waves and for the embedded extreme wave.

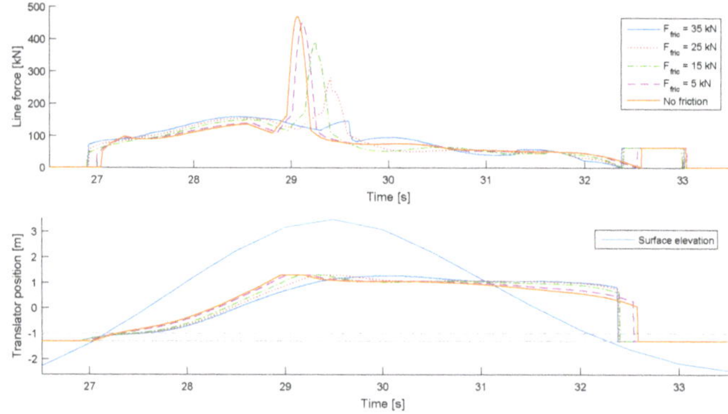

Figure 13. The line force and translator position for different levels of frictional damping. This figure shows the third wave peak of Figure 8. The generator damping factor is constant at $\gamma = 40$ kNs/m. The force of the endstop hit is reduced by an increased friction.

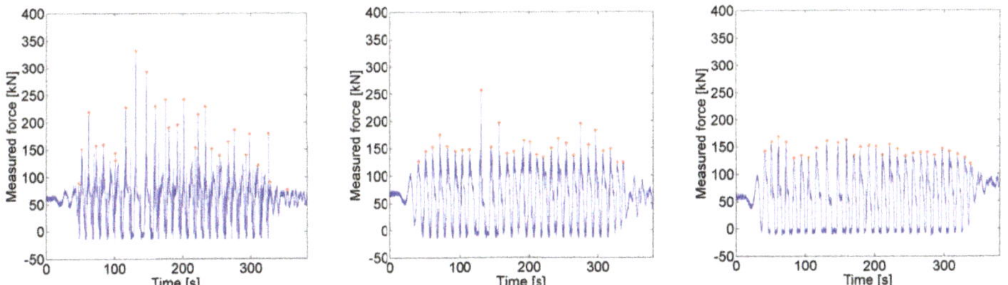

Figure 14. This figure is reprinted with permission from reference [24]. In a physical wave tank experiment, the line force of a WEC was studied when different levels of frictional damping were applied. The incident wave had a height of 5.7 m and an embedded focused extreme wave hit the WEC at t = 134 s. Three levels of friction was applied: 18 kN (**left**), 59 kN (**middle**), and 83 kN (**right**). The force in the endstop hits decrease with increased damping.

3.4. Influence of Line Length on Endstop Peak Force

Another possible factor for the survivability is the length of the connection line. An incorrect line length could be a consequence of the changing tide or incorrect calibrations during deployment. A tsunami event could also start with a withdrawal of the waterline and a decreased water level. In this section, the regular wave and the tsunami event were modeled for six different line lengths, comparing the magnitude of the endstop hits and the translator and buoy motion. The water depth was the same for all simulations, and the length of the connection line was altered.

In Figure 15, the line force and the translator position is shown together with the heave and surge motion of the buoy. It can be seen that a too short line (both the 0.5 m and the 1 m too short cases), increased the endstop force slightly. Studying the translator motion, it is seen that the endstop was hit earlier and that the endstop spring was fully compressed for a longer time with a shorter connection line. When the length of the connection line was instead increased, the force of the endstop hits decreased, and the duration of the endstop was shorter. For the line that was 2 m too long, the endstop was not hit at all. The heave motion of the buoy decreased with a shorter line, and increased with a longer line, since the translator motion was restricted by the upper endstop. The surge motion on the other hand did not increase with a longer line, but did decrease and got slightly delayed. The surge motion was influenced by the restricted heave motion; when the translator stayed in the upper endstop, the buoy could not move in positive heave direction, and was instead forced to move in surge and in negative heave direction.

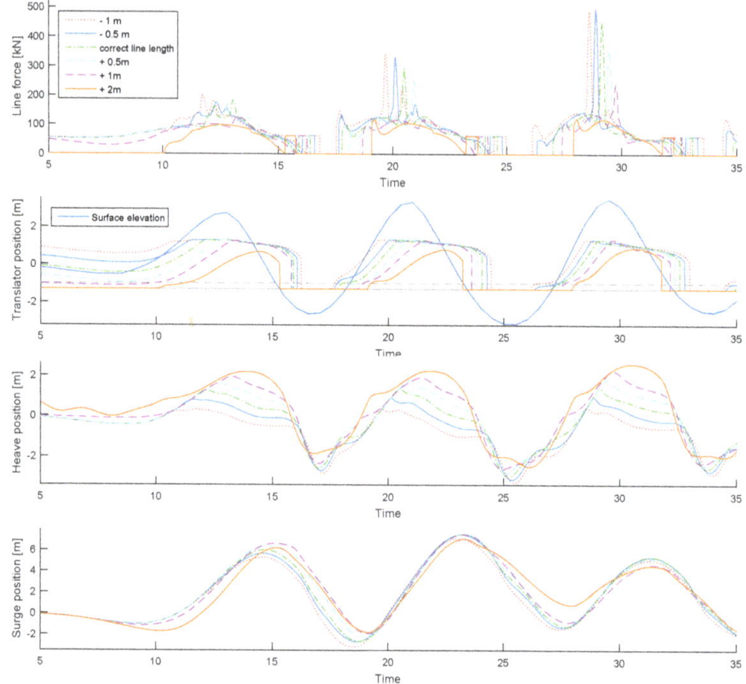

Figure 15. A regular wave was modeled, comparing the influence of different line lengths. The generator damping factor was $\gamma = 40$ kNs/m and the friction was set to $F_{fric} = 5$ kN.

The influence of line length when the WEC was hit by a tsunami wave event was also studied, and is shown in Figure 16. In contrast with the line length study with the regular wave, the magnitude of the peak forces increased with an increased line length. This can again be understood when comparing the different fluid velocity fields of the regular and tsunami waves, shown in Figure 17. Figure 17 shows the WEC with a 2 m to short line, for both the regular wave (left) and the tsunami wave (right), at the time of an endstop hit, t = 19 s for the regular wave and t = 7 s for the tsunami. Although the regular wave had the same wave height as the tsunami amplitude, the velocity fields differed significantly. The surface of the regular wave was smooth, while the tsunami wave had two stages; first a smooth surface elevation, followed by a turbulent bore. If the length of the connection line was correct or too short, the buoy got submerged by the smooth surface rise, and the turbulent bore propagated over the deeply submerged buoy. However, if the line length was too long, the buoy did not get submerged before the bore hit the WEC, and the buoy was hit by the bore and thus impacted by a higher force. The motion in heave followed the same pattern for the tsunami as for the regular wave; an increased line length increased the heave motion, which followed from the restricted translator motion. However, in contrast with the regular wave, the surge motion for the tsunami wave increased for an increased line length, which is explained by the difference between the velocity fields. The velocity of the water in a regular deep water wave follows a circular pattern which results in an oscillating surge motion of the buoy, as seen in Figure 15. The velocity field of a tsunami wave has a horizontal forward speed, and the current drags the buoy forward in the surge direction.

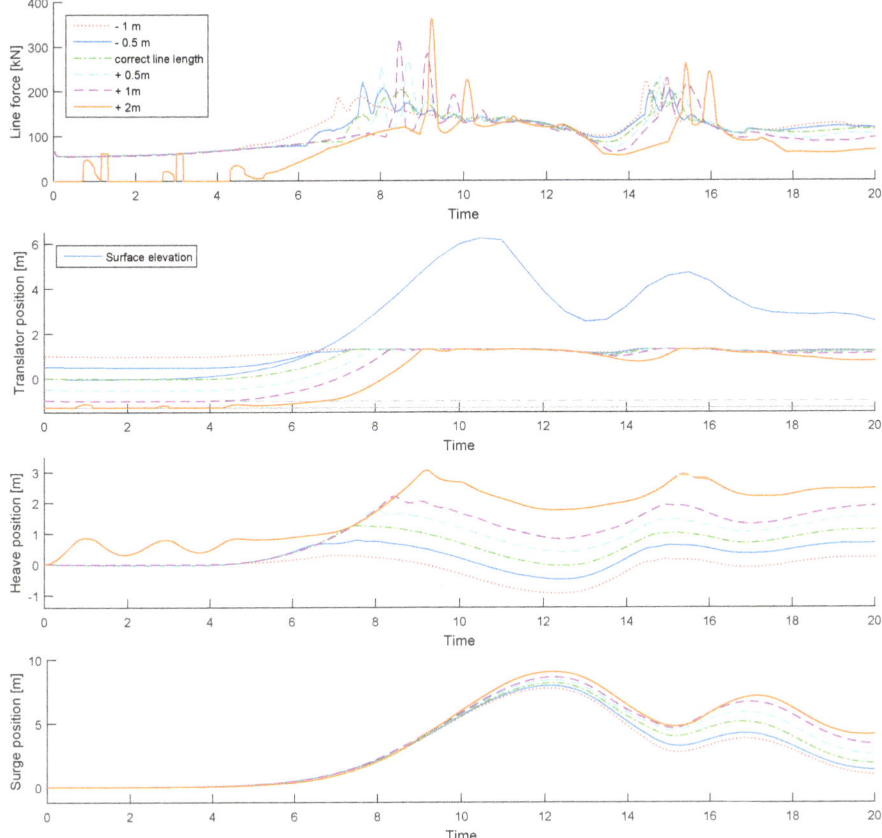

Figure 16. Line force, translator position and buoy position when the WEC is impacted by a tsunami wave. The influence of line length is compared.

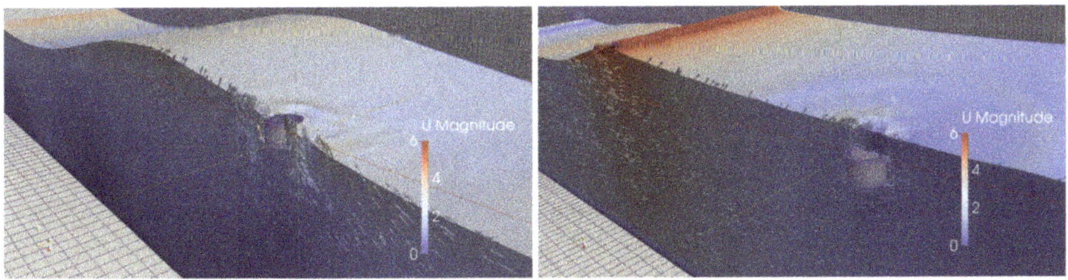

Figure 17. The WEC impacted by a regular wave (**left**), and by a tsunami wave (**right**). The incident waves are traveling from the left to the right in the figures. The connection line length is modelled as too short in those simulations. The origin marks the original position of the buoy in still water.

4. Discussion

The focus of this paper is the survivability of a point-absorbing WEC in extreme waves and tsunamis, using realistic parameters. The WEC parameters were chosen as suitable parameters for a WEC deployed at the Lysekil test site, Sweden [26], and the studied wave heights chosen to correspond to the wave climate at the Lysekil test site. A previously verified OpenFOAM model has been used to study the peak forces of endstop hits, and different parameters that can influence the survivability, namely: generator damping, friction damping, wave height, line length and wave type (high regular waves and tsunami waves). In addition, the paper has analyzed how overtopping waves and endstop hits affect the peak forces, which is of highest relevance for linear generators with limited stroke length. This study is an extension of the study in reference [25]. In reference [25], three levels of γ was used; no damping, a low damping at $\gamma = 10$ kNs/m and a medium damping at $\gamma = 30$ kNs/m. The endstop spring was longer and never got fully compressed, which it was in this paper. Reference [25] focused on the influence of overtopping waves, and it was seen that the overtopping did not influence the peak forces unless the generator damping was high enough to keep the translator from hitting the upper endstop spring. In this paper, a larger range of different and higher generator damping values γ was studied in higher waves, high enough to fully compress the upper endstop spring. It was seen that the force of the endstop hits, when the spring was fully compressed, increased with increasing wave heights and decreased with increased γ. It was also seen that if γ was high enough to prevent an endstop hit, the peak force was significantly lower, but then again increasing with further increased γ. However, it should be noted that this increased line force is expected and wanted as it can be converted to electricity in the generator, in contrast with the damaging peak force of the endstop hit.

In this paper, the decreasing PTO damping of a decreasing translator-stator overlap was included in the model. When the translator-stator overlap decreases, the PTO damping decreases and the translator can accelerate. It was suspected that this decreased damping when the stator moved closer to the endstop could affect the behavior with decreased peak forces with increased γ. However, this was not seen, and it was concluded that an increased γ will decrease the peak force of the endstop hit, regardless of the translator-stator overlap.

The influence of a constant damping, such as friction, was also studied, both for a regular extreme wave and for a tsunami wave. For periodic waves, peak three to seven in Figure 8, increased friction decreased the peak force of the endstop hit. This is confirmed by the physical wave tank experiments presented in reference [24] and reprinted in Figure 14. It is concluded that for periodic waves, increased damping decreases the peak forces of endstop hits. This is valid for both constant damping, $F_{fric} = constant$, as in Figures 13 and 14, and for linear generator damping, $F_{PTO} = \gamma \dot{r}$, as in Figure 12. However, if the waves are increasing in strength, like the two first wave peaks of Figure 8, and for the tsunami event in Figure 10, the friction can result in a latching effect and actually increase the force of the endstop hit instead of decreasing it. It is possible that this effect could occur for irregular waves as well, and should be considered in the design process of a WEC.

The influence of line length on survivability was also studied, and big differences were seen when the regular wave was compared with the tsunami wave. For the regular wave, an increased line length had a positive effect for the survivability; the peak forces of the endstop hits decreased, and for the longest line (2 m too long), the translator never hit the upper endstop. The surge motion decreased with increased line length, showing that for the correct or the too short line length, the surge motion is increased by the restricted translator motion; when the translator stands still at the upper endstop, the buoy is forced to move in the surge and negative heave direction. Since the 2 m too long line is the only studied length where the translator does not hit the upper endstop and the motion is unrestricted by the endstop, it can not be concluded how a further increased line length would affect the survivability. It was also seen that the WEC was more vulnerable to incorrect line length for the tsunami event. In contrast to the regular wave, the peak forces during the tsunami wave increased with increased line length, and the longest line resulted in significantly higher peak forces than a line of correct line length. The regular wave and the tsunami wave were comparable considering amplitude, but the velocity fields differed significantly. The surface of the regular wave was smooth, while the tsunami wave had two stages; an initial smooth surface elevation, followed by a turbulent bore. If the length of the connection line was correct or too short, the buoy got submerged by the initial smooth surface rise, and the turbulent bore propagated over the deeply submerged buoy. However, if the line was too long, the buoy was instead impacted by the turbulent bore, and the peak force increased. Although a 2 m or more too long line does not seem likely during normal operating conditions, it is possible that a tsunami wave event starts with a decreasing water level, and it is important to know that the consequences of the line length on survivability are significant.

This paper has only studied one wave period, and it should be noted that the response of a WEC will be dependent on wave period as well. The survivablility must be studied across all seastates at the deployment site, and the results in this paper should only be considered as a first step towards understanding the response of the studied WEC in extreme waves.

5. Conclusions

A previously verified OpenFOAM model has been used to study the survivability of a point-absorbing WEC in extreme waves and for a tsunami wave event, using realistic WEC parameters. It was concluded that:

1. For periodic waves, it was seen that both increased linear damping, $F_{PTO} = \gamma \dot{r}$, and increased constant damping $F_{fric} = constant$, decreased the force of the endstop hits. This corresponds well with established experimental results.
2. If the incident wave was not periodic, for the tsunami event or the transient waves at the front of the regular wave train, it was seen that increased friction could result in a latching effect and actually increase the force of the endstop hit instead of decreasing it. It is possible that this effect could also occur for irregular waves during normal operating conditions.
3. Due to the differences in fluid velocity fields, the WEC was more vulnerable to a too long line length when impacted by a tsunami wave than by a regular wave. For a regular wave, an increased line length resulted in lower endstop forces and decreased surge motion. For the tsunami wave on the other hand, an increased line length resulted in significantly higher endstop forces.

Acknowledgments: This research is supported by the Centre for Natural Disaster Science (CNDS), the Swedish Research Council (VR, grant Number 2015-04657). The computations were performed on resources provided by the Swedish National Infrastructure for Computing (SNIC) at UPPMAX.

Author Contributions: Linnea Sjökvist and Malin Göteman conceived and designed the numerical experiments. Linnea Sjökvist performed the numerical experiment, analyzed the data and wrote the paper.

Conflicts of Interest: The authors declare no conflict of interest.

Abbreviations

The following abbreviations are used in this manuscript:

WEC	Wave Energy Converter
CFD	Computational Fluid Dynamics
RANS	Reynolds Average Navier-Stokes
VOF	Volume of Fluid
PTO	Power Take Off

References

1. Wolgamot, H.; Fitzgerald, C. Nonlinear hydrodynamic and real fluid effects on wave energy converters. *Proc. IMechE Part A J. Power Energy* **2015**, *229*, 772–794.
2. Day, A.; Babarit, A.; Fontaine, A.; He, Y.; Kraskowski, M.; Murai, M.; Penesis, I.; Salvatore, F.; Shin, H. Hydrodynamic modeling of marine renewable energy devices: A state of the art review. *Ocean Eng.* **2015**, *108*, 46–69.
3. Palm, J.; Eskilsson, C.; Paredes, G.; Bergdahl, L. Coupled mooring Analysis for Floating Wave Energy Converters Using CFD: Formulation and Validation. *Int. J. Mar. Energy* **2016**, *16*, 83–99.
4. Buccino, M.; Vicinanza, D.; Salerno, D.; Banfi, D.; Calabrese, M. Nature and magnitude of wave loadings at Seawave Slotcone Generators. *Ocean Eng.* **2015**, *95*, 34–58.

5. Buccino, M.; Dentale, F.; Salerno, D.; Contestabile, P.; Calabrese, M. The use of CFD in the analysis of wave loadings acting on Seawave Slotcone Generators. *Sustainability* **2017**, *8*, 1255, doi:10.3390/su8121255.

6. Ransley, E. Survivability of Wave Energy Converter and Mooring Coupled 525 System Using CFD. Ph.D. Thesis, Plymouth University, Plymouth, UK, 2015.

7. Ransley, E.; Greaves, D.; Raby, A.; Simmonds, D.; Hann, M. Survivability of wave energy converters using CFD. *Renew. Energy* **2017**, *109*, 235–247.

8. Ransley, E.; Greaves, D.; Raby, A.; Simmonds, D.; Hann, M.; Jakobsen, M.; Kramer, M. RANS-VOF modelling of the wavestar point absorber. *Renew. Energy* **2017**, *109*, 49–65.

9. Chen, W.; Dolguntseva, I.; Savin, A.; Zhang, Y.; Li, W.; Svensson, O.; Leijon, M. Numerical modelling of a point-absorbing wave energy converter in irregular and extreme waves. *Appl. Ocean Res.* **2017**, *63*, 90–105.

10. Sjökvist, L.; Wu, J.; Ransley, E.; Engström, J.; Eriksson, M.; Göteman, M. Numerical models for the motion and forces of point-absorbing wave energy converters in extreme waves. *Ocean Eng.* **2017**, accepted.

11. Hann, M.; Greaves, D.; Raby, A. Snatch loading of a single taut moored floating wave energy converter due to focussed wave groups. *Ocean Eng.* **2015**, *96*, 258–271.

12. Waters, R.; Stålberg, M.; Danielsson, O.; Svensson, O.; Gustafsson, S.; Strömstedt, E.; Eriksson, M.; Leijon, M. Experimental Results from Sea Trials of an Offshore Wave Energy System. *Appl. Phys. Lett.* **2007**, *90*, 034105, doi:10.1063/1.2432168.

13. Engström, E. Hydrodynamic Modelling for a Point Absorbing Wave Energy Converter. Ph.D. Thesis, Uppsala University, Uppsala, Sweden, 2011.

14. Eriksson, M.; Isberg, J.; Leijon, M. Hydrodynamic modelling of a direct drive wave energy converter. *Int. J. Eng. Sci.* **2005**, *43*, 1377–1387.

15. Pastor, J.; Liu, Y. Power absorption modeling and optimization of a point absorbing wave energy converter using numerical method. *ASME J. Energy Resour. Technol.* **2014**, *136*, 021207.

16. Pastor, J.; Liu, Y. Wave climate resource analysis for deployment of wave energy conversion technology. *Sustainability* **2016**, *8*, 1321.

17. Waters, R. Energy from Ocean Waves. Ph.D. Thesis, Uppsala University, Uppsala, Sweden, 2008.

18. Rahm, M. Ocean Wave Energy. Ph.D. Thesis, Uppsala University, Uppsala, Sweden, 2010.

19. Ekström, R. Offshore Marine Substation for Grid-Connection of Wave Power Farms—An Experimental Approach. Ph.D. Thesis, Uppsala University, Uppsala, Sweden, 2010.

20. Stålberg, M.; Waters, R.; Danielsson, O.; Leijon, M. Influence of Generator Damping on Peak Power for a Direct Drive Wave Energy Converter. *J. Offshore Mech. Arct. Eng.* **2008**, *130*, doi:10.1115/1.2905032.

21. Sjökvist, L.; Krishna, R.; Castellucci, V.; Hagnestål, A.; Rahm, M.; Leijon, M. On the optimization of point absorber buoys. *J. Mar. Sci. Eng.* **2014**, *2*, 477–492.

22. Hong, Y. Numerical Modelling and Mechanical Studies on a Point Absorber Type Wave Energy Converter. Ph.D. Thesis, Uppsala University, Uppsala, Sweden, 2016.

23. Li, W.; Isberg, J.; Engström, J.; Waters, R.; Leijon, M. Parametric Study of tge Power Absorbtion for a Linear Generator Wave Energy Converter. *J. Ocean Wind Energy* **2015**, *2*, 248–252.

24. Göteman, M.; Engström, J.; Eriksson, M.; Leijon, M.; Hann, M.; Ransley, E.; Greaves, D. Wave loads on a point-absorbing wave energy device in extreme waves. *J. Ocean Wind Energy* **2015**, *2*, 176–181.

25. Sjökvist, L.; Göteman, M. The effect of overtopping waves on peak forces on a point absorbing WEC. In Proceedings of the Asian Wave and Tidal Energy Conference Series AWTEC, Singapore, 24–28 October 2016.

26. Ulvgård, L.; Sjökvist, L.; Göteman, M.; Leijon, M. Line Force and Damping at Full and Partial Stator Overlap in a Linear Generator for Wave Power. *J. Mar. Sci. Eng.* **2016**, *4*, 81.

27. Frost, A.; Ulvgård, L.; Sjökvist, L.; Leijon, M. Experimental study of generator damping at partial stator overlap in a linear generator for wave power. In Proceedings of the European Wave and Tidal Energy Conference Series EWTEC, Cork, Ireland, 27 August–1 September 2017.

28. Sjökvist, L.; Göteman, M. Peak Forces on a Point Absorbing Wave Energy Converter Induced by Tsunami Waves. *Renew. Energy* **2017**, under review.

29. Jacobsen, N.G.; Fuhrman, D.; Fredsøe, J. A wave generation toolbox for the open-source CFD library: OpenFOAM. *Int. J. Numer. Methods Fluids* **2012**, *70*, 1073–1088.

30. Waters, R.; Engström, J.; Isberg, J.; Leijon, M. Wave Climate off the Swedish West Coast. *Renew. Energy* **2009**, *34*, 1600–1606.

31. Mei, C.; Stiassnie, M.; Yue, D. *Theory and Applications of Ocean Surface Waves, Part 1: Linear Aspects*; World Scientific Publishing Co., Pte. Ltd.: Singapore, 2005.
32. Douglas, S.; Nistor, I. On the effect of bed condition on the development of tsunami-induces loading on structures using OpenFOAM. *Nat. Hazards* **2015**, *76*, 1335–1356.
33. Madsen, P.; Furman, D.; Shäffer, H. On the solitary wave paradigm for tsunamis. *J. Geophys. Res.* **2008**, *113*, doi:10.1029/2008JC004932.

MDPI

St. Alban-Anlage 66

4052 Basel

Switzerland

Tel. +41 61 683 77 34

Fax +41 61 302 89 18

www.mdpi.com

Energies Editorial Office

E-mail: energies@mdpi.com

www.mdpi.com/journal/energies